Advances in Experimental Medicine and Biology

Volume 941

More information about this series at http://www.springer.com/series/5584

Xiaojing Ma
Editor

Regulation of Cytokine Gene Expression in Immunity and Diseases

Springer

Editor
Xiaojing Ma
State Key Laboratory of Microbial Metabolism, Sheng Yushou Center of Cell Biology and Immunology, School of Life Science and Biotechnology
Shanghai Jiaotong University,
Shanghai, China

Department of Microbiology and Immunology
Weill Cornell Medical College
NY, USA

ISSN 0065-2598 ISSN 2214-8019 (electronic)
Advances in Experimental Medicine and Biology
ISBN 978-94-024-0919-2 ISBN 978-94-024-0921-5 (eBook)
DOI 10.1007/978-94-024-0921-5

Library of Congress Control Number: 2016955715

Printed on acid-free paper

This Springer imprint is published by Springer Nature
The registered company is Springer Science+Business Media B.V. Dordrecht

Contents

Contributors

Yi Ban Department of Microbiology and Immunology, Weill Cornell Medical College, New York, NY, USA

Wenjuan Dong Department of Microbiology and Immunology, Weill Cornell Medical College, New York, NY, USA

John Fenimore National Cancer Institute at Frederick, Cancer and Inflammation Program, Center for Cancer Research, NIH, Frederick, MD, USA

Kamran Ghoreschi Department of Dermatology, University of Tuebingen, Tuebingen, Germany

Stanislas Goriely WELBIO and Institute for Medical Immunology (IMI), Université Libre de Bruxelles, Charleroi, Belgium

I-Cheng Ho Brigham and Women's Hospital, Harvard Medical School, Boston, MA, USA

Tadamitsu Kishimoto Laboratory of Immune Regulation, World Premier International Immunology Frontier Research Center, Osaka University, Osaka, Japan

Jianguo Liu Division of Infectious Diseases, Allergy and Immunology, Department of Internal Medicine, Saint Louis University School of Medicine, St. Louis, MO, USA

Xiaojing Ma State Key Laboratory of Microbial Metabolism, Sheng Yushou Center of Cell Biology and Immunology, School of Life Science and Biotechnology, Shanghai Jiaotong University, Shanghai, China

Department of Microbiology and Immunology, Weill Cornell Medical College, New York, NY, USA

Kazuya Masuda Laboratory of Immune Regulation, World Premier International Immunology Frontier Research Center, Osaka University, Osaka, Japan

Shi-Chuen Miaw College of Medicine, National Taiwan University, Taipei, Taiwan

Masashi Narazaki Department of Respiratory Medicine, Allergy and Rheumatic Diseases, Osaka University Graduate School of Medicine, Osaka University, Osaka, Japan

Wenjun Ouyang Department of Immunology, Genentech, South San Francisco, CA, USA

Present Affiliation: Department of Inflammation and Oncology, Amgen, South San Francisco, CA, USA

Sascha Rutz Department of Cancer Immunology, Genentech, South San Francisco, CA, USA

Mei Song Department of Microbiology and Immunology, Weill Cornell Medical College, New York, NY, USA

Toshio Tanaka Department of Clinical Application of Biologics, Osaka University Graduate School of Medicine, Osaka University, Osaka, Japan

Qinghong Wang Division of Infectious Diseases, Allergy and Immunology, Department of Internal Medicine, Saint Louis University School of Medicine, St. Louis, MO, USA

Fang Wei State Key Laboratory of Microbial Metabolism, Sheng Yushou Center of Cell Biology and Immunology, School of Life Science and Biotechnology, Shanghai Jiaotong University, Shanghai, China

Iain Welsby WELBIO and Institute for Medical Immunology (IMI), Université Libre de Bruxelles, Charleroi, Belgium

Amir S. Yazdi Department of Dermatology, University of Tuebingen, Tuebingen, Germany

Howard A. Young National Cancer Institute at Frederick, Cancer and Inflammation Program, Center for Cancer Research, NIH, Frederick, MD, USA

Hua Zheng State Key Laboratory of Microbial Metabolism, Sheng Yushou Center of Cell Biology and Immunology, School of Life Science and Biotechnology, Shanghai Jiaotong University, Shanghai, China

Chapter 1
Regulation of IFN-γ Expression

John Fenimore and Howard A. Young

Abstract Interferon gamma, referred to here as IFN-γ, is a major component in immunological cell signaling and is a critical regulatory protein for overall immune system function. First discovered in 1965 (Wheelock Science 149: (3681)310–311, 1965), IFN-γ is the only Type II interferon identified. Its expression is both positively and negatively controlled by different factors. In this chapter, we will review the transcriptional and post-transcriptional control of IFN-γ expression. In the transcriptional control part, the regular activators and suppressors are summarized, we will also focus on the epigenetic control, such as chromosome access, DNA methylation, and histone acetylation. The more we learn about the control of this regulatory protein will allow us to apply this knowledge in the future to effectively manipulate IFN-γ expression for the treatment of infections, cancer, inflammation, and autoimmune diseases.

Keywords IFN-γ • Transcriptional control • Epigenetic control • LncRNA • MicroRNA • Activator • Suppressor

1.1 Introduction

Interferon gamma, referred to here as IFN-γ, is a major component in immunological cell signaling and is a critical regulatory protein for overall immune system function. First discovered in 1965 [1], IFN-γ is the only Type II interferon identified. Functionally it is a homodimer with an antiparallel interlocking structure, lacking beta sheets but possessing six alpha helixes per monomer. In humans the IFN-γ gene is found on chromosome 12q15 [2] and in mice on chromosome 10D2 [3]. The human gene for IFN-γ has four exon regions and three intron regions, covering 4.04 kbps [4]. The receptor complex for IFN-γ, IFNGR1 and IFNGR2, is almost

J. Fenimore (✉) • H.A. Young
National Cancer Institute at Frederick, Cancer and Inflammation Program, Center for Cancer Research, Frederick, MD 21702, USA
e-mail: john.fenimore@nih.gov

X. Ma (ed.), *Regulation of Cytokine Gene Expression in Immunity and Diseases*,
Advances in Experimental Medicine and Biology 941,
DOI 10.1007/978-94-024-0921-5_1

ubiquitous in all mammalian cells, save erythrocytes. In addition, the structure of IFN-γ is much conserved in studied jawed vertebrates [5].

IFN-γ's effects on cells are notable, having been shown to modulate the expression of over 2,300 human genes [6]. While of the functions of different interferons are often redundant, a lack of normal IFN-γ expression is linked with cases of heightened weaknesses to some diseases even in the presence of other interferons. As discussed by Kaufmann and Booty et al. [7, 8], when expression of IFN-γ is low, susceptibility to mycobacterial and fungal infections is common; this shows the need for IFN-γ expression to fight disease and depicts the nonredundant functions of IFN-γ's maintenance of host resistance [7–10].

Unlike other interferons, even though IFN-γ does possess an antiviral capacity, it is more notable for its broader ability to stimulate and modulate the immune system. IFN-γ triggers the immune responses that lead to phagocytosis and increased expression of the MHC receptors on the surface of cells. This is further highlighted by IFN-γ being used to treat chronic granulomatous diseases, a group of inherited immune disorders in which white blood cells fail to control bacterial infections, thus causing severe infections in the skin, liver, lungs, and bone as pathogens are contained in granules, but not consumed via phagocytosis [7]. In turn the presence of high amounts of IFN-γ has been linked to inflammation. In some diseases such as multiple sclerosis, an autoimmune disease involving inflammation in the brain and spinal column, IFN-γ was present in abnormally high levels in affected tissues [12]. IFN-γ has also been found to be involved in other chronic diseases such as Type 1 diabetes, where IFN-γ overexpression has been linked to autoimmune dysfunction, as infiltration by T cells expressing IFN-γ into islets containing pancreatic beta cells results in destruction of beta cells [13, 14].

The activity of IFN-γ is noted as being involved in the activation, growth, and differentiation of T cells, B cells, macrophages, natural killer (NK) cells, and other cell types such as endothelial cells and fibroblasts, thus making it vital to the inflammatory response and to cell-mediated immune responses [4]. The innate and adaptive immune systems rely on controlled IFN-γ expression to preserve the balance between an effective host immune response and the development of autoimmune disease. For this reason, it is very important to understand the pathways that regulate the expression of IFN-γ.

1.2 Transcriptional Controls

IFN-γ gene expression is controlled by a very complex system of regulation. IFN-γ has been noted to be normally expressed by cells from the immune system, such as the natural killer, or NK, cells and natural killer T, or NKT, cells which are both involved in the innate immune response. IFN-γ is also expressed during adaptive immune responses via $CD4^+$ Type 1 helper (Th1) T cell, or $CD4^+$ cells, and $CD8^+$ cytotoxic T lymphocyte effector cells, or $CD8^+$ cells. The expression of IFN-γ is one of the defining traits of CD4+ Th1 cell-type immune cells [11]. The differentiation of immune cells into cells that produce IFN-γ or inhibition of such differentiation is also inherent to the amount of IFN-γ that is expressed by a host undergoing an immune response. Production of IFN-γ is controlled at several levels, including epigenetic changes, chromosomal access, cell surface signaling, transcription factor binding to promoters and enhancers, mRNA stability, and long noncoding RNA (LncRNA) interactions with the IFN-γ locus. Given the importance of IFN-γ to host immune function, it is not surprising that this single gene is regulated at many different levels.

Some regions of the genome are noted as being important for the control of IFN-γ expression in specific cell types but not in others. For instance, the region 92–18 bp upstream of the start of transcription is noted as being vital for expression in human T cells and NKT cells, but not NK cells [15]. More work, illuminating the progress of our understanding of IFN-γ epigenetics, has demonstrated the importance of conformational changes to the structure of the IFN-γ locus and factors that affect access to the site [16–18]; this shows there are multiple mechanisms impacting the transcription of this gene.

1.2.1 Epigenetic Control

The role of access to and control of genetic information is an important factor when discussing the process of transcription. The IFN-γ gene is no exception, with several methods of epigenetic control affecting gene expression. However understanding the specific details of this control is a developing science. Specific areas of conserved noncoding sequences, or CNS, are vital for the expression of IFN-γ as relevant CNS regions are present before and after the transcription start site. Of great interest is the difference between their roles in different cell types. For example, it has been reported that CNS-30 can be deleted from NK cells without a loss of IFN-γ expression, while other cells that express IFN-γ require it for full transcription [15]. T cells require the presence of CNS-2, and all NKT and some T cells require the CNS present at +20 [17]. NK cell gene expression may also be enhanced by access to CNS-2 and CNS+20, but these regions are not required for basal IFN-γ gene expression [17]. Alternatively, CNS-16 has different effects depending upon the cell type; in NK cells it upregulates the expression of IFN-γ, but it is not required for transcription, while in the $CD4^+$ Th2 cells, suppression of IFN-γ expression requires this region [17].

1.2.1.1 Chromosomal Access/DNA Methylation/Histone Acetylation

Chromatin condensation, by preventing or allowing access to the promoter or other regulatory elements of a gene, is a first step in controlling gene expression. With respect to IFN-γ, expression is regulated first via chromatin modification. This is shown by the differences between nonactivated T cells and activated T cells with regard to the chromatin density around the IFN-γ gene [17]. The compression of the IFN-γ region has a marked difference when comparing Th1, Th2, and naïve T cells, as observed from a chromatin conformational capture assay [16]. This is also apparent in NK cells, which also have low chromatin density around the transcription start site indicating easier access to the promoter for binding of the required transcription factors and RNA polymerase [17].

Additional regulation of the expression of IFN-γ comes in the form of DNA methylation, which occurs when cytosine is converted to 5-methylcytosine at CpG nucleotides, and as a result prevents direct interactions with specific transcription factors to those regions of the DNA. The IFN-γ CNS-34, IFN-γ CNS-22, the IFN-γ promoter, and IFN-γ CNS+29 are methylated in naïve T helper cells, and methylation is lost as T helper cells progress through their differentiation pathways [19, 20]. The difference in methylation of DNA is notable in cells that do not express IFN-γ, where the locus is highly methylated, indicating that the methylation of specific transcription factor binding sites is an additional mechanism of transcriptional control. Contrasting this, hypomethylation, which is observed between the −200 and +1 positions of the IFN-γ promoter, contributes to the expression of the gene in activated Th1 and NK cells, through binding of transcription factors to those regulatory regions [20]. A target for methylation is a SnaBI restriction enzyme recognition site that is highly conserved in all species and is present at −52 bp in the *IFN-γ* promoter [17, 18]. Demethylation of this SnaBI site is rapid during the processes of antigen activation, differentiation, and proliferation of T cells [17]. Of importance NK cells do not have notable methylation at this site, and in NK cells, cytokine stimulation of IFN-γ expression (e.g., IL-12) results in the rapid demethylation of the broader IFN-γ locus [17].

Histone acetylation, another method of restricting transcription of a gene, is a reversible modification of DNA that regulates chromatin access. Histone acetylation of the IFN-γ locus is of interest because when naïve T cells are initially stimulated, the loci for the IFN-γ and IL-4 genes are unpacked by histones [17]. The process of histone acetylation changes after differentiation, with 50 kb of DNA both up- and downstream from the IFN-γ locus being closed off in non-IFN-γ expressing $CD4^+$ Th2 cells [17]. This can also be observed by depriving the activated Th1 cells of STAT4, thus suggesting that active binding of STAT4 driven by IL-12 is critical for maintaining an open chromatin conformation at the IFN-γ locus [22]. Additionally, in T cells it has been shown that chromatin confirmation is driven through an interplay of T-cell-specific T-box transcription factor (T-bet) and GATA3, which are transcription factors we will discuss later in this chapter. T-bet notably keeps the area where IFN-γ is located accessible via histone acetylation, while GATA3 seems to prevent the action of T-bet in keeping the IFN-γ locus accessible [17, 18]. This interaction between T-bet and GATA3 is notably absent in NK cells as the CNS regions that impact IFN-γ gene expression have, even in the absence of stimulation and active transcription, similar histone-acetylated domains as compared to that found in activated T cells [18].

1.2.1.2 Chromatin Positioning

Chromatin may form blocks to transcription or contrastingly cause specific areas of DNA to be more accessible; this is done not only via compression but also via moving elements away or toward enhancers. Chromatin may also cause movement in the three-dimensional space the gene occupies to prevent the association of transcription factors or to facilitate such binding, via chromatin looping. This accessibility is observed via a notable change in the conformation of the IFN-γ locus [16]. CCCTC-binding factor, or CTCF, is a zinc finger transcription factor that binds the core CCCTC sequence [21]. These proteins have been described as suppressors for which there are three binding sites across the IFN-γ gene locus. These sites have been defined as markers that have been used to denote the limits of this locus [22] and have also been depicted to create a loop in active T cells which has been hypothesized to be required for protection of the gene from heterochromatin silencing and to increase the efficiency of expression [17].

1.2.1.3 LncRNAs

Long noncoding RNAs, or lncRNAs, are noted as being inhibitors of transcription via interactions between sections of RNA greater than 200 bp that are transcribed alongside other regions of chromatin that link to and inhibit the action of proteins and RNA. The role of these lncRNAs in the enhancement or suppression of transcription is a developing field of study as their activity is becoming more defined. For example, the ncRNA repressor of the nuclear factor of activated T cells, or NRON, is a negative regulator of the transcription factor (NFAT), which interacts with the IFN-γ promoter [23]. This inhibition of NFAT by NRON is caused by the formation of protein and RNA complexes, resulting in sequestration of NFAT in the cytosol, away from the DNA [23]. It has also been reported that Theiler's murine encephalitis virus possible gene 1, or TMEVPG1, also named nettoie Theiler's pas Salmonella, or NeST, is the first identified lncRNA to regulate expression of IFN-γ [24]. NeST is located 170 kb downstream from the IFN-γ gene and transcribed from the antisense strand relative to IFN-γ; the 33 kb NeST gene is spliced into a 1.7 kb RNA transcript in human cells [24]. Studies have shown that NeST's expression in both mouse and human is Th1 T-cell selective and dependent on STAT4 and T-bet [24]. Due to this pattern of expression and its structural traits, it is believed that NeST associates with WD repeat-containing protein 5, or WDR5, to promote histone H3 lysine 4, drive methylation to enhance chromatin accessibility, and conversely decrease attraction of local histones at the nearby IFN-γ locus, thus promoting transcription [17].

1.2.1.4 miRNAs

MicroRNAs, or miRNAs, are between 18 and 25 nucleotides long and are noncoding RNAs that suppress gene expression by binding to the 3′ UTR of target genes, resulting in either translation inhibition or mRNA degradation. A broader role for miRNA control of IFN-γ expression was demonstrated by studies demonstrating that the elimination of the Dicer gene, vital for the processing of microRNAs, increased the expression of IFN-γ in T cells [25]. Interestingly this effect has been noted in $CD4^+$ TH2 cells indicating that there may be some changes to the epigenetics in the cells as a result of the inhibition of miRNA processing [25].

Recently studies have shown that the miRNA, miR-223, which is notably enhanced by estrogen, positively regulates the expression of IFN-γ via inhibition of anti-inflammatory elements such as IL-10 [26]. In other studies, microRNA-146a and microRNA-146b both suppress the activity of TRAF6 and IRAK1 protein expression, factors that are important for the function of the IFN-γ promoter, and thus cells expressing these miRNAs show decreased IFN-γ expression [27, 28]. Studies have also shown that miR-29 suppress IFN-γ production by directly targeting IFN-γ mRNA and enhancing its degradation [29]. However contrasting experimental evidence reports that miRNA-29 inhibition is driven via degradation of the T-box transcription factor T-bet mRNA instead of the IFN-γ mRNA [30]. Additional studies focused on miR-155 have shown that in NK cells this miRNA increases the expression of IFN-γ when NK cells are stimulated, notably by both IL-18 and IL-12 [31]. miR-155 is observed having its own expression increased when the NK cells are stimulated with IL-18 and IL-12 via downregulating SHIP1, a phosphatase, that in turn suppresses NF-kB activity, a transcription factor known to bind to the IFN-γ locus and stimulate transcription [31, 32].

1.3 Inducers of IFN-γ

IFN-γ is part of a system of biological signals that can induce substantial changes inside the cell and in the host overall. As such it must be regulated, since it is not passively expressed and different signals regulate expression either by enhancing or inhibiting gene expression. These signals come in many forms, but the primary methods are dependent upon cell activation through signaling via cell surface receptors. Regulation of IFN-γ expression is largely driven by activators like transcription factors such as T-bet and NFAT and specific inhibitors, like TGF-beta, in $CD4^+$ Type 1 helper T cells. Methods to manipulate this upregulation involve specific systems using phosphorylation of transcription factors, driven by activation of multimeric receptors of cytokines linked to Janus kinases (JAK) activation, while inhibitory signals act through inhibition of these signaling cascades and their receptors resulting in inhibition of the specific transcription factors required for IFN-γ transcription [33–37].

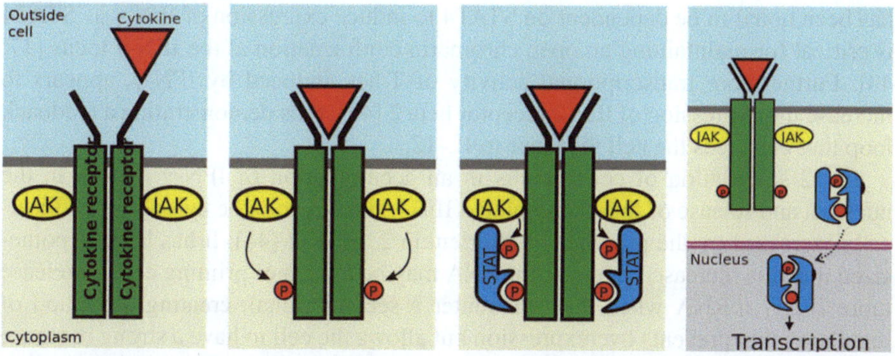

Fig. 1.1 The archetypal motif of the receptor and cytokine structure to stimulate STATs. The interaction of a cytokine causes the dimerization of surface receptors which in turn cause phosphorylation-driven changes which cascade to affect transcription

1.3.1 IL-2

IL-2 is a cytokine with three surface receptor subunits, alpha, beta, and a common gamma chain. The presences of these receptor subunits determine the level of affinity in each immune cell type for gene activation. For example, the beta and gamma chains together form a complex that binds IL-2 with intermediate affinity on NK cells, while all three receptor chains form a complex that binds IL-2 with high affinity on activated T cells [33]. The stimulated signaling pathways upregulate IFN-γ expression but also notably upregulate the expression of SOCS1 which functions to prevent overexpression of IFN-γ [38]. Suppressor of cytokine signaling, or SOCS1, is a suppressor of IFN-γ expression that interacts with the tyrosine kinase (TYK), Jak1, and inhibits expression downstream by reducing STAT phosphorylation [33].

IL-2 is also noted for triggering the binding of a STAT5 a/b heterodimer to a STAT5-binding element −3.6 kb from the transcription start site of the IFN-γ gene [39, 40] (Fig. 1.1).

1.3.2 IL-12/STAT4

Interleukin 12, or IL-12, is a cytokine heterodimer that can induce IFN-γ transcription and cause nuclear accumulation of IFN-γ mRNA in $CD4^+$ T helper cells and NK cells. At activation, two subunits of the IL-12 receptor which is a heterodimer comprised of the beta-1 and the beta-2 chains of the receptor come together to activate the receptor-associated Janus kinases, JAK2 and TYK2 [33, 40]. STAT4 is then phosphorylated by these tyrosine kinases and translocated into the nucleus to activate gene transcription. The pSTAT4 then interacts with IRF-1 and ERM to activate their binding to the IFN-γ promoter [14, 41, 42]. This activation also takes place in NK cells and is independent of most other forms of IFN-γ stimulation [23]. IL-12 has been noted to be dependent on STAT4 to induce expression of IFN-γ as STAT4 is critical for maintaining an open chromatin conformation at the IFN-γ locus [17, 43]. Furthermore transcriptional activity of T-bet, induced by IFN-γ, appears to increase the expression of IL-12 receptor beta 2 [43], thus demonstrating a feedback loop that enhances the cell response to IL-12.

IL-12 stimulation of cells results in an accumulation of IFN-γ mRNA in the nucleus, and release of this accumulated IFN-γ mRNA into the cytoplasm is drastically increased by the presence of interleukin 2, or IL-2 [44]. It has been hypothesized that this increased release of mRNA may be a form of priming cells to release more IFN-γ mRNA when they encounter a second signal, creating a method of regulation that prevents overexpression but allows the cell to have a strong response when needed. It has been shown that both IL-2 and IL-12 in combination promote a steady release of IFN-γ over time, notably more than either signal alone, thus demonstrating a synergistic response of the cell when multiple activating signals are encountered [44].

These previously described pathways of IFN-γ expression when $CD4^{+}$ cells are stimulated by IL-12 can occur only if the T-bet binding region is not blocked by the dual zinc finger transcription factor, GATA-binding protein 3 or GATA3 [45]. The GATA3 transcription factor does not affect the IFN-γ promoter region directly, but instead affects the downregulation of STAT4, thus preventing signaling needed to induce IFN-γ expression. GATA3 is in turn shown to be stimulated by exposure of a naïve T cell to interleukin 4 [45].

STAT3 has been reported to have variable effects in NK cells and their expression of IFN-γ. It has been noted that STAT3 regulates IFN-γ production by binding the IFN-γ promoter upon stimulation of IL-12. However STAT3 has not been shown to be required for IFN-γ expression by NK cells [46, 47].

It is important to note the synergistic effect of cytokines in regard to IFN-γ gene expression. A very strong synergy between IL-12 and IL-18 for IFN-γ expression is driven by the upregulation of receptors on cells that have been exposed to one or the other cytokines; the IL-12 beta-2 receptor chain has been noted to be upregulated by IL-18 and IL-12 can upregulate the IL-18 alpha receptor chain [48, 49], thus stimulating the target cell to be maximally responsive to these cytokines.

1.3.3 IL-15

Interleukin 15, or IL-15, is a heterodimeric cytokine that triggers IFN-γ gene expression through binding to a heterodimeric IL-15 receptor [35]. Signaling occurs through Janus kinases Jak 1 and Jak 3 activation which in turn phosphorylates STAT3 and STAT5. Studies have shown the responsiveness of $CD8^{+}$ cells to IL-15 results in the maintenance of CD8+ T-cell survival, thus affecting the host's overall production of IFN-γ [35, 41].

1.3.4 IL-18

A member of the IL-1 family, interleukin 18 (IL-18) (originally designated interferon gamma-inducing factor), induces the transcription of IFN-γ. The surface receptors of IL-18 are well characterized; the receptors are heterodimeric with a region for the binding of the ligand, the IL-1 receptor-related protein (IL-1Rrp), and the signal transducing subunit or the accessory protein-like subunit (AcPL) [36]. Once activated, this causes a signal cascade involving MYD88, the IL-1 receptor-associated kinase (IRAK), TNF receptor-associated factor 6, and the nuclear factor kappa-light-chain-enhancer of activated B cells (NF-kB) [34]. NF-kB is shown to bind and enhance activity of the IFN-γ promoter, increasing expression of IFN-γ [50]. NF-kB additionally undergoes a chain of interactions to stimulate the activation of STAT4 which stimulates the production of IFN-γ [52, 53].

Interleukin 10 (IL-10) is characterized as inhibiting the function of NF-kB and thus is considered an inhibitor of IFN-γ expression [54].

1.3.5 IL-27/STAT1/STAT3

Interleukin 27 (IL-27), a heterodimeric cytokine from the IL-12 family, enhances IFN-γ production by activated T cells and NK cells [33]. One such induction of expression is activated via IL-27's interactions with a dimeric receptor comprised of the interleukin-27 receptor alpha chain (IL27Ra) and glycoprotein 130 (gp130) [55]. Upon binding of IL-27 to this receptor, two arms of Janus kinases are activated: Jak1 linked to the IL27Ra and Jak2, Tyk2, and Jak1 linked to gp130 [55]. Triggering the phosphorylation of STAT1 and STAT3, this causes the transport of the pSTAT1/3 heterodimer to the nucleus where it activates the T-bet promoter, resulting in increased T-bet and subsequent IFN-γ expression [55, 56].

1.4 Positive Transcriptional Control Factors

IFN-γ production is highly variable and cells do not express IFN-γ in a resting state. Thus there are numerous methods to activate immune cells to express this cytokine and vary its levels of expression. These controls are driven by transcription factors that form complexes on specific regions of DNA to control the level of transcription via RNA polymerase 2 [57]. In this section we will discuss the transcription factors involved in regulating IFN-γ expression and briefly summarize our understanding of their activation and activity (Fig. 1.2).

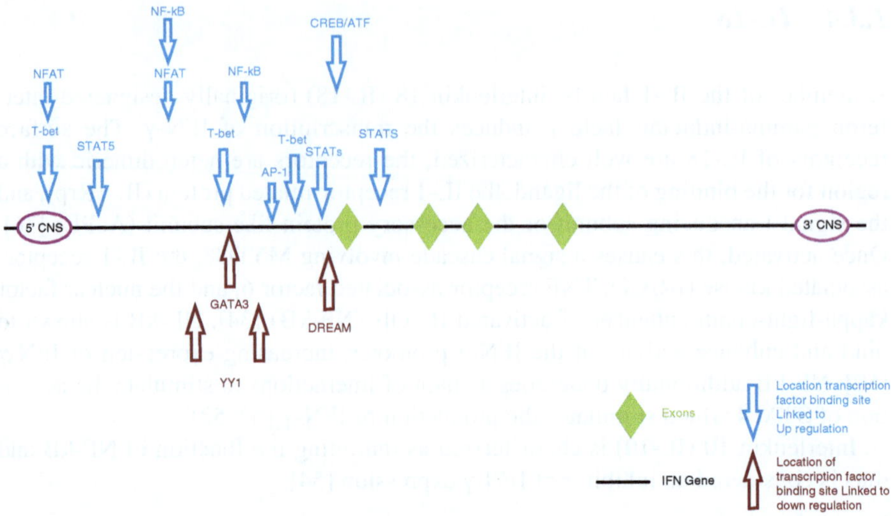

Fig. 1.2 The IFN-γ gene, depicting general structure and relative locations of protein-binding sites

1.4.1 T-bet

T-bet, or the T-cell-specific T-box transcription factor, is considered one of the most important controls for the expression of IFN-γ. T-bet has been defined as a key factor in T-cell differentiation and is expressed in many cell types in the immune system. However in many regards, it has been seen as the endpoint that other signal pathways must reach to stimulate the expression of IFN-γ or as the target that needs to be blocked to prevent expression [58]. Recent studies have shown T-bet is required for IFN-γ production and lineage commitment of $CD4^+$ T cells, but not of $CD8^+$ T cells [45]. Using T-bet KO mice, it was shown that, when stimulated, CD4+ T cells had a profound deficiency in IFN-γ expression [59]. Contrastingly it was determined there is little difference in IFN-γ expression in stimulated $CD8^+$ cells with or without T-bet [59]. Furthermore IL-12 has also been noted to induce T-bet in an IFN-γ receptor and STAT1-independent manner in $CD8^+$ T cells [60]. Induction of T-bet results in upregulation of IL-12RB2, a receptor essential for responsiveness to IL-12, and differentiation to a CD4+ Th1 cell type [34]. However studies have shown that T-bet represses the Th2 lineage commitment through a tyrosine kinase-mediated interaction that interferes with the binding of GATA3 to its target DNA [61]. Making T-bet both a suppressor and an enhancer depending on the perspective of Th1/Th2 T-cell determination indicates that T-bet has a broader influence on chromatin structure that directly and indirectly impacts IFN-γ gene expression [17].

1.4.2 Eomes

In CD8+ T cells, the primary driver of IFN-γ expression is via the transcription factor Eomesodermin (Eomes), which functions independent of T-bet [62]. However it has been reported that there is synergistic effect when both Eomes and T-bet transcription factors are expressed [63]. The activity of Eomes was also noted to increase phosphorylated STAT4 in CD8+ T cells, further promoting IFN-γ transcription [64]. Experiments with T-bet$^{-/-}$ CD8^{+} T cells have demonstrated normal IFN-γ expression, while Eomes −/− CD8+ cells did not express IFN-γ at notable levels, indicating a critical role for Eomes in CD8+ T-cell IFN-γ transcription [62, 63].

1.4.3 AP-1/CREB/ATF-2

Activating promoter 1, or AP-1, has been linked to the activity of the IFN-γ promoter [64]. AP-1 has been reported to enhance the activity of NFAT proteins through the formation of complexes with these transcription factors [42].

It has been shown that c-Jun, cAMP response element-binding protein (CREB), and ATF-2 are essential for activation-induced transcription of IFN-γ, and c-Jun binds preferentially to the IFN-γ proximal element at -52 as a heterodimer with ATF-2 [42]. The c-Fos and c-Jun transcription factors form a heterodimer known as activator protein 1 (AP-1), part of a subset of a large family of transcriptional control proteins, which in turn forms another complex with NFAT [65]. CREB or cAMP response element-binding protein is a known transcription factor closely associated with AP-1 that has been shown to bind to the proximal IFN-γ promoter −73 to −48 bp upstream of the transcription start site [64]. This strongly implicates the involvement of CREB proteins in the regulation of IFN-γ expression [65].

1.4.4 NFAT

Nuclear factor of activated T-cell (NFAT) binding sites have been identified at positions −280 bp to −270 bp and −160 bp to −155 bp upstream from the IFN-γ transcription start site, and studies show that they are required for full activity of the IFN-γ promoters in T cells [42, 66]. This effect is documented by experiments that demonstrated decreased levels of expression by full-length IFN-γ promoter-reported constructs with promoter regions that contain mutations in one or both of the NFAT-binding sites [66, 67].

1.5 Negative Transcriptional Controls

Overexpression of a specific signal can lead to drastic problems in all biological systems. In the immune system, such overexpression of a specific gene can cause autoimmune responses that can have far-reaching consequences to a host. IFN-γ is a powerful immunoregulator and the ability to turn the signal off is in many ways just as important as its activation. While not as fully understood as activating signals, characterization of the inhibitory signals leading to decreased expression of IFN-γ is proving vital to our understanding of the regulation of this gene.

1.5.1 Homeobox/Prox1

Prospero-related homeobox (Prox1) is a suppressor of the expression of IFN-γ in T cells; however, it does not typically function to suppress the genes as a homeobox gene would normally be expected to function. It has been reported that Prox1 does not bind to the promoter region of IFN-γ but that it uses a method of facilitating a conformational change with a protein intermediary to inhibit the synthesis of IFN-γ mRNA [68]. The nuclear receptor peroxisome proliferator-activated receptor gamma (PPARγ) has been shown to link to the Prox1 structure and inhibit IFN-γ expression [68]. The presence of Prox1 however is also notably downregulated as the cell differentiates into a $CD4^+$ T cell, but the mechanism is unknown [68]. PPARγ and PPAR delta have also been reported to inhibit IFN-γ gene expression in part by antagonizing the activities of the transcription factors AP-1, STAT, and NF-kB [69, 70]. However, once IFN-γ has been expressed, it has been demonstrated to in turn suppress PPARγ expression via upregulation of activated STAT1 [71], thus forming a regulatory cycle to prevent under- or overexpression.

IFN-γ has been linked to many sex-biased autoimmune diseases and the sex bias in PPARγ coincides with these phenomena [72]. Males have a notably higher level of expression of PPARγ than females, thus contributing to the finding that female mice and women express more IFN-γ than their male counterparts [72].

1.5.2 TGF-B

Transforming growth factor beta (TGF-B) is a superfamily of cytokines that, when activated, downregulate the expression of T-bet mRNA via direct interaction of SMAD proteins with the T-bet promoter [73]. TGF-β utilizes SMAD2, SMAD3, and SMAD4 to suppress IFN-γ via inhibiting the activity of T-bet by binding to the proximal T-bet site in the IFN-γ locus [74].

1.5.3 DREAM

Downstream regulatory element antagonist modulator (DREAM) can directly bind to the IFN-γ promoter and inhibit transcription. DREAM forms a complex with the related protein KChIP-2 around an area downstream of the TATA box in the IFN-γ promoter [75]. DREAM's binding takes place as a doublet with one direct and one inverted downstream regulatory element located at position +20 downstream from the transcription initiation site if IFN-γ, thus blocking transcription [75]. DREAM is constitutively expressed in unstimulated T lymphocytes, and it has been reported that DREAM is rapidly downregulated after stimulation of T cells [75]. This suggests DREAM may be important to prevent basal expression of IFN-γ when immune cells are not stimulated and are in a resting state.

1.5.4 GATA3

Transacting T-cell-specific transcription factor GATA3 is a potent suppressor of IFN-γ and has effects on T-cell differentiation. Members of the CREB/ATF, AP-1, octamer 1, and GATA families of transcription factors bind to the proximal −70 bp to −47 bp and distal −98 bp to −72 bp regions upstream from the start site of IFN-γ transcription [76]. It has been reported that GATA3 restricts access to the promoter regions of IFN-γ and T-bet, thus preventing interactions of activating transcription factors with the IFN-γ promoter [76]. In addition, Gata3 and octamer 1 enhance the expression of IL-4-driving cell differentiation away from a state where IFN-γ is expressed [77].

Published reports have promoted the idea that even though the binding of GATA3 can be detected at two GATA motifs, in positions −108 bp to −91-bp, of the human IFN-γ gene, it may be that the lack of expression of IFN-γ in the presence of GATA3 is more notably driven by the suppression of other transcription factors [76, 78]. Of note, there was a reduction in STAT4, involved in promoting IFN-γ transcription, in the presence of high levels of GATA3 [76]. Interestingly, these studies also showed a reduction of the amount of STAT1 in the cells expressing elevated levels of GATA3 [78]. In addition, runt-related transcription factor 3 (Runx3) enhances the binding of T-bet to the IFN-γ locus by binding to specific regions in the IFN-γ promoter; this in turn decreases the binding frequency of GATA3 [79]. The binding of T-bet acts to competitively suppress GATA3 binding, thus enhancing IFN-γ transcription [79, 80]. GATA3 has also been linked to changes in histone modulation, promoting access to some areas and restricting access to others in competition with T-bet [76].

1.5.5 Yin Yang 1

Ying Yang 1, or YY1, is a zinc finger transcription factor that belongs to the human GLI-Kruppel family of nuclear proteins and is an inhibitor of the IFN-γ promoter in two notable models [81]. There are two constitutive YY1-binding sites in the IFN-γ promoter at positions Y1 −199 bp to −203 bp and Y2 −217 bp to −221 bp upstream from the transcriptional initiation site [81]. These sites are both hypothesized to play a role in inhibiting the expression of IFN-γ by competition for DNA binding with AP-1 at the Y1 position [64] and by activation of an AP-2-like protein by YY1 binding at the Y2 position which then acts as a repressor of IFN-γ transcription [81].

1.6 Posttranscriptional Regulation

Posttranscriptional downregulation of IFN-γ is mostly characterized as being from processes degrading the IFN-γ mRNA. Removing the region in the 3′ UTR that contains the AUUA repeat elements (ARE) results in significantly higher levels of expression of IFN-γ due to stabilization of the mRNA [82]. A second method of control involves the 5′ UTR region of the mRNA. As the structure of the 5′UTR has revealed the presence of a semi-knot, this structure triggers a natural anti-dsRNA reaction by RNA-activated protein kinase (PKR) which helps downregulate the continued expression of IFN-γ via binding to the mRNA and facilitating degradation of the dsRNA present in the semi-knot [83, 84].

There are other forms of posttranscriptional controls for IFN-γ. A mechanism by which IFN-γ is posttranslationally modified, i.e., glycosylation at positions 25 and 97, stabilizes the protein and protects it from degradation [85]. In response to stress, the host cell can also change the protein's final structure, by downregulating the expression of the enzyme, Furin, which is needed to generate the IFN-γ protein into a functional state by cleaving the leader peptide [86]. This demonstrates that controlling functional expression of this cytokine is a continuum of mechanisms that extend beyond the domain of transcription alone.

1.7 Conclusions

Negative and positive controls of IFN-γ expression must remain in balance to prevent the promotion of autoimmune diseases or host deficiencies in fighting disease. The conserved nature and the wide-ranging effects of IFN-γ on the adaptive and innate immune systems ensure that the expression of IFN-γ is a potent response required to fight disease, but one that must be controlled in healthy hosts. This control is transcriptionally regulated at the level of chromatin access involving chromatin secondary and tertiary structure, DNA methylation, and histone acetylation. This process is in turn effecting and affected by transcription factors, which are both activators and inhibitors of gene expression. The specific host factors can be shaped by the process of transcription itself, as impacted by extracellular signals received by the target cells and RNA stability following gene induction. Understanding the labyrinthine system of controls may seem daunting, and the complete picture of IFN-γ control is still a point of active investigation. However, the more we learn about the control of this regulatory protein will allow us to apply this knowledge in the future to effectively manipulate IFN-γ expression for the treatment of infections, cancer, inflammation, and autoimmune diseases.

(The author would like to apologize for any articles or work that was missed in the collection of this work. The volume of data, space, and time constraints made it impossible for the author to create a complete list of the valuable research that has been done on this integral part of immunological research.)

References

1. Wheelock EF. Interferon-like virus-inhibitor induced in human leukocytes by phytohemagglutinin. Science. 1965;149(3681):310–1.
2. Naylor SL, Sakaguchi AY, Shows TB, Law ML, Goeddel DV, Gray PW. Human immune interferon gene is located on chromosome 12. J Exp Med. 1983;157(3):1020–7.
3. Naylor SL, Gray PW, Lalley PA. Mouse immune interferon (IFN-γ) gene is on chromosome 10. Somat Cell Mol Genet. 1984;10(5):531–4.
4. Ushio S, Namba M, Okura T, Hattori K, Nukada Y, Akita K, Torigoe K, et al. Cloning of the cDNA for human IFN-gamma-inducing factor, expression in Escherichia coli, and studies on the biologic activities of the protein. J Immunol. 1996;156(11):4274–9.
5. Savan R, Ravichandran S, Collins JR, Sakai M, Young HA. Structural conservation of interferon gamma among vertebrates. Cytokine Growth Factor Rev. 2009;20(2):115–24.
6. Rusinova I, Forster S, Yu S, Kannan A, Masse M, Cumming H, Chapman R, Hertzog PJ. INTERFEROME v2. 0: an updated database of annotated interferon-regulated genes. Nucleic Acids Research. 2013 January; 41 (database issue): D1040-D1046. Smith NL, Dennings DW Clinical implications of interferon gamma genetic and epigenetic variants. Immunology 2014.
7. Kaufmann SH. Immunopathology of mycobacterial diseases. In Seminars in immunopathology (pp. 1–4). 2016. Berlin/Heidelberg: Springer.
8. Booty MG, Nunes-Alves C, Carpenter SM, Jayaraman P, Behar SM. Multiple inflammatory cytokines converge to regulate CD8+ T cell expansion and function during tuberculosis. J Immunol. 2016;169(4):1822–31.
9. Marciano BE, Wesley R, Ellen S, Anderson VL, Barnhart LA, Darnell D, Holland SM, et al. Long-term interferon-γ therapy for patients with chronic granulomatous disease. Clin Infect Dis. 2004;39(5):692–9.
10. Schoenborn JR, Wilson CB. Regulation of interferon-gamma during innate and adaptive immune responses. Adv Immunol. 2007;96:41–101.
11. Janeway CA, Travers P, Walport M, Shlomchik MJ. ImmunoBiology. 6th ed. New York: Garland Science Taylor and Francis Group; 2005.
12. Hirsch RL, Panitch HS, Johnson KP. Lymphocytes from multiple sclerosis patients produce elevated levels of gamma interferon in vitro. J Clin Immunol. 1985;5(6):386–9.
13. Sarvetnick N, Shizuru J, Liggitt D, Martin L, McIntyre B, Gregory A, et al. Loss of pancreatic islet tolerance induced by beta-cell expression of interferon-gamma. Nature. 1990;346:844–7.

14. Azar ST, Tamim H, Beyhum HN, Habbal MZ, Almawi WY. Type I (insulin-dependent) diabetes is a Th1-and Th2-mediated autoimmune disease. Clin Diagn Lab Immunol. 1999;6(3):306–10.
15. Collins PL, Chang S, Henderson M, Soutto M, Davis GM, McLoed AG, Aune TM, et al. Distal regions of the human IFNG locus direct cell type-specific expression. J Immunol. 2010;185(3):1492–501.
16. Eivazova ER, Aune TM. Dynamic alterations in the conformation of the Ifng gene region during T helper cell differentiation. Proc Natl Acad Sci. 2004;101(1):251–6.
17. Aune TM, Collins PL, Collier SP, Henderson MA, Chang S. Epigenetic activation and silencing of the gene that encodes IFN-γ. Front Immunol. 2013;4:112.
18. Chang S, Aune TM. Histone hyperacetylated domains across the Ifng gene region in natural killer cells and T cells. Proc Natl Acad Sci U S A. 2005;102:17095–100.
19. Schoenborn JR, Dorschner MO, Sekimata M, Santer DM, Shnyreva M, Fitzpatrick DR, et al. Comprehensive epigenetic profiling identifies multiple distal regulatory elements directing transcription of the gene encoding interferon-gamma. Nat Immunol. 2007;8:732–42.
20. Falek PR, Ben-Sasson SZ, Ariel M. Correlation between DNA methylation and murine IFN-γ and IL-4 expression. Cytokine. 2000;12(3):198–206.
21. Phillips JE, Corces VG. CTCF: master weaver of the genome. Cell. 2009;137:1194–211.
22. Morinobu A, Gadina M, Strober W, Visconti R, Fornace A, Montagna C, O'Shea JJ, et al. STAT4 serine phosphorylation is critical for IL-12-induced IFN-γ production but not for cell proliferation. Proc Natl Acad Sci. 2002;99(19):12281–6.
23. Willingham AT, Orth AP, Batalov S, Peters EC, Wen BG, AzaBlanc P, et al. A strategy for probing the function of noncoding RNAs finds a repressor of NFAT. Science. 2005;309:1570–3.
24. Vigneau S, Rohrlich PS, Brahic M, Bureau JF. Tmevpg1, a candidate gene for the control of Theiler's virus persistence, could be implicated in the regulation of gamma interferon. J Virol. 2003;77:5632–8.
25. Muljo SA, Ansel KM, Kanellopoulou C, Livingston DM, Rao A, Rajewsky K. Aberrant T cell differentiation in the absence of Dicer. J Exp Med. 2005;202:261–9.
26. Dorhoi A, Iannaccone M, Farinacci M, Faé KC, Schreiber J, Moura-Alves P, … Kaufmann SHE. MicroRNA-223 controls susceptibility to tuberculosis by regulating lung neutrophil recruitment. The Journal of Clinical Investigation. 2013;123(11):4836–48. http://doi.org/10.1172/JCI67604.
27. Dai R, Phillips RA, Zhang Y, Khan D, Crasta O, Ahmed SA. Suppression of LPS-induced interferon-γ and nitric oxide in splenic lymphocytes by select estrogen-regulated microRNAs: a novel mechanism of immune modulation. Blood. 2008;112(12):4591–7.
28. Park H, Huang X, Lu C, Cairo MS, Zhou X. MicroRNA-146a and microRNA-146b regulate human dendritic cell apoptosis and cytokine production by targeting TRAF6 and IRAK1 proteins. J Biol Chem. 2015;290(5):2831–41.
29. Ma F, Xu S, Liu X, Zhang Q, Xu X, Liu M, Cao X, et al. The microRNA miR-29 controls innate and adaptive immune responses to intracellular bacterial infection by targeting interferon-[gamma]. Nat Immunol. 2011;12(9):861–9.
30. Steiner DF, Thomas MF, Hu JK, Yang Z, Babiarz JE, Allen CD, Ansel KM, et al. MicroRNA-29 regulates T-box transcription factors and interferon-γ production in helper T cells. Immunity. 2011;35(2):169–81.
31. Trotta R, Chen L, Ciarlariello D, Josyula S, Mao C, Costinean S, Caligiuri MA, et al. miR-155 regulates IFN-γ production in natural killer cells. Blood. 2012;119(15):3478–85.
32. Manning BD, Cantley LC. AKT/PKB signaling: navigating downstream. Cell. 2007;129(7):1261–74.
33. Kasahara TADASHI, Hooks JJ, Dougherty SF, Oppenheim JJ. Interleukin 2-mediated immune interferon (IFN-gamma) production by human T cells and T cell subsets. J Immunol. 1983;130(4):1784–9.

34. Hibbert L, Pflanz S, de Waal Malefyt R, Kastelein RA. IL-27 and IFN-α signal via Stat1 and Stat3 and induce T-Bet and IL-12R β 2 in naive T cells. J Interf Cytokine Res. 2003;23(9):513–22.
35. Calarota SA, Otero M, Hermanstayne K, Lewis M, Rosati M, Felber BK, Weiner DB, et al. Use of interleukin 15 to enhance interferon-γ production by antigen-specific stimulated lymphocytes from rhesus macaques. J Immunol Methods. 2003;279(1):55–67.
36. Biet F, Locht C, Kremer L. Immunoregulatory functions of interleukin 18 and its role in defense against bacterial pathogens. J Mol Med. 2002;80(3):147–62.
37. Bird JJ, Brown DR, Mullen AC, Moskowitz NH, Mahowald MA, Sider JR, Reiner SL, et al. Helper T cell differentiation is controlled by the cell cycle. Immunity. 1998;9(2):229–37.
38. Kandasamy K, Mohan SS, Raju R, Keerthikumar S, Kumar GSS, Venugopal AK, Gollapudi SK, et al. NetPath: a public resource of curated signal transduction pathways. Genome Biol. 2010;11(1):1–9.
39. Strebovsky J, Walker P, Lang R, Dalpke AH. Suppressor of cytokine signaling 1 (SOCS1) limits NFkB signaling by decreasing p65 stability within the cell nucleus. FASEB J. 2011;25(3):863–74.
40. Gonsky R, Deem RL, Bream J, Young HA, Targan SR. Enhancer role of STAT5 in CD2 activation of IFN-γ gene expression. J Immunol. 2004;173(10):6241–7.
41. Presky DH, Yang H, Minetti LJ, Chua AO, Nabavi N, Wu CY, Gately MK, GublerA U. functional interleukin 12 receptor complex is composed of two beta-type cytokine receptor subunits. Proc Natl Acad Sci U S A. 1996;93:14002–7.
42. Park WR, Nakahira M, Sugimoto N, Bian Y, Yashiro-Ohtani Y, Zhou XY, Fujiwara H, et al. A mechanism underlying STAT4-mediated up-regulation of IFN-y induction in TCR-triggered T cells. Int Immunol. 2004;16(2):295–302.
43. Ouyang W, Jacobson NG, Bhattacharya D, Gorham JD, Fenoglio D, Sha WC, Murphy KM, et al. The Ets transcription factor ERM is Th1-specific and induced by IL-12 through a Stat4-dependent pathway. Proc Natl Acad Sci U S A. 1999;96(7):3888–93.
44. Afkarian M, Sedy JR, Yang J, Jacobson NG, Cereb N, Yang SY, Murphy KM, et al. T-bet is a STAT1-induced regulator of IL-12R expression in naive CD4+ T cells. Nat Immunol. 2002;3(6):549–57.
45. Hodge DL, Martinez A, Julias JG, Taylor Lynn S, Young HA. Regulation of nuclear gamma interferon gene expression by interleukin 12 (IL-12) and IL-2 represents a novel form of post transcriptional control. Mol Cell Biol. 2002;22(6):1742–53.
46. Cook KD, Miller J. TCR-dependent translational control of GATA-3 enhances Th2 differentiation. J Immunol. 2010;185(6):3209–16.
47. Gotthardt D, Putz EM, Straka E, Kudweis P, Biaggio M, Poli V, Sexl V, et al. Loss of STAT3 in murine NK cells enhances NK cell–dependent tumor surveillance. Blood. 2014;124(15):2370–9.
48. Pensa S, Regis G, Boselli D, Novelli F, Poli V. STAT1 and STAT3 in tumorigenesis: two sides of the same coin? 2000.
49. Hoshino K, Tsutsui H, Kawai T, Takeda K, Nakanishi K, Takeda Y, Akira S. Cutting edge: generation of IL-18 receptor-deficient mice: evidence for IL-1 receptor-related protein as an essential IL-18 binding receptor. J Immunol. 1999;162(9):5041–4.
50. Kalina U, Kauschat D, Koyama N, Nuernberger H, Ballas K, Koschmieder S, Ottmann OG, et al. IL-18 activates STAT3 in the natural killer cell line 92, augments cytotoxic activity, and mediates IFN-γ production by the stress kinase p38 and by the extracellular regulated kinases p44erk-1 and p42erk-21. J Immunol. 2000;165(3):1307–13.
51. Zarling S, Berenzon D, Dalai S, Liepinsh D, Steers N, Krzych U. The survival of memory CD8 T cells that is mediated by IL-15 correlates with sustained protection against malaria. J Immunol. 2013;190(10):5128–41.
52. Kumar A, Takada Y, Boriek AM, Aggarwal BB. Nuclear factor-kB: its role in health and disease. J Mol Med. 2004;82(7):434–48.

53. Matikainen S, Paananen A, Miettinen M, Kurimoto M, Timonen T, Julkunen I, Sareneva T. IFN-α and IL-18 synergistically enhance IFN-y production in human NK cells: differential regulation of Stat4 activation and IFN-y gene expression by IFN-α and IL-12. Eur J Immunol. 2001;31(7):2236–45.
54. de Waal Malefyt R, Abrams J, Bennett B, Figdor CG, De Vries JE. Interleukin 10 (IL-10) inhibits cytokine synthesis by human monocytes: an autoregulatory role of IL-10 produced by monocytes. J Exp Med. 1991;174(5):1209–20.
55. Varma TK, Toliver-Kinsky TE, Lin CY, Koutrouvelis AP, Nichols JE, Sherwood ER. Cellular mechanisms that cause suppressed gamma interferon secretion in endotoxin-tolerant mice. Infect Immun. 2001;69(9):5249–63.
56. Egwuagu CE, Larkin III J. Therapeutic targeting of STAT pathways in CNS autoimmune diseases. JAK-STAT. 2013;2(1), e24134.
57. Schoenborn JR, Wilson CB. Regulation of interferon-y during innate and adaptive immune responses. Adv Immunol. 2007;96:41–101.
58. Eivazova ER, Markov SA, Pirozhkova I, Lipinski M, Vassetzky YS. Recruitment of RNA polymerase II in the Ifng gene promoter correlates with the nuclear matrix association in activated T helper cells. J Mol Biol. 2007;371(2):317–22.
59. Knox JJ, Cosma GL, Betts MR, McLane LM. Characterization of T-bet and eomes in peripheral human immune cells. Front Immunol. 2014;5:217.
60. Szabo SJ, Sullivan BM, Stemmann C, Satoskar AR, Sleckman BP, Glimcher LH. Distinct effects of T-bet in TH1 lineage commitment and IFN-gamma production in CD4 and CD8 T cells. Science (New York, NY). 2002;295(5553):338.
61. Yang Y, Ochando JC, Bromberg JS, Ding Y. Identification of a distant T-bet enhancer responsive to IL-12/Stat4 and IFNγ/Stat1 signals. Blood. 2007;110(7):2494–500.
62. Hamilton SE, Jameson SC. CD8+ T cell differentiation: choosing a path through T-bet. Immunity. 2007;27(2):180–2.
63. Gordon SM, Chaix J, Rupp LJ, Wu J, Madera S, Sun JC, Reiner SL, et al. The transcription factors T-bet and Eomes control key checkpoints of natural killer cell maturation. Immunity. 2012;36(1):55–67.
64. Samten B, Townsend JC, Weis SE, Bhoumik A, Klucar P, Shams H, Barnes PF. CREB, ATF, and AP-1 transcription factors regulate IFN-γ secretion by human T cells in response to mycobacterial antigen. J Immunol. 2008;181(3):2056–64.
65. MaciaÂn F, Lâopez-Rodrâiguez C, Rao A. Partners in transcription: NFAT and AP-1. Oncogene. 2001;20:2476–89.
66. Hatton RD, Harrington LE, Luther RJ, Wakefield T, Janowski KM, Oliver JR, Weaver CT, et al. A distal conserved sequence element controls Ifng gene expression by T cells and NK cells. Immunity. 2006;25(5):717–29.
67. Sica A, Dorman L, Viggiano V, Cippitelli M, Ghosh P, Rice N, Young HA. Interaction of NF-kB and NFAT with the interferon-γ promoter. J Biol Chem. 1997;272(48):30412–20.
68. Wang L, Zhu J, Shan S, Qin Y, Kong Y, Liu J, Xie Y, et al. Repression of interferon γ expression in T cells by Prospero-related homeobox protein. Cell Res. 2008;18(9):911–20.
69. Ricote M, Li AC, Willson TM, Kelly CJ, Glass CK. The peroxisome proliferator-activated receptor-γ is a negative regulator of macrophage activation. Nature. 1998;391(6662):79–82.
70. Cunard R, Eto Y, Muljadi JT, Glass CK, Kelly CJ, Ricote M. Repression of IFN-gamma expression by peroxisome proliferator-activated receptor gamma. J Immunol. 2004;172:7530–6.
71. Waite KJ, Floyd ZE, Arbour-Reily P, Stephens JM. Interferon-γ-induced regulation of peroxisome proliferator-activated receptor γ and STATs in adipocytes. J Biol Chem. 2001;276(10):7062–8.
72. Dunn SE, Ousman SS, Sobel RA, Zuniga L, Baranzini SE, Youssef S, Steinman L, et al. Peroxisome proliferator–activated receptor (PPAR) α expression in T cells mediates gender differences in development of T cell–mediated autoimmunity. J Exp Med. 2007;204(2):321–30.

73. Lin JT, Martin SL, Xia L, Gorham JD. TGF-β1 uses distinct mechanisms to inhibit IFN-γ expression in CD4+ T cells at priming and at recall: differential involvement of Stat4 and T-bet. J Immunol. 2005;174(10):5950–8.
74. Yu J, Wei M, Becknell B, Trotta R, Liu S, Boyd Z, Mao H, et al. Pro-and antiinflammatory cytokine signaling: reciprocal antagonism regulates interferon-gamma production by human natural killer cells. Immunity. 2006;24(5):575–90.
75. Savignac M, Pintado B, Gutierrez-Adan A, Palczewska M, Mellström B, Naranjo JR. Transcriptional repressor DREAM regulates T-lymphocyte proliferation and cytokine gene expression. EMBO J. 2005;24(20):3555–64.
76. Yagi R, Zhu J, Paul WE. An updated view on transcription factor GATA3-mediated regulation of Th1 and Th2 cell differentiation. Int Immunol. 2011;23(7):415–20.
77. Kim K, Kim N, Lee GR. Transcription factors Oct-1 and GATA-3 cooperatively regulate Th2 cytokine gene expression via the RHS5 within the Th2 locus control region. PLoS ONE. 2016;11(2), e0148576.
78. Kaminuma O, Kitamura F, Kitamura N, Miyagishi M, Taira K, Yamamoto K, Miyatake S, et al. GATA-3 suppresses IFN-γ promoter activity independently of binding to cis-regulatory elements. FEBS Lett. 2004;570(1):63–8.
79. Djuretic IM, Levanon D, Negreanu V, Groner Y, Rao A, Ansel KM. Transcription factors T-bet and Runx3 cooperate to activate Ifng and silence Il4 in T helper type 1 cells. Nat Immunol. 2007;8(2):145–53.
80. Penix L, Weaver WM, Pang Y, Young HA, Wilson CB. Two essential regulatory elements in the human interferon gamma promoter confer activation specific expression in T cells. J Exp Med. 1993;178(5):1483–96.
81. Ye J, Cippitelli M, Dorman L, Ortaldo JR, Young HA. The nuclear factor YY1 suppresses the human gamma interferon promoter through two mechanisms: inhibition of AP1 binding and activation of a silencer element. Mol Cell Biol. 1996;16(9):4744–53.
82. Hodge DL, Berthet C, Coppola V, Kastenmüller W, Buschman MD, Schaughency PM, Ortaldo JR, et al. IFN-gamma AU-rich element removal promotes chronic IFN-gamma expression and autoimmunity in mice. J Autoimmun. 2014;53:33–45.
83. Ben-Asouli Y, Banai Y, Pel-Or Y, Shir A, Kaempfer R. Human interferon-gamma mRNA autoregulates its translation through a pseudoknot that activates the interferon-inducible protein kinase PKR. Cell. 2002;108(2):221–32.
84. Williams BR. PKR; a sentinel kinase for cellular stress. Oncogene. 1999;18(45):6112–20.
85. Sareneva T, Cantell K, Pyhälä L, Pirhonen J, Julkunen I. Effect of carbohydrates on the pharmacokinetics of human interferon-γ. J Interf Res. 1993;13(4):267–9.
86. Pesu M, Muul L, Kanno Y, O'Shea JJ. Proprotein convertase furin is preferentially expressed in T helper 1 cells and regulates interferon gamma. Blood. 2006;108:983–5.

Chapter 2
The Interleukin-1 Family

Amir S. Yazdi and Kamran Ghoreschi

Abstract The interleukin-1 (IL-1) family consists of several pro- or anti-inflammatory proteins, with pro-inflammatory IL-1β being its best characterized member. IL-1β is one of the most prominent mediators of inflammation resulting in fever and immune activation via binding to IL-1 receptor 1. Due to its potency, its secretion is tightly regulated. First the transcription of the biologically inactive pro-form is induced by TLR activation, TNF, or IL-1 receptor activation by mature IL-1α or IL-1β. For the secretion of IL-1β, inflammasome activation as second stimulus is needed. Inflammasomes are cytosolic protein complexes whose activation results in the maturation of inflammatory caspases such as caspase-1. Caspase-1 then cleaves the inactive pro-IL-1β into its mature form which is then being secreted. While IL-1α and IL-1β are considered pro-inflammatory, IL-1Ra as a naturally occurring receptor antagonist acts as an inhibitor on IL-1 receptor signaling. Further members of the IL-1 family, such as IL-18, IL-33, or IL-36, are even involved in T-helper-cell differentiation and will also be discussed in this chapter.

Keywords Inflammasomes • Caspases • Autoinflammation • Nod-like receptor • IL-18 • IL-33 • IL-36

The innate immune system is the first line of the bodies' defense. Via unspecific, nondirected phagocytosis, complement activation, or the activation of innate immune receptors, the innate immune response is initiated. The innate immune system is highly conserved throughout evolution; even plants carry the most important immune receptors. These receptors do not detect particular peptides or antigens, but they rather react on structures, which are conserved in many pathogens or sterile exogenous or endogenous danger signals. Patterns of microbial origin are termed pathogen-associated molecular patterns (PAMPS), while DAMPs comprise danger-associated molecular patterns of nonmicrobial origin. After initial innate immune activation, adaptive immunity stars a more directed, specific response. Through the

A.S. Yazdi (✉) • K. Ghoreschi
Department of Dermatology, University of Tuebingen,
Liebermeisterstr. 25, 72076 Tuebingen, Germany
e-mail: amir.yazdi@med.uni-tuebingen.de

X. Ma (ed.), *Regulation of Cytokine Gene Expression in Immunity and Diseases*,
Advances in Experimental Medicine and Biology 941,
DOI 10.1007/978-94-024-0921-5_2

generation of target-specific T and B lymphocytes, an immunological memory can be established, which enables the organism an antigen-specific response upon second encounter. Supporting innate immune activation to generate an adaptive immune response is used upon other mechanisms in vaccination and immunization by the use of adjuvants [23], clearly demonstrating the close link and immunological overlap between innate and adaptive immune immunity.

The initial immune response is mediated both by resident cells and by invading immune cells. Not only resident tissue macrophages but also epithelial cells carry a plethora of immune receptors and are capable of producing innate cytokines to attract or support invading macrophages or neutrophils via cytokine or chemokine production [1].

Interleukin (IL)-1α, IL-1β, IL-6, IL-18, IL-33, and TNF [43] are the most prominent cytokines associated with the innate immune response. These cytokines can act both locally and systemically; therefore, their production and secretion need to be tightly regulated.

2.1 History of IL-1

Menkin and Beeson studied in the 1940s the pathogenesis of fever. Their studies showed that stimulated rabbit peritoneal exudate cells could release factors, which were able to induce fever. Thirty years later Waksman and several other investigators described a soluble factor which was released from myeloid cells to activate lymphocytes [10]. The term lymphocyte activation factor was coined, which was renamed interleukin in 1979. This term was chosen as the two first interleukins, namely, IL-1 and IL-2, were the first soluble mediators to be released from leukocytes, such as macrophages secreting IL-1 and lymphocytes secreting IL-2. In 1984 human and murine IL-1 cDNA were first isolated [3]. One year later IL-1α and IL-1β were first distinguished as different proteins [28].

2.2 The IL-1 Family

Currently, the IL-1 family consists of 11 members (IL-1α, IL-1β, IL-1RA, IL-18, IL-36Ra, IL-36α, IL-37, IL-36β, IL-36γ, IL-38, and IL-33). As depicted in Table 2.1, these cytokines officially are termed IL-1F1 (IL-1 family member) to IL-1F11 due to their first discovery, but more practically, the abovementioned nomenclature was reestablished [8]. In this chapter the best characterized IL-1 family members and their receptors will be introduced.

Table 2.1 The interleukin-1 family

Name	Synonym	Receptor	Active form
Pro-inflammatory			
IL-1α	IL-1F1	IL-1R1	Full length and cleaved
IL-1β	IL-F2	IL-1R1	Cleaved
IL-18	IL-F4	IL-18Rα	Cleaved
IL-36α	IL-F6	IL-36R	Cleaved
IL-36β	IL-F8	IL-36R	Cleaved
IL-36γ	IL-F9	IL-36R	Cleaved
IL-33	IL-F11	ST2	Full length and cleaved
Anti-inflammatory			
IL-1Ra	IL-F3	IL-1R1	No cleavage
IL-36Ra	IL-F5		Cleaved
IL-37	IL-F7	IL-18Rα	Only cleaved in the nucleus, full length and cleaved via receptor
IL-38	IL-F10	IL-36R? IL1R1?	Unclear

2.2.1 *IL-1α (IL-1F1) and IL-1β (IL-1F2)*

IL-1α and IL-1β are synthesized as precursor proteins, which are then proteolytically cleaved. In opposition to IL-1β, full-length IL-1α is already biologically active. Ca^{++}-dependent proteases such as calpains cleave IL-1α [26]; however, the reason and benefit of this cleavage are still debated.

IL-1β is yet much better characterized. Together with IL-1α it is expelled from the cell very early in the course of inflammation, inducing first localized inflammation and eventually systemic inflammation and fever. Both IL-1α and IL-1β share IL-1R1 as their common receptor. IL-1 receptor activation is a key event in the induction of fever. Dinarello et al. injected several nanograms of IL-1 intravenously into healthy humans, which responded with fever, leukocytosis, and hypotension [9]. In the last decade, however, we learned that IL-1β does not only mediate fever but is involved in many different organ systems, such as the vascular system in tumor angiogenesis [11], in rheumatoid arthritis, and in osteoarthritis [24] or even in diabetes, where IL-1 blockade serves as efficient treatment option [27].

Due to its high biological power and efficacy, transcription, synthesis, and secretion of both IL-1α and IL-1β are tightly regulated. While in particular resident cells such as keratinocytes IL-1α and IL-1β are present both on cDNA and on protein level at baseline [14, 15, 48], both cytokines need to be transcriptionally induced in myeloid cells such as macrophages, monocytes, and dendritic cells [16].

Inducers of pro-IL-1β and full-length IL-1α are TNF, TLR activation, phorbol-12-myristate-13-acetate (PMA), or IL-1 receptor activation by mature IL-1β or IL-1α themselves [5]. As IL-1α is present in the cell in a fully bioactive form, necrotic cell death results in the liberation of full-length IL-1α, which serves as an alarmin [7]. Alarmins such as S100A proteins or HMGB1 are preformed proteins, which are expelled from the cell after disruption of the cell membrane to induce inflammation. IL-1β cannot act as an alarmin as it needs to be processed to be active. As early as 1992, Thornberry identified caspase-1 as interleukin-1-converting enzyme (ICE) [44]. The exact mechanism of IL-1 activation and caspase-1 activation was elucidated 10 years later. Fabio Martinon and Jurg Tschopp introduced inflammasomes as cytosolic protein complexes responsible for the activation of inflammatory caspases [32]. These caspases can once being activated then cleave IL-1β and IL-18 and release IL-1α, IL-1β, and also IL-18 as pro-inflammatory mediators of innate immunity.

Inflammasomes structurally consist of a NOD-like receptor (NLR) protein such as NLRP1, NLRP3, or NLRC4 and the adaptor protein ASC, via which caspase-1 is recruited to the complex. NLRs can identify cytosolic danger signals. Particular NLR proteins and the non-NLR protein absent in melanoma 2 (AIM2) form inflammasomes upon activation. The NLR and AIM2, respectively, recruit ASC and caspase-1 after activation, most probably due to conformational changes enabling oligomerization. Most NLRs and AIM2 do not carry CARD and cannot directly bind to caspase-1. Therefore the pyrin domain (PYD) of ASC binds via PYD-PYD interactions to the NLR. ASC then via CARD-CARD interaction recruits caspase-1 to the complex leading to the formation of caspase-1 dimers or even oligomers. Caspases are known to autoactivate themselves. As not only caspase-1 itself but pro-IL-1β and pro-IL-18 are substrates of caspase-1, activated caspase-1 cleaves these mediators into their biologically active forms [32], which are then secreted.

The exact mechanism of protein secretion is still cryptic; however, Keller et al. identified caspase-1 being not only important in cytokine maturation but also in unconventional protein secretion [25]. In analogy to IL-1β, IL-1α also needs to be induced in myeloid cells. We could show recently that TLR ligation, exposure to heat-killed *Candida albicans*, and IL-1R1 activation serve as inducers of IL-1α cDNA expression [20]. This first signal does not only induce IL-1 cytokines but also NLR proteins such as NLRP3 [21], while ASC and caspase-1 are not transcriptionally regulated. For protein secretion, this first signal does not suffice. Here a second activator of the inflammasome is needed to induce the secretion of both IL-1α and IL-1β. Intrinsic activators of NLPR3 are ATP [29] and uric acid crystals which cause gout [33]. Uric acid crystals and ATP are released from dying or damaged cells and are important mediators of the inflammatory potential of necrotic cell death. Further important activators are bacterial toxins [29], UVB exposure [14], or metallic nanoparticles [49]. Besides in gout, the NLRP3 inflammasome seems to be crucial in further metabolic diseases. Cholesterol in crystals causing arteriosclerosis (Duewell et al. 2010) and putatively glucose [51] in diabetes mellitus also lead to IL-1 secretion in an NLRP3-dependent fashion. Vadanmagsar therefore considers NLRP3 as key sensor of metabolic stress [47].

Activators of NLRP3; NLRP1, such as anthrax-lethal toxin; or cytosolic dsDNA for AIM2 alone are not able to induce secretion of any IL-1 cytokine. For both cytokines, in myeloid cells two signals are needed; only the combination of transcriptional activation via TLR activation and consecutive inflammasome activation leads to protein secretion. After this two-step activation, we always detected co-secretion of both IL-1α and IL-1β after activation of all other inflammasomes tested (NLRP1, AIM2, and IPAF/NLRC4) [20]. Depending on the stimulus, either the full length or the cleaved form of IL-1α is secreted, but the functional relevance of processing is still unclear.

One important difference in IL-1α and IL-1β processing is the dependency of caspase-1. IL-1β secretion entirely depends on active caspase-1, no matter which inflammasome is activated. IL-1α secretion on the other hand demands dependency on the activator caspase-1 or not. Nigericin, ATP, *C. albicans*, *Salmonella*, dsDNA, or anthrax-lethal toxin results in a caspase-1-dependent secretion of IL-1α, while caspase-1-deficient cells are still able to secrete significant amounts of IL-1α after uric acid, alum, or *Clostridium* toxin B exposure. NLRP3 activation is regulated by the depletion of reactive oxygen species (ROS) [12, 52] and a diminished potassium gradient from intracellular to extracellular potassium [38]. Only the NLRP3 and caspase-1-dependent activators impair IL-1α secretion after increasing extracellular potassium concentration or scavenging ROS [20], hinting toward a caspase-1-dependent and caspase-1-independent but calcium-dependent route of IL-1α processing and secretion. The calcium-ionophore ionomycin, for instance, only leads to a secretion and cleavage of IL-1α, but not of caspase-1 or IL-1β.

2.3 IL-1 Receptors

After release from the cell, both IL-1α and IL-1β bind to IL-1 receptor 1 (IL-1R1). IL-1R1 belongs to the large family of Toll-like/IL-1 receptor (TIR) superfamily receptors. Similar to Toll-like receptors (TLRs), IL-1R1 signals via the adaptor protein MyD88 then lead to the activation of prostaglandins and the transcription factor NFkB to initiate inflammation. IL-1 receptor 2 (IL-1R2) is a soluble or membrane-bound receptor and was initially considered as endogenous antagonist of IL-1R1, therefore inhibiting IL-1 signaling [40]. Very recently, Zheng et al. identified an intracellular form of IL-1R2. This form binds full-length IL-1α, but not IL-1β and prevents the cleavage of IL-1α. Once caspase-1 is activated, IL-1R2 is released from IL-1α. IL-1α can then be cleaved and secreted [50]. IL-1R2 hereby seems to determine the different activation of IL-1α in necrosis versus inflammasome activation, explaining the partial dependence of IL-1α secretion on functional inflammasomes [17, 20]. IL-1 in concert with other cytokines like IL-6, IL-23, and TGF-β is also critically implicated in the generation of T helper 17 cells (Th17) [19], which are important for the defense of extracellular pathogens and in the pathogenesis of inflammatory autoimmune diseases such as psoriasis and rheumatoid arthritis [18].

2.3.1 IL-18 (IL-1F4)

Identical to IL-1β IL-18 needs to be cleaved for activation. This cleavage is mediated by inflammatory caspases such as caspase-1, which converts inactive full-length Il-18 to its 17 kDa mature form to be secreted. In opposition to IL-1β, IL-18 gene expression does not have to be induced. IL-18 cDNA is present in myeloid cells, mouse spleen, epithelial cells, and many epithelial cells such as keratinocytes [37]. A TLR-mediated induction of gene expression is therefore not needed. After being activated by inflammasomes and inflammatory caspases, IL-18 is unconventionally secreted mainly from myeloid cells. After binding to its own receptor, the IL-18 receptor, which similar to IL-1R1 carries a TIR domain, IL-18 induces an increase in interferon (IFN)-γ production. In autoimmune diseases provoked by interferons, such as lupus erythematosus, psoriasis, or inflammatory bowel disease, IL-18 is locally increased in inflamed tissue [11]. The IFN production promotes the activation of T lymphocytes toward a Th1 phenotype [13].

Interestingly, IL-18-deficient mice are not immunocompromised, but they are more prone to metabolic syndrome with hyperlipidemia, diabetes, and arteriosclerosis, supporting the relevance of IL-1 family members as key proteins in metabolism [35].

2.4 IL-33 (IL-1F11)

While inflammatory caspases are known to proteolytically activate IL-1β and IL-18, IL-33 is not activated by caspase-1. IL-33 is constitutively expressed in certain cell types, such as endothelial cells [34], and can be induced in other tissues by TLR activation [39], similar to IL-1α and IL-1β. In analogy to IL-1α, IL-33 is a dual-function cytokine being able to shuttle from the cytosol to the nucleus. As ST-2, the membrane-bound receptor of IL-33, can be activated by full-length IL-33 present in many cell types, IL-33 is suggested to act as an alarmin. The function of the proteolytic processing of the cytokine via calpains, cathepsins, or apoptotic caspase-3 and caspase-7 as reviewed in [31] is still cryptic. Functionally, intranuclear IL-33 can bind to chromatin [6] and to NFkB and modulates its function [2]. Once being released from the cell, IL-33 can bind to ST-2. Originally, this binding was associated with a T-helper-cell-2 (Th2) response [42]. This year Guo et al. demonstrated the dependency of eosinophilic inflammation after helminth infection on IL-33 [22], while Schiering et al. postulate that the balance between IL-23 and IL-33 determines regulatory T-cell functions in the intestine [41]. In conclusion, both the processing and the function of IL-33 remain partially elusive.

2.5 IL-36 (IL-1F5, IL-1F6, IL-1F8, IL-1F9) and IL-37 (IL-1F7)

The IL-1 cytokines IL-36α, IL-36β, and IL-36γ and IL-36Ra belong to the recent group of IL-36 cytokines. IL-36Ra is the receptor antagonist and therefore anti-inflammatory, while the other three cytokines exert pro-inflammatory properties. Similar to IL-18 and IL-1β, IL-36 must be cleaved to be activated [46], but the protease responsible for its processing has not yet been identified. The highest expression of IL-36 mRNA is in keratinocytes; therefore, IL-36 might impact psoriasis [30] via the activation of dendritic cells [45].

IL-37 can be a ligand for the IL-18 receptor being an inhibitor protein of IL-18 [36]. Similar to IL-33 and IL-1α, IL-37 can translocate to the nucleus and bind to a membrane-bound receptor. As for other IL-1 family members, caspase-1 is of importance, but not for the release of IL-37, rather for the nuclear translocation of intracellular IL-37 [4].

2.6 Conclusion

In summary, cytokines of the IL-1 family are key players in innate immunity. Their expression, processing, and secretion are tightly regulated. Inflammatory caspases, which can be activated by inflammasomes, are of central importance in many IL-1 members. Although part of the innate immune response, IL-1 cytokines also impact T-cell functions, therefore being important linkers between early innate immune reactions and the generation of an adaptive immune response.

Acknowledgments KG and ASY are supported by the Deutsche Forschungsgemeinschaft (DFG) SFB685 and SFB/TRR 156 to KG and ASY.

References

1. Akira S, Uematsu S, Takeuchi O. Pathogen recognition and innate immunity. Cell. 2006;124:783–801.
2. Ali S, Mohs A, Thomas M, et al. The dual function cytokine IL-33 interacts with the transcription factor NF-kappaB to dampen NF-kappaB-stimulated gene transcription. J Immunol. 2011;187:1609–16.
3. Auron P, Webb AC, Rosenwasser LJ, et al. Nucleotide sequence of human monocyte interleukin 1 precursor cDNA. Proc Natl Acad Sci U S A. 1984;81:7907–11.
4. Bulau AM, Nold MF, Li S, et al. Role of caspase-1 in nuclear translocation of IL-37, release of the cytokine, and IL-37 inhibition of innate immune responses. Proc Natl Acad Sci U S A. 2014;111:2650–5.
5. Burns K, Martinon F, Tschopp J. New insights into the mechanism of IL-1beta maturation. Curr Opin Immunol. 2003;15:26–30.

6. Carriere V, Roussel L, Ortega N, et al. IL-33, the IL-1-like cytokine ligand for ST2 receptor, is a chromatin-associated nuclear factor in vivo. Proc Natl Acad Sci U S A. 2007;104:282–7.
7. Chen CJ, Kono H, Golenbock D, et al. Identification of a key pathway required for the sterile inflammatory response triggered by dying cells. Nat Med. 2007;13:851–6.
8. Dinarello C, Arend W, Sims J et al. IL-1 family nomenclature. Nat Immunol. 2010; 11:973.
9. Dinarello CA. Biologic basis for interleukin-1 in disease. Blood. 1996;87:2095–147.
10. Dinarello CA. The IL-1 family and inflammatory diseases. Clin Exp Rheumatol. 2002;20:S1–13.
11. Dinarello CA. Immunological and inflammatory functions of the interleukin-1 family. Annu Rev Immunol. 2009;27:519–50.
12. Dostert C, Petrilli V, Van Bruggen R, et al. Innate immune activation through Nalp3 inflammasome sensing of asbestos and silica. Science. 2008;320:674–7.
13. Fantuzzi G, Puren AJ, Harding MW, et al. Interleukin-18 regulation of interferon gamma production and cell proliferation as shown in interleukin-1beta-converting enzyme (caspase-1)-deficient mice. Blood. 1998;91:2118–25.
14. Feldmeyer L, Keller M, Niklaus G, et al. The inflammasome mediates UVB-induced activation and secretion of interleukin-1beta by keratinocytes. Curr Biol. 2007;17:1140–5.
15. Feldmeyer L, Werner S, Le F, et al. Interleukin-1, inflammasomes and the skin. Eur J Cell Biol. 2010;89:638–44.
16. Fenton MJ. Review: transcriptional and post-transcriptional regulation of interleukin 1 gene expression. Int J Immunopharmacol. 1992;14:401–11.
17. Fettelschoss A, Kistowska M, Leibundgut-Landmann S, et al. Inflammasome activation and IL-1beta target IL-1alpha for secretion as opposed to surface expression. Proc Natl Acad Sci U S A. 2011;108:18055–60.
18. Ghoreschi K, Laurence A, Yang X, et al. T helper 17 cell heterogeneity and pathogenicity in autoimmune disease. Trends Immunol. 2011;32:395–401.
19. Ghoreschi K, Laurence A, Yang X, et al. Generation of pathogenic T(H)17 cells in the absence of TGF-beta signalling. Nature. 2010;467:967–71.
20. Gross O, Yazdi AS, Thomas CJ, et al. Inflammasome activators induce interleukin-1alpha secretion via distinct pathways with differential requirement for the protease function of caspase-1. Immunity. 2012;36:388–400.
21. Guarda G, Zenger M, Yazdi AS, et al. Differential expression of NLRP3 among hematopoietic cells. J Immunol. 2011;186:2529–34.
22. Guo L, Huang Y, Chen X, et al. Innate immunological function of TH2 cells in vivo. Nat Immunol. 2015;16:1051–9.
23. Janeway Jr CA, Medzhitov R. Innate immune recognition. Annu Rev Immunol. 2002;20:197–216.
24. Kapoor M, Martel-Pelletier J, Lajeunesse D, et al. Role of proinflammatory cytokines in the pathophysiology of osteoarthritis. Nat Rev Rheumatol. 2011;7:33–42.
25. Keller M, Ruegg A, Werner S, et al. Active caspase-1 is a regulator of unconventional protein secretion. Cell. 2008;132:818–31.
26. Kobayashi Y, Yamamoto K, Saido T, et al. Identification of calcium-activated neutral protease as a processing enzyme of human interleukin 1 alpha. Proc Natl Acad Sci U S A. 1990;87:5548–52.
27. Larsen CM, Faulenbach M, Vaag A, et al. Interleukin-1-receptor antagonist in type 2 diabetes mellitus. N Engl J Med. 2007;356:1517–26.
28. March CJ, Mosley B, Larsen A, et al. Cloning, sequence and expression of two distinct human interleukin-1 complementary DNAs. Nature. 1985;315:641–7.
29. Mariathasan S, Weiss DS, Newton K, et al. Cryopyrin activates the inflammasome in response to toxins and ATP. Nature. 2006;440:228–32.
30. Marrakchi S, Guigue P, Renshaw BR, et al. Interleukin-36-receptor antagonist deficiency and generalized pustular psoriasis. N Engl J Med. 2011;365:620–8.

31. Martin MU. Special aspects of interleukin-33 and the IL-33 receptor complex. Semin Immunol. 2013;25:449–57.
32. Martinon F, Burns K, Tschopp J. The inflammasome: a molecular platform triggering activation of inflammatory caspases and processing of proIL-beta. Mol Cell. 2002;10:417–26.
33. Martinon F, Petrilli V, Mayor A, et al. Gout-associated uric acid crystals activate the NALP3 inflammasome. Nature. 2006;440:237–41.
34. Moussion C, Ortega N, Girard JP. The IL-1-like cytokine IL-33 is constitutively expressed in the nucleus of endothelial cells and epithelial cells in vivo: a novel 'alarmin'? PLoS ONE. 2008;3, e3331.
35. Netea MG, Joosten LA, Lewis E, et al. Deficiency of interleukin-18 in mice leads to hyperphagia, obesity and insulin resistance. Nat Med. 2006;12:650–6.
36. Nold-Petry CA, Lo CY, Rudloff I, et al. IL-37 requires the receptors IL-18Ralpha and IL-1R8 (SIGIRR) to carry out its multifaceted anti-inflammatory program upon innate signal transduction. Nat Immunol. 2015;16:354–65.
37. Novick D, Kim S, Kaplanski G et al. Interleukin-18, more than a Th1 cytokine. Semin Immunol. 2013; 25:439–48.
38. Petrilli V, Papin S, Dostert C, et al. Activation of the NALP3 inflammasome is triggered by low intracellular potassium concentration. Cell Death Differ. 2007;14:1583–9.
39. Polumuri Sk, Jayakar Gg, Shirey Ka et al. Transcriptional regulation of murine IL-33 by TLR and non-TLR agonists. J Immunol. 2009; 189:50–60.
40. Re F, Muzio M, De Rossi M, et al. The type II "receptor" as a decoy target for interleukin 1 in polymorphonuclear leukocytes: characterization of induction by dexamethasone and ligand binding properties of the released decoy receptor. J Exp Med. 1994;179:739–43.
41. Schiering C, Krausgruber T, Chomka A, et al. The alarmin IL-33 promotes regulatory T-cell function in the intestine. Nature. 2014;513:564–8.
42. Schmitz J, Owyang A, Oldham E, et al. IL-33, an interleukin-1-like cytokine that signals via the IL-1 receptor-related protein ST2 and induces T helper type 2-associated cytokines. Immunity. 2005;23:479–90.
43. Schroder K, Tschopp J. The inflammasomes. Cell. 2010;140:821–32.
44. Thornberry NA, Bull H, Calaycay JR, et al. A novel heterodimeric cysteine protease is required for interleukin-1 beta processing in monocytes. Nature. 1992;356:768–74.
45. Tortola L, Rosenwald E, Abel B, et al. Psoriasiform dermatitis is driven by IL-36-mediated DC-keratinocyte crosstalk. J Clin Invest. 2012;122:3965–76.
46. Towne JE, Renshaw BR, Douangpanya J, et al. Interleukin-36 (IL-36) ligands require processing for full agonist (IL-36alpha, IL-36beta, and IL-36gamma) or antagonist (IL-36Ra) activity. J Biol Chem. 2011;286:42594–602.
47. Vandanmagsar B, Youm Y, Ravussin A, et al. The NLRP3 inflammasome instigates obesity-induced inflammation and insulin resistance. Nat Med. 2011;17:179–88.
48. Yazdi AS, Drexler SK, Tschopp J. The role of the inflammasome in nonmyeloid cells. J Clin Immunol. 2010;30:623–7.
49. Yazdi AS, Guarda G, Riteau N, et al. Nanoparticles activate the NLR pyrin domain containing 3 (Nlrp3) inflammasome and cause pulmonary inflammation through release of IL-1alpha and IL-1beta. Proc Natl Acad Sci U S A. 2010;107:19449–54.
50. Zheng Y, Humphry M, Maguire Jj et al. Intracellular interleukin-1 receptor 2 binding prevents cleavage and activity of interleukin-1alpha, controlling necrosis-induced sterile inflammation. Immunity. 2013; 38:285–95.
51. Zhou R, Tardivel A, Thorens B, et al. Thioredoxin-interacting protein links oxidative stress to inflammasome activation. Nat Immunol. 2010;11:136–40.
52. Zhou R, Yazdi AS, Menu P, et al. A role for mitochondria in NLRP3 inflammasome activation. Nature. 2011;469:221–5.

Chapter 3
Regulation of IL-4 Expression in Immunity and Diseases

I-Cheng Ho and Shi-Chuen Miaw

Abstract IL-4 was first identified as a T cell-derived growth factor for B cells. Studies over the past several decades have markedly expanded our understanding of its cellular sources and function. In addition to T cells, IL-4 is produced by innate lymphocytes, such as NTK cells, and myeloid cells, such as basophils and mast cells. It is a signature cytokine of type 2 immune response but also has a nonimmune function. Its expression is tightly regulated at several levels, including signaling pathways, transcription factors, epigenetic modifications, microRNA, and long non-coding RNA. This chapter will review in detail the molecular mechanism regulating the cell type-specific expression of IL-4 in physiological and pathological type 2 immune responses.

Keywords IL-4 • Cytokine • Type 2 immune response • Transcription • Epigenetic

3.1 IL-4-Producing Cells

IL-4 is a pleiotropic glycoprotein. It was first identified in 1982 as a T cell-derived soluble factor that stimulates the proliferation of B cells [1, 2]; its cDNA was later successfully cloned in 1986 [3, 4]. The cDNA of human IL-4 encodes 153 amino acid residues but yields a secreted protein of 129 amino acid residues. Depending on the addition of N-linked oligosaccharides, secreted IL-4 has a molecular weight of 15, 18, or 19 kDa. Its molecular structure features a globular hydrophobic core, held together by three disulfide bonds [5] and four alpha-helixes, which are arranged in a left-handed bundle with two overhand connections [6–9].

I.-C. Ho (✉)
Brigham and Women's Hospital, Harvard Medical School, Boston, MA, USA
e-mail: iho@partners.org

S.-C. Miaw
College of Medicine, National Taiwan University, Taipei, Taiwan

X. Ma (ed.), *Regulation of Cytokine Gene Expression in Immunity and Diseases*, Advances in Experimental Medicine and Biology 941,
DOI 10.1007/978-94-024-0921-5_3

IL-4 has become the signature cytokine of type 2 helper T (Th2) cells. After encountering antigens, naive CD4+ Th cells can differentiate into several distinct functional subsets [10], which are defined by the expression of the so-called master transcription factors and subset-specific cytokines. For example, Th1 cells express T-bet and produce IFN-γ; Th17 cells express RORγt and produce IL-17. Th2 cells are defined by the expression of the transcription factor GATA3 and the production of IL-4, IL-5, and IL-13, which are collectively called type 2 cytokines. Given their ability to expand into large numbers in response to antigens, Th2 cells are one of the major sources, if not the major source, of IL-4. Follicular helper Th (Tfh) cells [11], which are strategically located in the B cell follicles of lymph nodes, also produce a high level of IL-4 but little IL-5 or IL-13. Unlike Th2 cells, they express a low level of GATA3 but a high level of transcription factor Bcl6 [12–14], which is essential for their differentiation. A unique subset of T cells called NKT cells that recognize lipid antigens presented by CD1d molecules are also capable of producing IL-4. In addition, several types of innate immune cells, including mast cells, basophils, and eosinophils, are known as IL-4-producing cells.

The development of several strains of IL-4 reporter mice has markedly increased the sensitivity of detecting IL-4-producing cells in vivo and ex vivo [15–18]. Studies using the IL-4 reporter mice indicate that the ability to produce IL-4 protein is acquired through a two-step process [18]. In the first step, the *Il4* gene is rendered accessible through chromatin remodeling. Most of the IL-4-competent innate or innate-like cells, such as mast cells, basophils, and NKT cells, obtain this competence during their differentiation [19]. In contrast, naive Th cells require antigen priming under conditions favoring Th2 differentiation to open up their *Il4* locus [18]. This priming of naive Th cells takes place in the paracortex of draining lymph nodes. Once primed, the IL-4-competent Th cells quickly disseminate to peripheral tissues [20]. In the second step, IL-4-competent cells rapidly produce IL-4 protein in response to appropriate stimulation.

3.2 Function of IL-4

3.2.1 IL-4 Receptors and Signaling Pathways

IL-4 is capable of binding to two types of receptor [21–24]. The type I receptor is comprised of IL-4Rα and the common γ_c subunit, which is also a subunit of IL-2, IL-7, IL-9, IL-15, and IL-21 receptor complexes. The type II receptor consists of IL-4Rα and IL-13Rα1 subunits. The type I receptor binds exclusively to IL-4, whereas the type II receptor binds to both IL-4 and IL-13. Ligand binding to IL-4 receptors activates tyrosine kinases Jak1, Jak3, and Tyk2 through IL-4Rα, γ_c, and IL-13Rα1, respectively. Phosphorylation of the cytoplasmic domain of IL-4Rα results in the recruitment and phosphorylation of Stat6, a signaling and transcription factor [25]; thus, the majority of IL-4 function is carried out via Stat6. However, type I receptors can also activate the insulin receptor substrate (IRS) 2 and subsequently PI3K and Akt [26], eventually leading to the activation of mTOR [27]. Type I receptors are expressed mainly in hematopoietic cells and are nearly the only IL-4 receptors seen in T cells, basophils, mast cells, and mouse B cells. In contrast, the expression of type II receptors is limited to non-hematopoietic cells, which do not express the common γ_c subunit. The differential expression of IL-4Rα, γ_c, and IL-13Rα1 determines the sensitivity of each type of cell to IL-4 and IL-13 [23].

3.2.2 Immunological Function of IL-4

IL-4 is a potent growth and survival factor for B cells, whose effects are mediated by Stat6. It also promotes immunoglobulin isotype switching to IgE and IgG1 and augments the expression of CD23 and MHC class II in B cells [28]. Although mouse B cells express only IL-4 type I receptors, human B cells express both type I and type II receptors and can respond to both IL-4 and IL-13 [29, 30].

IL-4 also plays a critical role in regulating the function of macrophages. Two major functional subsets of macrophages have been identified [31, 32]. Classically activated macrophages (M1) are induced by IFN-γ and TLR ligands. M1 macrophages express a high level of inflammatory cytokines, such as IL-12, IL-23, and IL-1β, and nitric oxide (NO) and can contribute to inflammatory processes. In contrast, alternatively activated macrophages (M2) express arginase 1, mannose receptor, and chitinase, produce IL-10, and display anti-inflammatory activity. IL-4 and IL-13 are the main inducers of M2 macrophages [33, 34].

IL-4 is also the cytokine that drives the differentiation of Th2 cells, which secrete IL-4, thereby forming a positive feedback loop to reinforce Th2 differentiation. By activating Stat6 and inducing the expression of GATA3, IL-4 can also inhibit the differentiation of other Th subsets, such as Th1 and Th17 cells. The impact of the IL-4/Stat6/GATA3 axis on the differentiation of Th2 cells will be further elaborated later.

Although IL-4 is not required for the development of mast cells [35], it critically regulates their function [36]. IL-4 promotes the proliferation, survival, adhesion, and chemotaxis of mast cells. It augments the expression of FcεRI, allowing the binding of IgE and IgE-induced degranulation to take place [37, 38]. Its effect on mast cells is mediated mainly by Stat6, but also by several other signaling pathways including MEK and p38.

IL-4 along with GM-CSF has been used to drive the differentiation of dendritic cells from myeloid precursors in vitro. However, its effect on the survival, function, and trafficking of dendritic cells is complicated and varies according to experimental settings. IL-4 is not required for the generation and homeostasis of NK cells but can inhibit the activity of NK cells partly by suppressing their expression of NKG2D [39]. Innate CD8+ T cells are a recently discovered lineage of T cells. They are generated in the thymus, express transcription factor Eomes, display memory phenotype, and rapidly produce IFN-γ in response to TCR stimulation [40]. PLZF+

NKT cell-derived IL-4 is required for the differentiation of the innate CD8+ T cells through the induction of Eomes [41–44].

3.2.3 IL-4 and Type 2 Immunity

A type 2 immune response is characterized by the involvement of Th2 cells, mast cells, eosinophils, and group 2 innate lymphoid cells and by the production of IL-4, IL-5, IL-13 and IgE [45]. Type 2 immune responses defend against helminths but can also cause allergic inflammation such as asthma and atopic dermatitis. Exogenous IL-4 facilitates the clearance of helminths and induces airway hyperresponsiveness, serving as the key mediator of type 2 immune responses [46, 47]. Reciprocally, genetic ablation of IL-4 or blocking IL-4 attenuates the clearance of some helminths, such as *Trichuris muris* and *Heligmosomoides polygyrus*, and reduces airway inflammation in animal models of asthma [48–51]. It has become clear that IL-4 and IL-13 have unique and overlapping functions in type 2 immune responses given the differential expression of type I and type II IL-4 receptors. As T and B cells express type I receptors almost exclusively, IL-4 is responsible for the initiation of allergic response and the production of IgE. In contrast, IL-13 is more potent than IL-4 in eradicating *Nippostrongylus brasiliensis* and inducing hypertrophy of airway smooth muscles and hyperplasia of goblet cells [47, 52–54].

3.2.4 Nonimmunological Function of IL-4 and Type 2 Cytokines

IL-4 and type 2 cytokines can also act directly or indirectly on nonimmune cells. IL-4 and IL-13 act directly on lymphatic endothelial cells and inhibit their proliferation and tube formation in vitro as well as lymphangiogenesis in vivo [55, 56]. IL-4 derived from T cells can promote regrowth of neurons, which express IL-4Rα, in animal models of CNS injury [57].

Recent animal studies have also strongly suggested that IL-4/IL-13 is indispensible in coping with environmental stresses. In particular, IL-4/IL-13 is essential in combating hypothermia by promoting the differentiation of alternatively activated macrophages in brown adipose tissue. Cold exposure triggers sympathetic discharge, which induces the release of noradrenaline in mostly brown adipose tissue, resulting in the upregulation of thermogenic genes and lipolysis [58]. Interestingly, cold exposure is also associated with the upregulation of genes characteristic of alternatively activated macrophages, which are responsible for the browning of white adipose tissue, an event that is dependent on IL-4/IL-13 and Stat6 [59]. Indeed, mice deficient in IL-4/IL-13 or Stat6 are more susceptible to hypothermia in cold environments.

In addition to promoting the differentiation of alternatively activated macrophages in adipose tissue, IL-4 acts directly on adipose tissue and facilitates lipolysis [60]. By using IL-4/GFP reporter mice, Wu et al. identified eosinophils as the main source of IL-4 in adipose tissue [61]. Genetic ablation of eosinophils markedly attenuates the differentiation of alternatively activated macrophages and blunts the cold-induced expression of thermogenesis and the development of functional beige fat [61, 62]. Rao et al. further demonstrated that one of the signals inducing the production of IL-4 by adipose eosinophils is meteorin-like [63], a hormone secreted by adipose tissue upon cold exposure. These elegant studies establish a thermogenic circuit involving meteorin-like, eosinophils, IL-4/IL-13, and alternative activated macrophages to counteract hypothermia. How meteorin-like induces the expression of IL-4 in eosinophils is currently unknown.

3.3 Mechanisms Regulating the Cell Type-Specific Expression of IL-4

IL-4 has been a model gene for studying the molecular mechanisms regulating the expression of Th2 cell-specific genes. Earlier studies focusing on the *Il4* gene promoter have identified five binding sites for the NFAT/AP-1 complex, namely, P0, P1, P2, PRE-I/P4, and P5, forming the backbone of the *Il4* promoter (Fig. 3.1). This approximately 400 bp promoter also contains binding sites for other positive transcription factors, such as C/EBP, NF-kB, and IRF4, and negative transcription factors such as IRF1 and CIITA. These studies have been summarized in an elegant review and will not be reiterated here [64]. However, none of these *cis*-acting elements or cognate transcription factors can explain the cell type-specific expression of IL-4. Recent studies have identified additional *cis*-acting elements, signaling pathways, and transcription factors that are critical for Th2-specific expression of IL-4 (Fig. 3.2). These signaling pathways and transcription factors form several positive feedback loops to further enforce the expression of IL-4 and the differentiation of Th2 cells (Fig. 3.3). Epigenetic modifications further stabilize the ability of Th2 cells to express IL-4. However, as the expression of IL-4 is tightly linked to the differentiation of Th2 cells, it is often impossible to separate these two processes. In addition, most studies are conducted using Th cells and the conclusions thus generated are not necessarily applicable to other IL-4-producing cells. In the following sections, we will discuss the signaling pathways, trans-acting factors, *cis*-acting elements, and epigenetic modifications that are important for the expression of IL-4 in Th2 cells and highlight the difference between Th2 cells and other IL-4-producing cells.

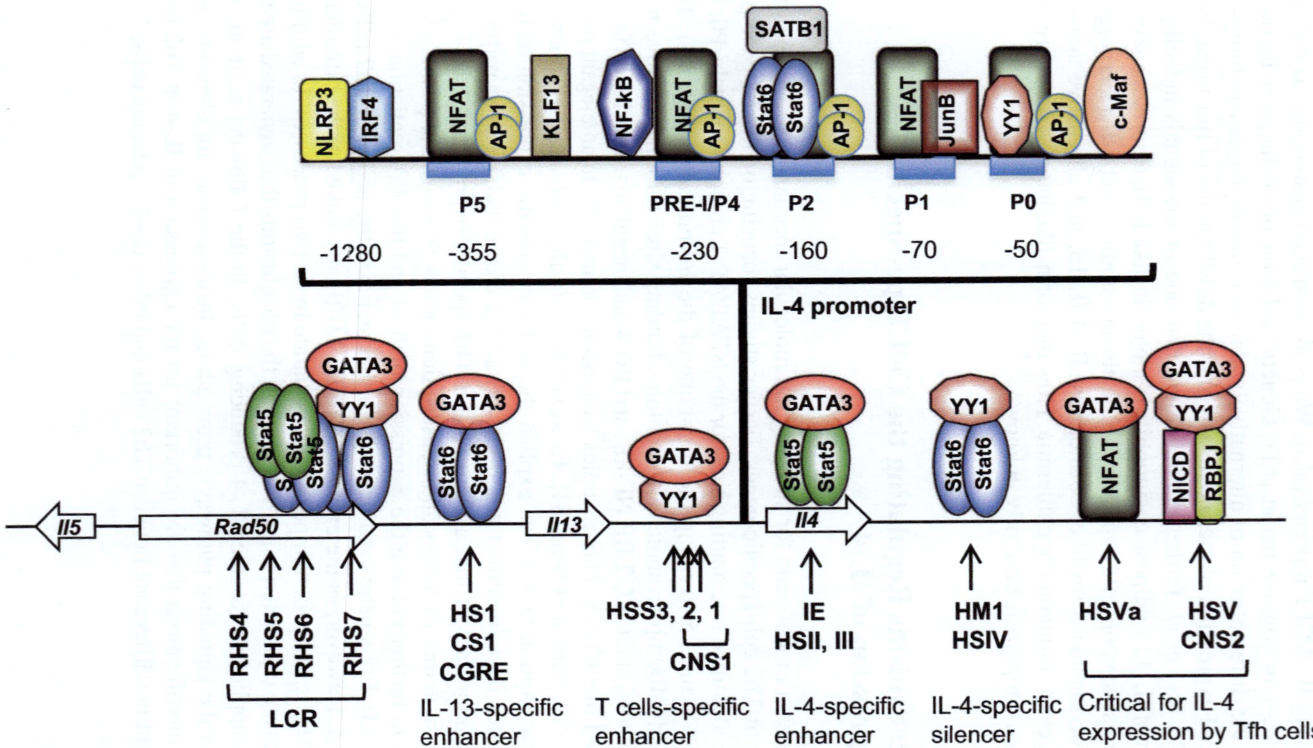

Fig. 3.1 *Cis*-acting elements and their cognate transcription factors within the proximal *Il4* promoter (the *top panel*) and the Th2 locus (the *bottom panel*) that are critical for the expression of IL-4 and the differentiation of Th2 cells. The diagrams are not on scale

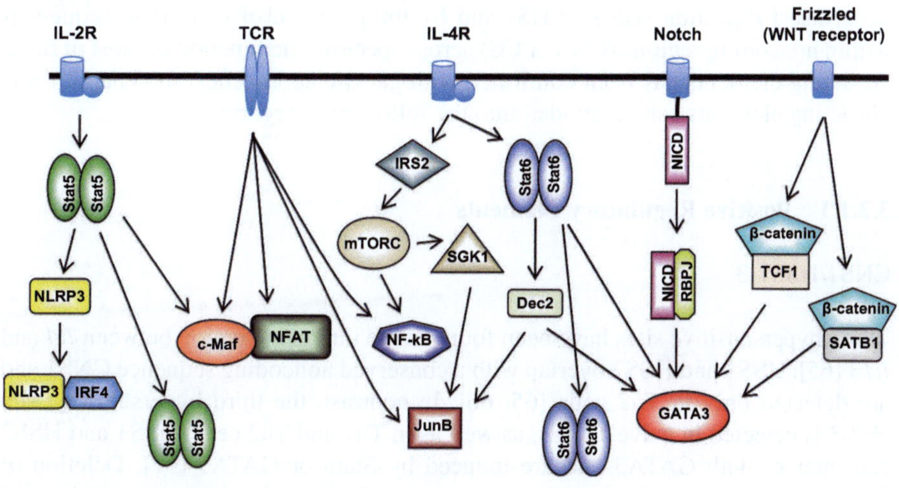

Fig. 3.2 Major signaling pathways and transcription factors that are critical for the expression of IL-4 and the differentiation of Th2 cells

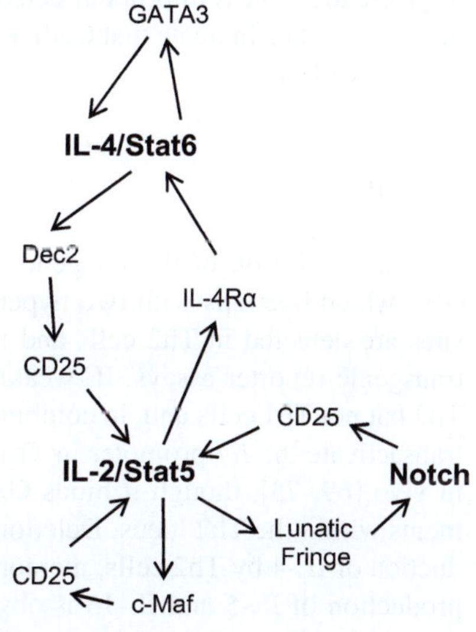

Fig. 3.3 Positive feedback circuits that enforce the differentiation of Th2 cells

3.3.1 *The* Il4 *Gene and Th2 Locus*

The *Il4* gene contains four exons and is located in a genomic locus at human chromosome 5 or mouse chromosome 11, tentatively called the Th2 locus, which also contains *Il13*, *Rad50*, and *Il5*. This arrangement is relatively conserved in mammals. As IL-4 is often co-expressed with IL-5 and IL-13, the expression of IL-4 is also subjected to regulation by additional *cis*-acting elements spreading across the Th2 locus (Fig. 3.1). These *cis*-acting elements were identified by their sensitivity to DNase I digestion (HSS or HS) and by the presence of conserved sequences within noncoding regions (CNS or CS) across species. The function of most of these *cis*-acting elements has been confirmed with genetic approaches. In general, these *cis*-acting elements can be divided into the following categories.

3.3.1.1 Positive Regulatory Elements

CNS1/HSS1-3

Three hypersensitive sites have been found at the intergenic region between *Il4* and *Il13* [65]. HSS1 and HSS2 overlap with a conserved noncoding sequence CNS1 and are detected only in Th2 cells [65, 66]. In contrast, the third hypersensitive site HSS3 is detected in naive T cells, as well as in Th1 and Th2 cells. HSS1 and HSS2 can interact with GATA3 and are induced by Stat6 or GATA3 [67]. Deletion of CNS1 from the mouse genome or a yeast artificial chromosome containing the human Th2 locus attenuates the production of not only IL-4 but also IL-5 and IL-13 in Th2 cells [66, 68, 69]. Interestingly, the production of IL-4 by mast cells in response to ionomycin is not affected by the deletion of CNS1 [68]. Taken together, the existing data indicate that CNS1/HSS1-2 is a T cell-specific coordinate enhancer of Th2 cytokines.

IE/HSII

The second intron of the *Il4* gene contains an enhancer called intronic enhancer (IE), which overlaps with two hypersensitive sites, HSII and HSIII [70, 71]. Both sites are detected in Th2 cells and mast cells but not in Th1 or NIH3T3 cells. In transgenic reporter assays, IE weakly enhances the activity of the *Il4* promoter in Th2 but not Th1 cells and, in combination with CNS1, allows exogenous GATA3 to transactivate the *Il4* promoter in Th1 cells [72]. HSII binds both Stat5 and GATA3 in vivo [69, 73], though it binds GATA3 most strongly among all *cis*-acting elements within the Th2 locus. Deletion of HSII leads to a profound defect in the production of IL-4 by Th2 cells, memory Th cells, and NKT cells, but no defect in the production of IL-5 and IL-13 is observed [69]. Deletion of HSII also renders differentiating Th1 cells unable to produce IL-4 in response to exogenous GATA3, but the GATA3-driven production of IL-5 and IL-13 is not affected. Mice missing HSII are resistant to allergic airway inflammation and mount only a weak antigen-specific IgE response after immunization. Taken together, these data suggest that IE/HSII is an IL-4-specific enhancer, but the role of HSIII remains to be determined.

CNS2/HSV and HSVa

Hypersensitive site HSV, located at the 3′ end of *Il4*, is detected only in Th2 cells and overlaps with the conserved noncoding sequence CNS2 [71]. Immediately 5′ of HSV is another hypersensitive site, HSVa, that is detected only in stimulated Th2 cells and binds NFAT and GATA3 [74]. In transient transfection assays, HSVa enhances the activity of the *Il4* promoter [74]. HSVa very likely dictates the probability of IL-4 expression among differentiated Th2 cells, along with the *Il4* promoter, by serving as the binding site of NFAT-containing protein complexes [75, 76]. HSV+HSVa augments *Il4* promoter activity in transgenic reporter assays, and its activity is dependent on RBPJ, a critical mediator of Notch signals, but not Stat6 [72, 77]. Deletion of HSV+HSVa or HSV results in modest reduction in the expression of IL-4 and IL-13 in in vitro differentiated Th2 cells [78–80]; a marked reduction in the production of IL-4 and, to a lesser degree, IL-13 is also observed in mast cells but not basophils [78, 81]. While deletion of HSV+HSVa or HSV results in a similar defect in the production of IL-4 and IL-13, impaired IL-5 expression is observed only in mice with a deletion of HSV+HSVa. These observations suggest that HSVa cannot fully compensate for the loss of HSV.

Additionally, recent studies suggest that HSV and HSVa play a more important role in Tfh than in Th2 cells. In transgenic reporter mice, HSV+HSVa is active in NKT cells and CD44+ Th cells, which express a low level of GATA3 and are very likely Tfh cells, but not in differentiating Th2 cells. Deletion of HSV+HSVa affects the expression of IL-4 more profoundly in Tfh than in Th2 cells. The lack of IL-4 production by Tfh cells in mice carrying a deletion of HSV+HSVa or HSV may explain the marked reduction in the level of antigen-specific IgE after immunization [79, 80].

HS1/CS1/CGRE

A hypersensitive site matching a conserved GATA3-responsive element (CGRE) maps to the promoter region of *Il13*. Deletion of HS1/CS1/CGRE substantially reduces the expression of IL-13 but not IL-4 in differentiating Th2 cells, suggesting that it is an IL-13-specific enhancer [82].

Locus Control Region

A locus control region (LCR) is a *cis*-acting element that confers a high-level tissue-specific expression. It also serves as an insulator and blocks the influence of neighboring chromatin. In transgenic reporter assays, LCRs exhibit copy number-dependent and tissue-specific enhancer activity that is site independent. Given the coordinated expression of IL-4, IL-5, and IL-13, the presence of an LCR within the Th2 locus was strongly suspected. However, none of the aforementioned positive *cis*-acting elements meets the definition of an LCR. By establishing a series of transgenic mouse strains carrying a bacteria artificial chromosome with various regions of the Th2 locus driving an IL-4 promoter/luciferase cassette, Lee et al. identified an LCR 3′ of the *Rad50* gene [83]. Deletion of the LCR renders Th cells unable to produce IL-4, IL-5, and IL-13, even when differentiated under Th2 polarizing conditions [84], whereas the production of IFN-γ by Th1 cells is unaffected. Accordingly, LCR-deficient mice are resistant to airway allergic inflammation.

This LCR contains at least four hypersensitive sites: RHS4, RHS5/RAD50-O, RHS6/RAD50-(A+B), and RHS7/RAD50-C [85, 86]. RHS6 is detected in naive Th, Th1, and Th2 cells, whereas RHS4 and RHS7 are Th2 specific. The appearance of RHS7 is dependent on Stat6 but not GATA3 [86]. RHS5, RHS6, and RHS7 are also conserved across species, and these three sites together recapitulate the function of the LCR in transgenic reporter assays [85]. RHS7 is required for interactions between the LCR and the *Il4*, *Il5*, and *Il13* promoters in Th cells; chromosome conformation capture assays demonstrate that the LCR is in direct contact with these promoters, thereby keeping the Th2 cytokine genes in a "poised" position [87]. The establishment and maintenance of this intrachromosomal interaction requires GATA3 and Stat6 and is disrupted in the absence of RHS7 [87, 88].

The function of each of the four hypersensitive sites has also been examined in genetically engineered mice [88, 89]. Deletion of RHS4+5 results in no apparent defect in the production of type 2 cytokines. Deletion of RHS7 leads to a partial defect in the expression of IL-4 and IL-13 in Th2 cells and mast cells but no apparent defect in IL-5. In contrast, deletion of RHS6 deprives Th2 cells of the ability to produce type 2 cytokines. Thus, each of the hypersensitive sites within the LCR has both unique and redundant functions in regulating the expression of IL-4, IL-5, and IL-13.

3.3.1.2 Negative Regulatory Element

HSIV/HM1

A silencer has been identified downstream of the 3′ UTR of *Il4* [90]. This silencer matches to the hypersensitive site HSIV, which is detected in naive Th, Th1, and Th2 cells, though its silencing effect is Th1 specific [71]. Its sequence is not fully conserved across species, but it suppresses the activity of the *Il4* promoter in both transient transfection assays and transgenic reporter assays [72, 90]. Mutation of a consensus Stat6 binding site within the silencer or blocking IL-4/Stat6 signaling activates this silencer even in Th2 cells [90]. Another silencer has also been identified as the conserved noncoding sequence HM1, located downstream of the nonconserved HSIV but upstream of HSV; HM1 shows increased accessibility to restriction enzymes and is also named HSIV (thus called HM1/HSIV) [72, 91]. HM1/HSIV is decorated with the transcriptionally active mark H3K4me2 in naive T, Th1, and Th2 cells. HM1/HSIV also displays potent silencer activity in transgenic reporter assays [72]. Deletion of HM1/HSIV has little impact on the differentiation and cytokine production by Th2 cells but results in derepression of IL-4 in Th1 cells and overproduction of IL-4 in mast cells, with rather modest, if any, effects on IL-13 and IL-5 [69, 91]. HM1/HSIV silences the expression of IL-4 by serving as a binding site for Runx3/T-bet complex. In the absence of HM1/HSIV, Runx3 and/or T-bets are less efficient at suppressing IL-4 expression in differentiating Th2 cells [82]. Thus HSIV and HM1/HSIV function as IL-4-specific silencers and suppress IL-4 expression in Th1 cells.

3.3.2 Signaling Pathways Promoting Th2 Differentiation and IL-4 Production

Several signaling pathways have been shown to regulate the expression of IL-4 in Th cells. However, the expression of IL-4 is tightly linked to the differentiation of Th2 cells. A signal pathway may promote the differentiation of Th2 cells, thereby enhancing the expression of all Th2 cytokines, but not directly augment the transcription of the *Il4* gene. Here we will review a few major signaling pathways that are critical for the differentiation of Th2 cells and the expression of IL-4.

3.3.2.1 TCR Signals

Signals emanating from TCR are potent and essential triggers of IL-4 expression in Th cells. The effects of TCR signals on IL-4 expression are mediated mainly through NFAT proteins, activated by calcium-dependent phosphatase calcineurin, and AP-1, which is induced through the RAS-RAF-ERK pathway and forms a dimer with NFAT. Inhibition of calcineurin with cyclosporine prevents T cells from expressing IL-4. TCR signals also activate NF-kB through the PKCθ-CARMA1 pathway. CARMA1 also augments the expression of JunB and GATA3 [92]. The strength of TCR signals can influence the Th1/Th2 fate decision of differentiating Th cells. Stimulating naive Th cells with suboptimal concentrations of antigens or weaker antigens favors the production of IL-4 and the differentiation of Th2 cells [93–95]. It was subsequently shown that weaker TCR signals result in weaker ERK activity and shift AP-1 complexes from c-Fos/JunB heterodimers to JunB/JunB homodimers [96].

3.3.2.2 IL-4/Stat6 Pathway

The most important signaling pathway for the differentiation of Th2 cells is IL-4/Stat6. Stat6 is a member of the signal transducer and activator of transcription (Stat) family of proteins. Stat6 monomers tether to the intracellular tail of IL-4Rα and are phosphorylated by Jaks upon binding of IL-4 to IL-4 receptors [25]. Phosphorylated Stat6 monomers then form dimers through their SH2 domains and translocate to the nucleus. Stat6 dimers preferentially bind to DNA sequences of $TTC(N)_4GAA$ and transactivate gene expression. In differentiating human Th2 cells, Stat6 is responsible for the expression of up to 80 % of IL-4-inducible genes [97], but surprisingly, Stat6 is not necessary for IL-4 production by NKT cells, basophils, and eosinophils [98–100], suggesting that Stat6 does not directly transactivate the *Il4* gene. Stat6-deficient (Stat6KO) Th cells are unable to respond to IL-4 and display a profound defect in the differentiation of Th2 cells in vitro [101–103]. Accordingly, Stat6KO mice in general are unable to mount type 2 immune responses in vivo. Reciprocally, ectopic expression of activated Stat6 is sufficient to redirect differentiating Th1 cells to IL-4-producing Th2 cells [104].

The phosphorylation of Stat6 is reversed by SHP1 [105], a protein tyrosine phosphatase. Mice carrying SHP1-deficiency only in T cells have an elevated level of serum IgE, as their Th cells have prolonged Stat6 phosphorylation after IL-4 stimulation and express more IL-4 [106]. The strength of IL-4/Stat6 signaling is also negatively regulated by Grail, a type 1 transmembrane E3 ubiquitin ligase. Grail ubiquitinates Stat6, resulting in its degradation. Accordingly, Grail deficiency leads to enhanced Th2 differentiation, and Grail-deficient mice are more susceptible to allergic airway inflammation. Interestingly, the expression of Grail is induced by IL-4/Stat6 and is highest in Th2 cells among various Th subsets, thereby establishing a negative feedback mechanism [107].

Published data indicate that Stat6 promotes IL-4 expression in Th cells partly by inducing the expression of GATA3. Indeed, exogenous GATA3 is sufficient to restore Th2 differentiation in Stat6KO Th cells [108]. GATA3 can be transcribed through two promoters, 1a and 1b, separated by 10 kb apart [109, 110]. Transcripts from both promoters are detected in the first 2 days of in vitro Th2 differentiation, but promoter 1a is exclusively used after day 3. Stat6 is recruited to both promoters; however, although Stat6 contributes to the expression of GATA3 through promoter 1b, it is not essential for the activity of this proximal promoter. In contrast, very little promoter 1a activity is detected in the absence of Stat6. Interestingly, both promoters are active in already differentiated Th2 cells and are induced by TCR signals but not IL-4.

Stat6 is also recruited to several sites within the Th2 locus in Th2 cells [111]. While the functional significance of this observation is still not fully established, Stat6 very likely contributes to the folding and epigenetic modifications of the Th2 locus. For example, the appearance of RHS7 and its demethylation are induced in differentiating Th2 cells in a Stat6-dependent manner, but are not dependent on GATA3 [86, 112]. Stat6 also directly binds to more than 4000 genes in Th2 cells, and approximately 60 % of Stat6 binding sites co-localize with H3K4me4 marks [113]. These observations indicate that Stat6 can also promote Th2 differentiation and IL-4 expression independently of GATA3. In addition, Stat6 is required for the migration of Th2 cells, Th2 memory, and stable expression of GATA3 in differentiated Th2 cells [114–116]; thus, Stat6 is also critical for the maintenance of Th2 phenotype and type 2 immune responses.

3.3.2.3 Non-Stat6 Pathways

Despite the critical role of the IL-4/Stat6 pathway in promoting the differentiation of Th2 cells, residual Th2 differentiation can still be detected in the absence of Stat6 or IL-4Rα, indicating the presence of IL-4/Stat6-independent pathways [114, 117, 118].

IL-2/Stat5 Pathway

The IL-2/Stat5 pathway works parallel to the IL-4/Stat6 pathway in driving Th2 differentiation and IL-4 production. Compared to differentiating Th1 cells, differentiating Th2 cells express a higher level of CD25, the receptor of IL-2, and are more responsive to exogenous IL-2 [119]. A timely blockade of IL-2 inhibits the production of IL-4 and Th2 differentiation in both human and mouse Th cells [119–121]. Moreover, a deficiency in either Stat5a or Stat5b, the signaling/transcription factor downstream of IL-2, results in a marked defect in the Th2 immune response in vivo [122]. Reciprocally, the expression of a constitutively active Stat5a restores IL-4 production in the absence of IL-2 [73]. The IL-4-promoting effect of IL-2/Stat5 is still observed in differentiating Th2 cells deficient in Stat6 or IL-4Rα [73], indicating that the effect of IL-2/Stat5 is independent of the IL-4/Stat6 pathway. The IL-2/Stat5 pathway probably accounts for the residual Th2 differentiation of Stat6KO Th cells, because a deficiency in both Stat5a and Stat6 nearly completely blocks in vitro Th2 differentiation and in vivo Th2 cell-dependent allergic airway inflammation [123]. IL-2/Stat5 does not induce the expression of GATA3, and forced expression of GATA3 does not rescue IL-4 production by differentiating Th cells in the presence of IL-2 blockade [121]. In addition, constitutively active Stat5 is unable to induce IL-4 production in the absence of GATA3 [124]. Thus, both GATA3 and IL-2/Stat5 are essential for optimal Th2 differentiation.

IL-2/Stat5a promotes the expression of IL-4 and Th2 differentiation by several mechanisms:

(a) Stat5a is recruited to the promoter of *Maf*, which encodes c-Maf, in an IL-2-dependent manner [125]. The induction of c-Maf by TCRs in developing Th2 cells is completely blocked by daclizumab, a humanized anti-CD25 antibody. c-Maf in turn enhances the induction of CD25 in developing Th2 cells by an IL-4-independent mechanism [119]. Thus, IL-2, Stat5, c-Maf, and CD25 form another positive feedback loop, parallel to the IL-4/Stat6/GATA3 loop during Th2 differentiation (Fig. 3.3).

(b) IL-2 induces the expression of IL-4Rα independently of IL-4. Endogenous IL-2 is also essential for TCR-induced expression of IL-4Rα [126], providing a connection between the IL-2/Stat5 and IL-4/Stat6 loops.

(c) Stat5a is also recruited to several principal *cis*-acting elements of the Th2 locus, including HSII, HSIII, and LCR, but not HSVa sites, and alters the epigenetic landscape of the Th2 locus in a manner different from GATA3 [73, 126].

(d) In addition, Stat5 augments the expression of Lunatic Fringe, which enhances Notch signals in T cells, and promotes IL-4 production and Th2 differentiation in the presence of Notch ligands, such as DLL4 [127]. Reciprocally, Notch signals potentiate the expression of CD25 and the responsiveness of Th cells to IL-2 [128], suggesting cross talk between the IL-2/Stat5 and Notch pathways.

(e) Stat5a directly transactivates SOCS3, which is a potent inhibitor of IL-12/Stat4 signals [129]. Thus, IL2/Stat5 passively promotes Th2 differentiation by inhibiting IL-12-induced Th1 differentiation.

(f) Recently, the nucleotide-binding oligomerization domain (NOD)-like receptor pyrin domain containing family of gene 3 (NLRP3), the key component of NLRP3 inflammasome, was found to be expressed preferentially in Th2 cells over Th1 cells [130]. Its expression is induced by IL-2/Stat5. It can form a dimer with IRF4 and binds to the promoter of *Il4*, thereby augmenting the expression of IL-4 independently of inflammasome activity.

Notch Signaling Pathway

The Notch signaling pathway plays a role of facilitator in the expression of IL-4 and the differentiation of Th2 cells. The mammalian Notch signaling pathway consists of four Notch receptors (Notch 1–4) and five ligands of the Jagged and Delta-like family (Jagged 1, Jagged 2, Delta-like ligand [DLL] 1, DLL2, and DLL3) [131]. Upon ligand binding, Notch receptors undergo a series of cleavages involving the ADMA family of metalloproteinases and the γ-secretase activity of a presenilin-containing protein complex, leading to the release of the Notch intracellular domain (NICD). NICD then translocates to the nucleus and forms a protein complex with the transcription factor recombination signal binding protein for immunoglobulin kappa J region (RBPJ). The NICD/RBPJ complex further recruits coactivators, such as mastermind proteins (MAML) and p300, to transactivate Notch target genes. Mice deficient in RBPJ in T cells have a marked reduction in the basal level of serum IgE and have impaired Th2 responses after immunization [132, 133]. Moreover, combined deficiency of Notch 1 and Notch 2 also attenuates Th2 differentiation [133]. Th2 cells derived from these mice express less IL-4 and GATA3 when differentiated under Th2 polarizing conditions [132–134]. Similarly, inhibition of Notch signals by expression of a dominant negative MAML (DNMAML) attenuates Th2 differentiation in vitro and renders mice susceptible to infection with *T. muris* [135, 136], but reciprocally enhances in vitro Th1 differentiation [136].

Forced expression of NICD1 but not NICD2 enhances IL-4 production through its synergy with GATA3 even in the absence of Stat6 [134, 135], indicating that Notch signals promote IL-4 expression and Th2 differentiation independently of Stat6. Instead, the Th2-promoting effect of Notch signals is mediated mainly through induced expression of GATA3 [133, 135]; RBPJ binds to a conserved site approximately 10 kb upstream of the exon1 of *Gata3* and transactivates the *Gata3* 1a promoter independently of IL-4 [133, 135]. NICD also enhances the transcription of the *Il4* gene through the 3′ HSV, which contains a conserved RBPJ binding site; mutation of the RBPJ site ablates the transactivation by NICD [134].

Notch signals are also important for the differentiation of Th1 cells. LPS induces the expression of DLL4 in APCs [134, 137], and subsequently interactions between DLL and Notch promote Th1 differentiation [134, 138, 139]. In contrast, Jagged 2 expression in APCs is induced by PGE2 and cholera toxin, and the interaction between Jagged 2 and Notch enhances Th2 differentiation [134]. These observations lead to the hypothesis that Notch signals, depending on the ligands, instruct Th1 or Th2 fate. However, this model has been challenged by a few recent studies. Neither DLL1 nor Jagged 1 expression in APCs induces Th1 or Th2 fate in the absence of polarizing cytokines, nor do they redirect cytokine-driven Th1 or Th2 differentiation [140, 141]. Blocking Notch signaling, either with γ-secretase inhibitors or DNMAML, inhibits both Th1 and Th2 genetic programs regardless of cytokine milieu [142]. Recruitment of NICD1 to *Il4* and *Gata3* is detected even in Th1 cells, as well as recruitment to the *Tbx21* locus, which encodes T-bet, in Th2 cells, suggesting that Notch signals "globally" facilitate, but do not instruct, the differentiation of Th cells in the presence of polarizing cytokines.

WNT/β-Catenin Pathway

WNTs are secreted glycoproteins [143]. Binding of WNTs to their receptors, called Frizzled, results in the stabilization of β-catenin, which translocates to the nucleus and forms a complex with transcription factor TCF-1 or LEF-1, members of the high-mobility group (HMG) domain-containing proteins. This complex then binds to WNT-responsive elements and recruits CBP and p300, resulting in gene expression. Naive human Th cells express a low level of WNT and exhibit a basal TCF-1 activity even in the absence of exogenous WNT. Exogenous Dkk1, a WNT inhibitor, or siRNA against β-catenin suppress the expression of GATA3 and IL-4 in differentiating human Th2 cells [144]. In agreement with this result, genetic ablation of β-catenin also leads to a profound defect in Th2 differentiation, whereas forced expression of β-catenin strongly enhances Th2 differentiation. This Th2-promoting effect of WNT/β-catenin is in part mediated by TCF-1, which is also required for the optimal differentiation of Th2 cells [145].

Both β-catenin and TCF-1 bind to the promoter 1b of *Gata3* and drive the transcription of *Gata3* in response to TCR stimulation [145], independently of IL-4/Stat6 or Notch signaling. Interestingly, during the differentiation of Th2 cells, the expression of TCF-1 is suppressed by IL-4, despite TCF-1's role in promoting Th2 differentiation [146]. It turns out that TCF-1 has a short alternative spliced form, which lacks the N-terminal β-catenin-binding domain and is a constitutive suppressor. IL-4/Stat6 signals preferentially suppress the expression of the short isoform, increasing the proportion of full-length TCF-1.

While both TCF-1 and LEF-1 can interact with β-catenin, the Th2-promoting effect of WNT/β-catenin appears to be mediated by TCF-1 but not LEF-1, and LEF-1 can in fact inhibit Th2 differentiation. In contrast to TCF-1, forced expression of LEF-1 in differentiating Th2 cells results in attenuated expression of IL-4, IL-5, and IL-13 [147]. Reciprocally, knocking down LEF-1 with siRNA in Jurkat cells increases the level of IL-4 transcript [148]. LEF-1 can inhibit Th2 differentiation by two mechanisms. It can prevent GATA3 from binding to DNA by forming a complex with GATA3 [147], an interaction mediated by LEF-1's HMG box and the zinc fingers of GATA3. In addition, LEF-1 can directly bind to a negative regulatory element within the *Il4* promoter [148]. Consistent with its role as a negative regulator of IL-4 expression and Th2 differentiation, the expression of LEF-1 is downregulated when naive Th cells differentiate into Th2 cells [146, 148]. This downregulation of LEF-1 requires signals from IL-4/Stat6 and TCR.

mTOR

A serine-threonine kinase, mammalian target of rapamycin (mTOR), is the core component of two distinct signaling complexes, mTORC1 and mTORC2 [149]. mTORC1 contains the scaffold protein Raptor and several other subunits including mLST8, PRAs40, and Deptor. mTORC1 is activated sequentially by PI3K, Akt, and Rheb, a RAS-like GTPase, and in turn promotes the phosphorylation of S6K1 and 4E-BP1. In contrast, mTORC2 consists of the scaffold proteins Rictor, mLST8, mSIN1, and Protor, and it promotes the phosphorylation of Ser473 of Akt. The activation of mTORC2 is less understood, but its activation by PI3K-triggered ribosome binding has been reported [150]. mTOR is also activated by IL-4 in Th cells through the IRS2-PI3K-Akt pathway [27]. mTOR complexes can sense growth factors and nutrients in the environment and regulate cellular metabolism.

The role of mTOR complexes in influencing the differentiation of Th cells was first reported by Delgoffe et al. [151]. They found that mTOR is essential for the differentiation of effector Th cells, including Th1, Th2, and Th17. Th cells deficient in mTOR default into Treg cells upon stimulation. They subsequently found that Rictor, the scaffold protein of mTORC2, is essential for the differentiation of Th2 cells, whereas a deficiency of Rheb, an activator of mTORC1, specifically interferes with the differentiation of Th1 [152]. However, this mTORC1-Th1 and mTORC2-Th2 dichotomy has been challenged by two reports. Lee et al. showed that Rictor is required for the differentiation of both Th1 and Th2 cells [153], while Yang et al. demonstrated that a deficiency of Raptor also leads to a profound defect in the differentiation of Th2 cells [154]. Thus, both mTORC1 and mTORC2 contribute to the differentiation of Th2 cells.

How mTORC promotes the differentiation of Th2 cells is also controversial. A reduction in the level of GATA3 is a universal finding in all experiments using mice or cells with impaired mTORC activity. However, other mechanisms have been suggested:

(a) Attenuation of Stat6 signaling through SOCS5: Delgoffe et al. showed that Rictor-deficient Th cells express a higher level of SOCS5, an inhibitor of Stat6, and a lower level of p-Stat6 [152]. However, this observation cannot be reproduced by Lee et al., who found no significant difference in the level of p-Stat6 between WT and Rictor-deficient Th cells [153].
(b) Attenuation of p-PKCθ and NF-kB activity: Instead, Lee et al. found that Rictor-deficient Th cells have a reduced level of p-PKCθ and nuclear NF-kB, including RelA, RelB, and p50. Forced expression of a constitutively active p-PKCθ rescues Th2 differentiation [153].
(c) Reduction in the level of JunB protein: mTORC2 also phosphorylates and activates SGK1 (serum- and glucocorticoid-regulated kinase 1), a member of the AGC family of serine-threonine kinase. Once activated, SGK1 inhibits E3 ligase Nedd4-2. Heikamp et al. recently showed that a deficiency of SGK1 results in marked impairment in the differentiation of Th2 cells [155]. SGK1-deficient Th2 cells are unable to express GATA3, IL-4, IL-5, and IL-13. This defect is attributed to a reduced protein level of JunB caused by hyperactive Nedd4-2, which facilitates the ubiquitination and degradation of JunB.

Nucleic Acid Sensing Pathway

A recent paper by Imanishe et al. showed that Th cells can sense non-CpG DNA and self-DNA, which elicits co-stimulation signals independently of MyD88/Trif or other known nucleotide sensors such as STING and ZBP1 [156]. This DNA-induced co-stimulation promotes the expression of GATA3 and IL-4 and reciprocally suppresses the expression of T-bet and IFN-γ, eventually leading to the polarization of Th2 cells even in nonpolarizing conditions. This induction of GATA3 and IL-4 is dependent on the IL-4/Stat6 pathway. However, how sensing of non-CpG or self-DNA leads to the induction of GATA3 and IL-4 is still unknown.

3.3.3 Transcription Factors

3.3.3.1 GATA3

A member of the GATA family of transcription factors, GATA3, was originally cloned as a T cell-specific transcription factor binding to the enhancers of the TCR α and δ genes [157, 158]. GATA3 is essential for several critical steps during the differentiation of T cells, including T cell fate determination, thymic β-selection, and CD4 lineage specification [159–161]. It also regulates the homeostasis and activation of NKT cells and CD8+ T cells [162–164]. Although naive Th cells express GATA3, its expression is further induced during Th2 differentiation, driven by the IL-4/Stat6 pathway – but not IL-2/Stat5 – [165, 166] and by the Notch signaling pathway. Forced expression of GATA3 is sufficient to drive the differentiation of Th2 cells in the absence of Stat6 and to convert differentiating Th1 cells into IL-4-producing cells. Studies using conditional GATA3-deficient mice have demonstrated that GATA3 is essential for the differentiation of Th2 cells driven by IL-4/Stat6, Notch, or IL-2/Stat5 and for IL-4 production by NKT cells [124, 133, 162, 167].

Existing data have demonstrated that GATA3 promotes the expression of IL-4 and the differentiation of Th2 cells through several mechanisms:

GATA3 as a chromatin-remodeling factor of the Th2 locus: GATA3 can bind to almost all DNase I hypersensitive sites within the Th2 locus, including multiple sites at LCR, CGRE, HSII, and HSV. Forced expression of GATA3 induces the appearance of HSII, HSIII, and HSV even in the absence of Stat6. GATA3 together with Stat6 induces long-range interactions among LCR and the *Il4* and *Il13* promoters [87, 108, 168], strongly suggesting that GATA3 is capable of remodeling the chromatin conformation of the Th2 locus.

GATA3 as a transcription factor of type 2 cytokines: Despite GATA3's essential role in Th2 differentiation, GATA3 deficiency in already differentiated Th2 cells has a minimal impact on the expression of IL-4, but results in a profound defect in IL-5 and IL-13 expression [124], suggesting that GATA3 serves as a direct transcription factor of IL-5 and IL-13, but not IL-4. Indeed, GATA3 can bind to and transactivate the promoters of IL-5 and IL-13 [169, 170], very likely through different mechanisms. For example, a positively charged amino acid residue at position 261, either arginine or lysine, is essential for GATA3 to support IL-5 expression but dispensable for its induction of IL-4 and IL-13 in Th2 cells or suppression of IFN-γ production in differentiating Th1 cells [171].

GATA3 as an inhibitor of Th1 cells: In addition to promoting the differentiation of Th2 cells, GATA3 also actively suppresses the differentiation of Th1 cells through several mechanisms. It suppresses the upregulation of IL-12Rβ-2 and Stat4, the key signal transduction pathway for the induction of Th1 differentiation, neutralizes the function of Runx3, and epigenetically inhibits the expression of T-bet and IFN-γ [161, 172]. Since the differentiation of naive Th cells to Th1 or Th2 cells is often mutually exclusive, by suppressing Th1 differentiation,

GATA3 further enforces the expression of IL-4 and the differentiation of Th2 cells.

Several genome-wide ChIP-seq analyses have identified potential target genes of GATA3 in Th2 cells [173–175]. The data strongly suggest that GATA3 also regulates the expression and maintenance of many Th2-specific genes. One of the GATA3 target genes is *Gata3* itself. This finding is consistent with the observation that GATA3 can induce its own expression in Th cells [108], thereby forming a positive feedback loop. However, the role of this positive feedback loop in the differentiation of Th2 cells and expression of IL-4 has yet to be confirmed.

3.3.3.2 c-Maf

c-Maf is a leucine zipper transcription factor that is preferentially expressed in Th2 cells over Th1 cells [176]. Its expression in differentiating Th2 cells is induced by TCR/CD28, ICOS signaling, and IL-2/Stat5 [125, 176–178]. c-Maf physically interacts with NFAT and binds to a half MARE (Maf recognition element) site located at approximately −40 position, immediately next to a NFAT binding site within the *Il4* promoter. Forced expression of c-Maf and NFAT is sufficient to drive endogenous IL-4 production in non-IL-4 producers, such as M12 B cells. This effect is further enhanced by the NFAT-interacting protein NIP45 [179], which is required for the optimal production of type 2 cytokines by Th2 cells [180]. c-Maf transgenic mice have a higher level of serum IgE and IgG1, and their Th cells skew toward a Th2 fate in an IL-4-dependent manner when activated under nonpolarizing conditions [181]. Reciprocally, c-Maf-deficient Th cells display a profound defect in differentiating into Th2 cells under nonpolarizing conditions but are able to become Th2 cells in the presence of exogenous IL-4 [182]. c-Maf-deficient Th2 cells derived in this manner nevertheless still express a low level of IL-4, albeit producing normal levels of IL-5 and IL-13. Thus, transactivation by c-Maf is specific to IL-4 among type 2 cytokines. However, unlike GATA3, c-Maf is unable to force established Th1 cells to produce IL-4 when at a level comparable to that of Th2 cells [181]. The activity of c-Maf is regulated by several posttranslational modifications; SUMOylation at Lys33 of c-Maf interferes with its recruitment to the *Il4* promoter [183, 184], whereas Tec kinase-mediated phosphorylation of Tyr21/92/131 enhances its transactivation of the *Il4* gene [185, 186].

3.3.3.3 JunB

JunB is a member of the AP-1 family of transcription factors. It was first reported that Th2 cells contain more AP-1 activity, which correlates with the level of JunB-containing complexes, than Th1 cells [187]. It was subsequently found that Th2 cells express more JunB than Th1 cells [188]. It binds to the P1 element within the *Il4* promoter and synergizes with c-Maf to transactivate the promoter. Deficiency of

JunB results in a significant reduction in the expression of IL-4 and IL-5, but not IL-13, and an increase in the level of IFN-γ and T-bet in Th2 cells [189]. Accordingly, mice deficient in JunB only in T cells are resistant to allergic airway inflammation. The expression of JunB in Th cells is induced by TCR signals through CARMA1 and ERK1 [92, 190]. Its activity is positively regulated by JNK-mediated phosphorylation of Thr102 and Thr104 as well as SUMOylation at K237, K267, and K301 [188, 191]. Its degradation is mediated by Itch E3 ubiquitin ligase [192], which requires Ndfip1 (Nedd4 family interacting protein-1) to function [193]. Deficiency of Itch or Ndfip1 results in a prolonged half-life of JunB after T cell activation, excessive production of IL-4 by Th2 cells, and type 2 inflammation.

3.3.3.4 Dec2

The helix-loop-helix transcription factor Dec2 is preferentially expressed in Th2 cells over Th1 and Th17 cells [194, 195]. Its expression is induced by the IL-4/Stat6 pathway and is further enhanced by ICOS or IL-25. Interestingly, GATA3 is not essential for its initial induction during the differentiation of Th2 cells but is necessary for sustaining its expression in established Th2 cells. Dec2-deficient mice have a significant defect in mounting a Th2 immune response in vivo, and T cell-specific Dec2 transgenic mice are more susceptible to allergic airway inflammation. Dec2 deficiency results in a marked reduction in the level of type 2 cytokines, GATA3, and JunB in Th2 cells, but has little impact on the differentiation of Th1 or Th17 cells. Although Dec2 augments IL-4-driven Th2 differentiation, it has a negligible effect in the absence of IL-4/Stat6 signals. The impairment of Th2 differentiation in the absence of Dec2 is rescued by forced expression of JunB or GATA3. This observation is consistent with the finding that Dec2 is recruited to the promoters of *Gata3* and *Junb*, suggesting that Dec2 directly transactivates *Gata3* and *Junb*. In addition, Dec2 augments the expression of CD25 in a Stat6-dependent manner, thereby enhancing the sensitivity of differentiating Th2 cells to IL-2.

3.3.3.5 Other Positive Transcription Factors

In addition to the aforementioned transcription factors that are preferentially expressed in Th2 cells, optimal expression of IL-4 also requires several other transcription factors that are not restricted to Th2 cells. Here we will discuss those acting directly on *Il4* or the Th2 locus.

Kruppel-like factor (KLF)13: KLF13 can bind directly to a CACCC box located at the -338 position of the *Il4* gene. A weak transactivator by itself, KLF13 interacts with c-Maf and potentiates the transcriptional activity of c-Maf on IL-4 [196].

Yin-Yang 1 (YY1): YY1 is a zinc finger DNA-binding protein that can function as a transactivator or repressor. It can bind to several sites within the Th2 locus, including RHS7, HSV, HSIV, CNS1, and the P0 element located at -53 to -59 of the *Il4* promoter [197, 198]. It not only directly transactivates the *Il4* promoter but also facilitates the recruitment of GATA3 to the Th2 locus. Haploinsufficiency or knockdown of YY1 results in a significant defect in the production of IL-4, IL-5, and IL-13 [198, 199].

Special AT-rich sequence-binding protein 1 (SATB1): SATB1 is a "chromatin organizer" as well as a transcription factor. It binds to base-unpairing regions (BURs), which are enriched with stretches of DNA sequences containing a mixture of adenine, thymidine, and cytosine (but not guanine) on one strand, and recruits chromatin-remodeling complexes [200, 201]. The expression of SATB1 is induced in activated Th cells, and its induction during the differentiation of Th2 cells is dependent on Stat6 but not GATA3 [202]. SATB1 binds to nine BURs within 200 kb of the Th2 locus and is also recruited to CNS1 and CNS2, which do not contain conserved BURs [111]. The anchoring of SATB1 to these sites allows the formation of densely packed chromatin loops, bringing the promoters of *Il4*, *Il5*, and *Il13* close to one another and to the LCR of the Th2 locus. Knocking down SATB1 in Th2 cells thus results in a marked reduction in the expression of IL-4, IL-5, and IL-13 [111, 144]. In addition to facilitating the chromatin looping of the Th2 locus, SATB1 can directly bind to the P2 element of the *Il4* promoter and, together with JunB, synergistically transactivate IL-4 [203]. SATB1 is also required for the induction of c-Maf in Th2 cells in response to stimulation and functions as a mediator of WNT signals to augment the expression of GATA3 [111, 144]. Upon activation with WNT, SATB1 recruits β-catenin and p300 complex to a region upstream of exon 1a of *Gata3*. It is still unclear whether SATB1 can also bind and recruit β-catenin to promoter 1b as TCF-1 does.

3.3.3.6 Negative Transcription Factors

The expression of IL-4 and the differentiation of Th2 cells are also subjected to negative regulation. IRF1 can bind to the *Il4* promoter and mediate the suppressive effect of IFN-γ on IL-4 expression [204]. A recent paper further demonstrated that IL-1α and IL-1β, elicited during *Listeria* infection, induce the stabilization and nuclear translocation of IRF1, which bound to the 3′UTR region instead of the promoter of *Il4*, thus suppressing IL-4 expression [205]. This effect appears to be specific to IL-4 because the expression of IL-5 and IL-13 is relatively unaffected. Transcription factors like T-bet and RORγt that promote the differentiation of other functional subsets of Th cells can inhibit the differentiation of Th2 cells, thereby inhibiting the expression of IL-4. For example, the IL-12/Stat4 pathway induces T-bet, the master transcription factor of Th1, which then induces the expression of Runx3. T-bet and Runx3 form a complex, which binds to the HSIV and silences the expression of IL-4. Accordingly, Runx3-deficient Th1 cells aberrantly produce IL-4 [82, 206]. Runx3 can also interact with GATA3 and attenuate its activity [207]. Other transcription factors, such as PU.1, Bcl6, and Sox4, also inhibit the expression of IL-4 and/or the differentiation of Th2 cells by inhibiting the function or expression of GATA3 [208–211].

3.3.4 Epigenetic Regulation of IL-4

Epigenetic modifications allow transcription factors to access DNA to initiate gene expression. Once an expression profile is established, additional modifications are needed to maintain the expression of transcriptionally active genes while keeping silenced genes transcriptionally inactive. Epigenetic modifications can occur on both DNA and histones. For example, demethylation of CpG islands facilitates gene expression, while methylation of lysine 4 of histone H3 (H3K4me3) is associated with active transcription. In contrast, methylation of lysine 27 of histone H3 (H3K27me3) or deacetylation of histones marks transcriptionally silenced loci. These modifications are not just epiphenomenon of transcriptional activity; instead, they can actively regulate gene expression.

3.3.4.1 Epigenetic Landscape of *Il4* and the Th2 Locus in Th2 Cells

During the differentiation of Th2 cells, the *Il4* gene and the Th2 locus undergo numerous changes in their epigenetic landscape. In murine naive Th cells, the entire *Il4* gene, including the promoter and the first intron, and CNS1 are CpG hypermethylated. However, the degree of methylation in the promoter region is slightly less than in other regions. Such a design may enable the *Il4* promoter to stay poised to respond to Th2 polarizing stimulation[212, 213]. When Th cells are activated in the presence of IL-4, the *Il4* promoter becomes demethylated within 2–4 days after stimulation in an IL-4/Stat6- and cell cycle-dependent manner, with partial dependence on IL-2 and CD28 co-stimulation [112]. The demethylation process then extends several kilobases from the *Il4* promoter into the *Il4* gene. This extension of demethylation into the gene body correlates with a high expression level of IL-4. In addition, the *Il4*/*Il13* intergenic regions, corresponding to HSS3 and CNS1, and RHS7 within the LCR are also demethylated during Th2 differentiation [85, 112]. In contrast, CNS2 is hypomethylated in naive Th cells and Th2 cells but becomes methylated during Th1 differentiation [212]. However, these changes in methylation in mouse Th cells are not identical to those observed in human Th cells. CpG demethylation of *Il4* and the Th2 locus in human Th2 cells is less global, limited to several Th2-specific DNase I hypersensitive sites (DHS), including the second intron of the *Il4* gene, the *Il13* promoter, and the region between CNS1 and *Il13*. CpG demethylation is not detected in the human *Il4* promoter [214].

Surprisingly, in both humans and mice, the status of demethylation of the *Il4* promoter is not different between Th1 and Th2 cells. In addition, ectopic expression of GATA3 in Th1 cells induces the expression of IL-4 and the appearance of Th2-specific DHS but does not lead to demethylation [215]. These observations indicate that the status of DNA methylation alone is not enough to dictate the expression of IL-4. Histone modifications also play a critical role in stably repressing or maintaining the expression of IL-4. The differentiation of Th2 cells is associated with the appearance of permissive histone modifications not only in the *Il4* gene but also over the entire Th2 locus [216]. For example, the HSII within the second intron of the *Il4* gene is decorated with nonpermissive H3K27me3 in naive Th cells but is highly enriched with permissive marks such as H3K4me3, H3K9Ac, and H3K14Ac in differentiated Th2 cells [69, 217]. This is consistent with the observation that HSII is bound by GATA3 and Stat5 and is responsible for GATA3-mediated expression of IL-4 [69]. Seumois et al. used ChIP to compare global H3K4me marks among human naive, CCR4+ (Th2-like), and CCR4- (Th1-like) memory cells and identified many differentially enriched regions (DERs) [218]. They found several Th2 DERs corresponding to LCR-O, LCR-A, HSII, and HSV of the mouse Th2 locus, whereas a human Th1 DER overlapped with mouse HSIV. In addition, mouse HSS3 appears to be a DER shared by both human Th1 and Th2 cells. Surprisingly, there was no DER matching to mouse LCR-B (part of RHS6), LCR-C (RHS7), CGRE, CNS1, or HSIII.

3.3.4.2 Epigenetic Modifiers

Despite the detailed characterization of epigenetic modifications within the Th2 locus, the functional role of each specific modification at a given site is still poorly understood due to the lack of molecular tools that can alter epigenetic modifications in a site-specific manner. However, studies using Th cells rendered deficient in epigenetic modifiers clearly demonstrate that epigenetic modifications play a critical role in regulating the expression of IL-4 and are not just consequences of active or inactive transcription of the *Il4* gene.

Epigenetic Modifiers that Suppress IL-4 Expression

1. *DNA (cytosine-5)-methyltransferase 3a (DNMT3a) and 1 (DNMT1):* DNMT3a is responsible for de novo DNA methylation and is recruited to CNS1 in single-positive (SP) thymocytes [213]. It is not required for basal DNA methylation of the *Il4* gene in naive T cells [219]. However, DNMT3a-deficient Th cells activated in vitro in the absence of polarizing cytokines, a condition that leads to the production of IFN-γ but not IL-4 in wild type (WT) cells, start to express IL-4 and lose methylation of the *Il4* gene. The production of IL-4 and IFN-γ is otherwise normal in DNMT3a-deficient Th cells differentiated under Th2 and Th1 conditions, respectively [219–221]. DNMT1, which is responsible for maintaining and spreading DNA methylation, is recruited to CNS1, the promoter, and the second intron of the *Il4* gene in activated SP thymocytes [213]. Deficiency of DNMT1 leads to a global decrease in the level of CpG methylation in the *Il4*/*Il13* locus and derepression of Th2 cytokines [213, 222]. Consequently, there is a high level of Th2 cytokines in CD8 T and Th1 cells, which normally do not express these cytokines. These changes are associated with an increase in the degree of H3K4me2 but not the level of GATA3.

2. *Methyl-CpG-binding domain protein 2 (MBD2):* MBD2 binds to methylated CpGs and recruits nucleosome-remodeling and deacetylase (NuRD) repressive complexes, thereby acting as a link between DNA methylation and repressive histone modifications. It is recruited to CNS1 and the second intron of the *Il4* gene in Th1 cells [215]. Its recruitment to these regions is inhibited by ectopic expression of GATA3. Deficiency of MBD2 results in derepression of IL-4 in Th1 cells and overexpression of IL-4 in Th2 cells, though this effect is not due to an increase in the level of GATA3.
3. *Enhancer of zeste homologue 2 (EZH2):* EZH2 is a histone methyltransferase and a component of the polycomb repressive complex 2 (PRC2). It is responsible for maintaining H3K27me3 of the Th2 locus in naive and nonpolarized Th cells [223, 224]. Deficiency of EZH2 leads to augmented expression of IL-4 in Th2 cells and derepression of IL-4 in Th1 cells.
4. *Histone deacetylase 1 (HDAC1):* HDAC1 binds to CNS2 and the second intron of the *Il4* gene in naive Th cells [225]. The expression of IL-4 and other type 2 cytokines in Th2 cells is enhanced in the absence of HDAC1, but no derepression of IL-4 in Th1 cells is observed.
5. *Jumonji domain containing 3 (Jmjd3)*: Jmjd3 is a histone demethylase that is responsible for the demethylation of H3K27me2/3. Deficiency of Jmjd3 skews unpolarized Th cells into a Th2 fate and results in derepression of IL-4 in Th1 cells, but has little impact on the differentiation of Th2 cells [226]. The status of H3K27me3 in the Th2 locus has yet to be examined in Jmjd3-deficient Th cells. This effect of Jmjd3 deficiency is very likely mediated by reduced expression of T-bet.

Epigenetic Modifiers that Promote IL-4 Expression

1. *Mixed lineage leukemia (MLL):* MLL is a member of the Trithorax group protein family and is responsible for H3K4me3 permissive modification. It is highly expressed in naive Th cells and memory Th2 cells, but not in memory Th1, effector Th1 or Th2 cells [227]. It also forms a complex with GATA3 and c-Myc in human memory Th cells [228]. Haploinsufficiency of MLL results in a reduction in the level of H3K4me3 and H3K9ac in the Th2 locus [227]. MLL+/− naive Th cells are capable of differentiating into Th2 cells; however, MLL+/− memory Th2 cells are unable to maintain the expression of IL-4, IL-5, and IL-13. This defect is associated with a reduction in the level of GATA3 in memory Th2 cells.
2. *Mel-18 and G9a:* Although enzymes mediating repressive modifications are expected to suppress IL-4 expression, there are still exceptions. Mel-18 is a component of polycomb repressive complex 1, which usually suppresses gene expression. Unexpectedly, deficiency of Mel-18 results in impairment in the differentiation of Th2 but not Th1 cells. This defect is associated with attenuated expression of GATA3 and decreased demethylation of the *Il4* gene and is thus

rescued by exogenous GATA3 [229]. The abnormal demethylation and impaired Th2 differentiation are therefore very likely secondary to attenuated expression of GATA3. Another example is histone lysine methyltransferase G9a. G9a is responsible for H3K9me2 modifications, a repressive histone mark, in CNS1 and the *Il4* gene of unpolarized Th and Th1 cells [230]. Surprisingly, deficiency of G9a attenuates Th2 differentiation but not Th1 differentiation, though there is no alteration in the expression of Stat6 or GATA3.

3.3.5 Regulation of IL-4 Expression by microRNA and lncRNA

Several classes of noncoding RNA, including microRNA (miR) and long noncoding RNA (lncRNA), can also regulate gene expression [231]. Several miRs have been shown to regulate the differentiation of mouse Th1/Th2 cells either in vitro or in animal models of allergic inflammation. However, these data must be interpreted very carefully for two reasons. Firstly, most of these miRs are expressed in other types of cells in addition to Th cell. Thus, germ-line deficiency of any given miR can have impacts beyond just Th cells. This may lead to conflicting results between in vivo and in vitro experiments. Secondly, the sequences of miRs and their targets are not as conserved as coding exons between mouse and human. It remains questionable whether their function is also conserved across species.

3.3.5.1 miR-21

A high level of miR-21 is associated with an excessive Th2 immune response and can be detected in the lung of IL-4 or IL-13 transgenic mice [232], in the esophageal tissue from patients with eosinophilic esophagitis [233], and in the lesional skin from patients with allergic contact dermatitis to diphenylcyclopropenone [234]. miR-21 expression in Th cells can be inhibited by Bcl6 [235], a transcriptional suppressor of Th2 differentiation [209, 210]. Forced expression of miR-21 also enhances the expression of IL-4 and GATA3 in Th cells differentiated under neutral conditions. miR-21 mimic increases the transcript level of *Gata3* but the mechanism of this effect is still unclear [236].

3.3.5.2 miR-155

The effect of miR-155 on Th2 cells is discordant between in vitro and in vivo experiments. In vitro differentiated Th2 cells deficient in miR-155 produce more IL-4, IL-5, and IL-10 compared to their WT counterparts, suggesting that miR-155 suppresses IL-4 expression. One of the targets of miR-155 is c-Maf [237, 238].

However, a study looking at global miR expression in the skin found that miR-155 is upregulated in patients with atopic dermatitis compared to healthy controls [239] and that the main source of miR-155 is T cells. This observation is consistent with the finding that miR155-deficient mice are actually more resistant to ovalbumin-induced allergic airway inflammation [240]. In addition, Th cells obtained from the peribronchial lymph nodes of immunized/challenged miR155-deficient mice produce far less type 2 cytokines, though they retain a normal transcript level of *Gata3* and *Maf*, but show an increase in PU.1 expression.

3.3.5.3 miR-19

miR-19 promotes Th2 proliferation by suppressing the expression of Pten. miR-19a is a member of the miR-17~92 cluster, which contains six miRs transcribed in one polycistronic pri-miR. In a study by Simpson et al. comparing the expression profile of 190 miRs between bronchial lavage CD4+ T cells from asthmatic patients and those from healthy controls [241], only miR-19a was consistently elevated in asthmatic patients, and this elevation was not affected by steroid treatment. Simpson et al. further demonstrated that deletion of the miR-17~92 cluster results in impaired expression of IL-4, IL-5, and IL-13, but not GATA3 by Th2 cells. This impairment in cytokine production is rescued by exogenous miR-19a and miR-19b but not by other miR from this cluster. Inhibiting or mimicking miR-19 in human polarizing Th2 cells modestly attenuates or enhances the expression of IL-13, but not IL-4, respectively. Pten, one of the targets of miR-19, inhibits the proliferation and the survival of T cells and is upregulated in Th2 cells deficient in miR-17~92. Haploinsufficiency of Pten partially rescues the production of IL-4 and IL-13.

3.3.5.4 Long Noncoding RNA (lncRNA)

Linc-Maf-4 Inhibits the Expression of c-Maf

A recent profiling of lncRNA from 13 subsets of human T and B lymphocytes has identified lncRNAs that are preferentially expressed in various functional subsets of Th cells [242]. One of these lncRNAs is linc-MAF-4, which is enriched in Th1 cells. Located 139.5 kb upstream of *Maf*, its transcription level is inversely correlated with that of c-Maf but not of other neighboring genes. Knocking down linc-MAF-4 with siRNA in activated T cells results in an increase in the level of c-Maf, IL-4, and GATA3. Chromosome conformation capture analysis further indicates the presence of *in cis* chromatin-looping conformation between linc-MAF-4 and the *Maf* promoter. This physical approximation enables the linc-MAF-4 transcript to act as a scaffold, recruiting chromatin modifiers EZH2 and LSD1, which inhibit the transcription of *Maf*.

Th2-LCR lnRNA Maintains H3K4me3 of the Th2 Locus

Spurlock et al. compared lncRNAs among human Th1, Th2, and Th17 cells and identified a cluster of four alternatively spliced Th2-specific lncRNAs encoded in reverse direction within the last intron and 3′ region of *Rad50*, partly overlapping with the LCR [243]. This cluster of Th2-LCR lncRNA is associated with hsRNPs and WDR5, a component of the H3K4 methyltransferase complex. Knocking down the Th2-LCR lncRNA with siRNA reduces the expression of IL-4, IL-5, and IL-13, but not RAD50, in human primary Th cells. There is also a marked reduction in the level of H3K4me3 in *Il4* and *Il13*. Thus, one possible function of the Th2-LCR lncRNA is to facilitate the formation of H3K4me3 at the Th2 locus.

3.4 Regulation of IL-4 Expression in Non-Th2 Cells

Tfh cells are considered a separate lineage of CD4+T cells [11]. They produce a high level of IL-4 but little IL-5 and IL-13 [116], and their ability to produce IL-4 is dependent on the *cis*-acting elements HSV and HSVa within the Th2 locus and the SLAM/SAP signaling pathway [79, 80, 244]. Unlike Th2 cells, Tfh cells express a low level of GATA3, and their differentiation and function are negatively regulated by Stat5 [116, 245, 246]. Tfh differentiation also requires Bcl6, ICOS, and IL-21. In addition, ICOS signaling induces the expression of c-Maf [177, 178], which works collaboratively with Bcl6 to activate the Tfh gene expression program and very likely directly transactivates *Il4* in Tfh cells [247].

Innate immune cells, such as mast cells, basophils, and eosinophils, do not express GATA3 or c-Maf [248]; their ability to produce IL-4 is independent of the IL-4/Stat6 signaling pathway [99, 100, 249]. Instead, they express two other GATA members, GATA1 and GATA2. Both GATA proteins bind to CNS1 and HSVa within the Th2 locus in bone marrow-derived mast cells [250]. A recent study also demonstrated that Stat5 is recruited to the *Gata2* gene in bone marrow-derived mast cells [251]. Thus, the Stat5/GATA2 or GATA1 pathway probably plays a role in innate immune cells, similar to the involvement of Stat6/GATA3 in Th2 cells.

Knocking down GATA1 in mature basophils results in a marked reduction in the expression of IL-4 in response to IgE/antigen [252]. Similarly, ablation of the *Gata2* gene in differentiated basophils leads to diminished expression of IL-4 and IL-13 in response to PMA/ionomycin or IgE cross-linking [253]. These results indicate that both GATA1 and GATA2 are essential for optimal production of IL-4 in basophils. How GATA1 and GATA2 promote the expression of IL-4 in basophils is still not fully understood. The absolute dependence on GATA1 and GATA2 for IL-4 expression in differentiated basophils is different from the dispensable role of GATA3 in differentiated Th2 cells, suggesting that GATA1 and GATA2 act as direct transcription factors of IL-4 in innate immune cells.

In addition to GATA1 and GATA2, other transcription factors, such as PU.1, CEBPα, and NFAT2, have been implicated in positively regulating the expression of type 2 cytokines or the chromatin accessibility of the Th2 locus in mast cells and basophils [254–256]. In basophils, FcεRI-induced nuclear translocation of NFAT and IL-4 production is partly dependent on the presence of zinc-binding proteins metallothioneins I and II [257]. However, the actual role of these positive regulatory factors – besides metallothioneins I and II – has yet to be confirmed with knockout cells.

3.5 Expression of IL-4 in Protective and Pathogenic Type 2 Immune Responses

IL-4 and type 2 immune responses mediate immune defenses against helminth infection, but they also play an important role in the pathogenesis of atopic diseases, such as asthma and atopic dermatitis. Numerous animal studies have demonstrated that alterations in the aforementioned signaling pathways, transcription factors, and epigenetic modifications can attenuate or aggravate IL-4 expression and type 2 inflammation in vivo. However, how protective and pathological type 2 immune responses are triggered in human diseases is still not fully understood. Recently, genomic and epigenomic studies have been deployed to discover genetic variations that are associated with a higher risk of human atopic diseases or genes that are dysregulated in these diseases. These studies have identified both expected and unexpected pathways.

3.5.1 Genome-Wide Association Studies

Genetic variations in several hundred genetic loci have been found to be associated with a higher risk of asthma or a high level of serum IgE [258, 259]. As expected, genetic variations at HLA-DR and HLA-DQ loci are identified in some of these studies [260, 261]. This finding is consistent with the presence of IL-4-producing Th cells in the bronchial lavage and airways of asthmatic patients [262, 263] and lends additional evidence of a critical role of Th2 cells in the development of type 2 inflammation. Several of the genetic variations thus identified fall within the *Il4* gene, the Th2 locus, or loci encoding known regulators of IL-4 and type 2 immunity, such as Stat6 and Rad50. However, it is still unclear whether these genetic variations indeed influence the expression of IL-4 and/or the development of type 2 immunity and, if so, how. For example, a single nucleotide polymorphism (SNP), rs2240032, located within RHS7 of the LCR alters the binding of SMAD3 to RHS7 and the CpG methylation status of the *Il13* promoter. This SNP is also associated with reduced expression of IL-4 but not IL-13 [264]. However, it is unclear how the alteration in SMAD3 binding leads to attenuated expression of IL-4.

3.5.2 Epigenomic Studies

Seumois et al. compared global H3K4me marks of human naive, CCR4+ (Th2-like), and CCR4- (Th1-like) memory cells between asthmatic and healthy individuals. Although many differences across the entire genome were identified, there was no major difference at the Th2 locus [218]. Similarly, no major changes in the status of DNA methylation at the Th2 locus were identified in the PBMC from asthmatic or atopic patients [265, 266]. However, a recent epigenomic association study demonstrated that in peripheral blood leukocytes, the degree of methylation of a CpG site upstream of the transcriptional start site of the *Il4* gene is inversely correlated with serum IgE level [267]. This inverse correlation cannot be explained by any SNP within the *Il4* gene. Another study found that *Il4*, *Il13*, and *Runx3* are hypomethylated in the PBMC of asthmatic children and that the degree of methylation is inversely correlated with the level of expression of these genes [268]. Runx3 is known to suppress IL-4 expression and the differentiation of Th2 cells, so it is unclear why its expression is higher in the PBMC of asthmatic individuals than in healthy individuals.

3.5.3 Roles of Epithelium-Derived Cytokines and Group 2 Innate Lymphoid Cells

While these genomic and epigenomic studies have yet to conclusively identify what triggers asthma or hyper-IgE, one picture has emerged from these studies, as well as from clinical studies of asthma: pathological type 2 inflammation can be achieved through several mechanisms, involving not only Th2 cells but also innate or innate-like cells. Many genome-wide association studies have found that several epithelium-derived cytokine genes, such as *Il33* and *Tslp* and *Il1rl1*, which encodes the subunit of the IL-33 receptor T1/ST2, are strongly associated with asthma [260, 269–271]. These findings are consistent with heightened expression of TSLP and IL-33 in the airways of asthmatic patients [272–275]. In addition, IL-25 expression in airway epithelial cells is also higher in asthmatic patients, and this increased level can be induced by allergen challenge or rhinoviral infection [276–278].

These three epithelium-derived cytokines are known to activate group 2 innate lymphoid cells (ILC2s), a newly discovered lineage of lymphocytes [279, 280], whose differentiation and function require GATA3. Though they do not express the known lineage markers, they express the common γ-chain, CD127, CD25, Sca-1, c-Kit, and Thy1, as well as the receptors for IL-33, TSLP, and IL-25. They also produce a high level of IL-13 and IL-5 in response to epithelium-derived cytokines. Thus, an emerging hypothesis of type 2 inflammation posits that allergens, irritants, and viruses induce epithelial cells to secrete IL-33, TSLP, and/or IL-25, which then act on ILC2s to produce type 2 cytokines and trigger type 2 inflammation. Indeed, ILC2s have been demonstrated to have a critical role in animal models of asthma [280]; an increased number of ILC2-like cells have been found in the sputum and bronchial lavage of asthmatic patients when compared to controls [281, 282]. As ILC2s produce a high level of IL-13 and IL-5, this model can explain some clinical features seen in asthma, such as hypertrophy of airway smooth muscle, eosinophilia, and overproduction of mucus. However, this ILC2-centered hypothesis does not involve adaptive immune cells and very likely operates mainly in "nonallergic" type 2 inflammation.

It is still controversial as to whether ILC2s are capable of producing a meaningful level of IL-4, as IL-33, TSLP, and/or IL-25 does not induce IL-4 production in ILC2 cells. However, stimulation of human blood or nasal polyp ILC2s with IL-2 and IL-33, human skin ILC2s with prostaglandin D2, or mouse pulmonary ILC2s with leukotriene D4 can induce the secretion of IL-4 protein [283–285]. Interestingly, the receptors for IL-33, TSLP, and IL-25 are also expressed in differentiated and memory Th2 cells; stimulation of differentiated and memory Th2 cells with IL-33 induces the expression of IL-13 and/or IL-5 but not IL-4 [286, 287]. However, both TSLP and IL-25, though not as potent as IL-4, can promote the differentiation of GATA3-expressing/IL-4-producing Th2 cells in an IL-4- and Stat6-dependent manner [288, 289]. Since TSLP augments the proliferation of differentiated Th2 cells and IL-4 further enhances the expression of TSLPR [290, 291], there is a resulting positive feedback loop. These results provide a link between epithelium-derived cytokines and IL-4-producing Th2 cells.

3.5.4 Other Cellular Sources of IL-4

IL-4 has a number of other cellular sources besides just Th2. Existing data have indicated that different sources of IL-4 have unique and shared functions. For example, in the context of eradiating worm infection, innate cell-derived IL-4/IL-13 is at least as important as IL-4/IL-13 derived from T cells. Basophils are one of the major sources of innate cell-derived IL-4 after infection with helminths [115, 292], collaborating with DCs to initiate type 2 immunity during worm infection. Mice deficient in IL-4/IL-13 in both T cells and basophils have a marked defect in clearing *N. brasiliensis*, whereas deficiency of IL-4/IL-13 in only T cells or basophils has no effect [293, 294]. In addition, mice deficient in IL-4/IL-13 only in basophils have a reduced number of Th2 cells in mesenteric lymph nodes after reinfection with *H. polygyrus* and are unable to clear the infection [295].

One explanation for the critical role of basophils in defense against worm infection is their ability to respond directly to protease activity. For example, active cysteine protease papain directly induces basophil production of IL-4 and IL-13 [296]. This effect requires IL-3, FcRγ, calcium flux, and activation of PI3K and NFAT [297]. Accordingly, immunization with active papain induces a strong Th2 response in a basophil-dependent manner. Helminths secrete a variety of proteases to facilitate their migration through tissues. These proteases and the protease Der p1, a potent allergen of house dust mite, can induce basophils to secrete IL-4 and IL-13 in vitro [298].

Basophils and their IL-4 also participate in other models of type 2 inflammation. In a basophil-dependent animal model of allergic dermatitis, basophil-derived IL-4 converted infiltrating monocytes into M2-like macrophages [299], which have anti-inflammatory effects and attenuate inflammation. They can also respond to TSLP, which induces basophilia in mice in an IL-3-independent manner [300]. Compared to IL-3-elicted basophils, TSLP-elicited basophils produce significantly more IL-4 in response to IL-3, IL-18, or IL-33.

Studies using IL-4 reporter mice also demonstrate that after helminth infection or immunization with helminth antigens, the majority of IL-4-secreting cells in draining lymph nodes are Tfh cells [116, 301, 302]. Given its strategic location, IL-4 derived from Tfh cells is probably responsible for the production of IgE and IgG1 by B cells during type 2 immune response.

The role of NKT cells in allergic asthma or immune defense against helminths is still controversial [303–305]. Asperamide B, a glycolipid derived from *Aspergillus fumigatus*, can directly activate both mouse and human NKT cells to produce IL-4 and IL-13 [306], suggesting that NKT cells and their IL-4 potentially play a role in the pathogenesis of allergic reactions to *A. fumigatus*.

Thus, the initial source of IL-4 and type 2 cytokines probably depends on the nature of the trigger of an immune response: ILC2 for irritants or viruses, basophils for proteases or helminths, and NKT cells for glycolipid antigens. Although eosinophils and mast cells are capable of secreting IL-4 and are also important players in type 2 inflammation [36, 307], the role of their IL-4 in allergy and worm infection has yet to be established. Eosinophils are essential for the induction of peanut food allergies. However, both IL-4-sufficient and IL-4-deficient eosinophils are capable of restoring Th2 priming and allergic reaction in eosinophil-deficient mice [308]. In addition, IL-4 derived from eosinophils is dispensable for the induction of allergic respiratory inflammation in mice [309]. Thus far, the protective and pathogenic role of IL-4 derived from mast cells has yet to be elucidated.

3.6 Conclusions and Perspectives

While the efficacy of recombinant IL-4Rα and anti-IL-4 in asthma has been disappointing [310, 311], anti-IL-4Rα, which blocks both type I and type II IL-4 receptors, has yielded promising results in patients with asthma or atopic dermatitis [312, 313]. In addition, GATA3-specific DNAzyme, which reduces the expression of GATA3, was recently shown to attenuate asthmatic responses after allergen provocation in patients with allergic asthma [314]. These encouraging results demonstrate the therapeutic potential of targeting molecules regulating the expression and/or signaling of IL-4 in pathological type 2 inflammation. Although the molecular mechanisms regulating the expression of IL-4 in vitro and in animal models of type 2 inflammation have been characterized, we still know very little about how IL-4 is dysregulated in human diseases. Clinical studies of asthma have clearly indicated that type 2 inflammation, such as in asthma, has a very heterogeneous pathogenesis and its clinical picture is heavily influenced by both genetic and environmental factors. A better understanding of the cellular source of IL-4 and how its expression is disturbed in various clinical settings will help us identify the most rewarding targets for drug development. Investigation into the functional significance of genetic variations that are associated with a higher risk of disease will eventually lead to the development of personalized treatments. In addition, the functional role of IL-4 in nonimmune cells, such as lymphatic endothelial cells, neurons, and adipocytes, is expanding and yet its cellular source and regulation in these contexts are poorly understood. Further investigation very likely will identify therapeutic targets of many other diseases.

References

1. Howard M, et al. Identification of a T cell-derived b cell growth factor distinct from interleukin 2. J Exp Med. 1982;155:914–23.
2. Isakson PC, Pure E, Vitetta ES, Krammer PH. T cell-derived B cell differentiation factor(s). Effect on the isotype switch of murine B cells. J Exp Med. 1982;155:734–48.
3. Lee F, et al. Isolation and characterization of a mouse interleukin cDNA clone that expresses B-cell stimulatory factor 1 activities and T-cell- and mast-cell-stimulating activities. Proc Natl Acad Sci U S A. 1986;83:2061–5.
4. Noma Y, et al. Cloning of cDNA encoding the murine IgG1 induction factor by a novel strategy using SP6 promoter. Nature. 1986;319:640–6. doi:10.1038/319640a0.
5. Carr C, Aykent S, Kimack NM, Levine AD. Disulfide assignments in recombinant mouse and human interleukin 4. Biochemistry. 1991;30:1515–23.
6. Walter MR, et al. Crystal structure of recombinant human interleukin-4. J Biol Chem. 1992;267:20371–6.
7. Wlodawer A, Pavlovsky A, Gustchina A. Crystal structure of human recombinant interleukin-4 at 2.25 A resolution. FEBS Lett. 1992;309:59–64.
8. Smith LJ, et al. Human interleukin 4. The solution structure of a four-helix bundle protein. J Mol Biol. 1992;224:899–904.

9. Powers R, et al. Three-dimensional solution structure of human interleukin-4 by multidimensional heteronuclear magnetic resonance spectroscopy. Science. 1992;256:1673–7.
10. Schmitt N, Ueno H. Regulation of human helper T cell subset differentiation by cytokines. Curr Opin Immunol. 2015;34:130–6. doi:10.1016/j.coi.2015.03.007.
11. Crotty S. T follicular helper cell differentiation, function, and roles in disease. Immunity. 2014;41:529–42. doi:10.1016/j.immuni.2014.10.004.
12. Johnston RJ, et al. Bcl6 and Blimp-1 are reciprocal and antagonistic regulators of T follicular helper cell differentiation. Science. 2009;325:1006–10. doi:10.1126/science.1175870.
13. Yu D, et al. The transcriptional repressor Bcl-6 directs T follicular helper cell lineage commitment. Immunity. 2009;31:457–68. doi:10.1016/j.immuni.2009.07.002.
14. Nurieva RI, et al. Bcl6 mediates the development of T follicular helper cells. Science. 2009;325:1001–5. doi:10.1126/science.1176676.
15. Riviere I, Sunshine MJ, Littman DR. Regulation of IL-4 expression by activation of individual alleles. Immunity. 1998;9:217–28.
16. Ho IC, Kaplan MH, Jackson-Grusby L, Glimcher LH, Grusby MJ. Marking IL-4-producing cells by knock-in of the IL-4 gene. Int Immunol. 1999;11:243–7.
17. Hu-Li J, et al. Regulation of expression of IL-4 alleles: analysis using a chimeric GFP/IL-4 gene. Immunity. 2001;14:1–11.
18. Mohrs K, Wakil AE, Killeen N, Locksley RM, Mohrs M. A two-step process for cytokine production revealed by IL-4 dual-reporter mice. Immunity. 2005;23:419–29. doi:10.1016/j.immuni.2005.09.006.
19. Gessner A, Mohrs K, Mohrs M. Mast cells, basophils, and eosinophils acquire constitutive IL-4 and IL-13 transcripts during lineage differentiation that are sufficient for rapid cytokine production. J Immunol. 2005;174:1063–72.
20. Mohrs K, Harris DP, Lund FE, Mohrs M. Systemic dissemination and persistence of Th2 and type 2 cells in response to infection with a strictly enteric nematode parasite. J Immunol. 2005;175:5306–13.
21. Wang T, Secombes CJ. The evolution of IL-4 and IL-13 and their receptor subunits. Cytokine. 2015. doi:10.1016/j.cyto.2015.04.012.
22. Munitz A, Brandt EB, Mingler M, Finkelman FD, Rothenberg ME. Distinct roles for IL-13 and IL-4 via IL-13 receptor alpha1 and the type II IL-4 receptor in asthma pathogenesis. Proc Natl Acad Sci U S A. 2008;105:7240–5. doi:10.1073/pnas.0802465105.
23. Junttila IS, et al. Tuning sensitivity to IL-4 and IL-13: differential expression of IL-4Ralpha, IL-13Ralpha1, and gammac regulates relative cytokine sensitivity. J Exp Med. 2008;205:2595–608. doi:10.1084/jem.20080452.
24. Ramalingam TR, et al. Unique functions of the type II interleukin 4 receptor identified in mice lacking the interleukin 13 receptor alpha1 chain. Nat Immunol. 2008;9:25–33. doi:10.1038/ni1544.
25. Goenka S, Kaplan MH. Transcriptional regulation by STAT6. Immunol Res. 2011;50:87–96. doi:10.1007/s12026-011-8205-2.
26. Heller NM, et al. Type I IL-4Rs selectively activate IRS-2 to induce target gene expression in macrophages. Sci Signal. 2008;1:ra17. doi:10.1126/scisignal.1164795.
27. Stephenson LM, Park DS, Mora AL, Goenka S, Boothby M. Sequence motifs in IL-4R alpha mediating cell-cycle progression of primary lymphocytes. J Immunol. 2005;175:5178–85.
28. Vazquez MI, Catalan-Dibene J, Zlotnik A. B cells responses and cytokine production are regulated by their immune microenvironment. Cytokine. 2015;74:318–26. doi:10.1016/j.cyto.2015.02.007.
29. Gauchat JF, et al. A novel 4-kb interleukin-13 receptor alpha mRNA expressed in human B, T, and endothelial cells encoding an alternate type-II interleukin-4/interleukin-13 receptor. Eur J Immunol. 1997;27:971–8. doi:10.1002/eji.1830270425.
30. Zurawski G, de Vries JE. Interleukin 13, an interleukin 4-like cytokine that acts on monocytes and B cells, but not on T cells. Immunol Today. 1994;15:19–26. doi:10.1016/0167-5699(94)90021-3.

31. Sica A, Mantovani A. Macrophage plasticity and polarization: in vivo veritas. J Clin Invest. 2012;122:787–95.
32. Mosser DM, Edwards JP. Exploring the full spectrum of macrophage activation. Nat Rev. 2008;8:958–69.
33. Stein M, Keshav S, Harris N, Gordon S. Interleukin 4 potently enhances murine macrophage mannose receptor activity: a marker of alternative immunologic macrophage activation. J Exp Med. 1992;176:287–92.
34. Doherty TM, Kastelein R, Menon S, Andrade S, Coffman RL. Modulation of murine macrophage function by IL-13. J Immunol. 1993;151:7151–60.
35. Banks EM, Coleman JW. A comparative study of peritoneal mast cells from mutant IL-4 deficient and normal mice: evidence that IL-4 is not essential for mast cell development but enhances secretion via control of IgE binding and passive sensitization. Cytokine. 1996;8:190–6. doi:10.1006/cyto.1996.0027.
36. McLeod JJ, Baker B, Ryan JJ. Mast cell production and response to IL-4 and IL-13. Cytokine. 2015. doi:10.1016/j.cyto.2015.05.019.
37. Chen XJ, Lycke N, Enerback L. Surface and gene expression of immunoglobulin E receptors on mast cells and mast-cell numbers in interleukin-4-gene knockout mice. Immunology. 1999;96:544–50.
38. Yamaguchi M, et al. IgE enhances Fc epsilon receptor I expression and IgE-dependent release of histamine and lipid mediators from human umbilical cord blood-derived mast cells: synergistic effect of IL-4 and IgE on human mast cell Fc epsilon receptor I expression and mediator release. J Immunol. 1999;162:5455–65.
39. Marcais A, et al. Regulation of mouse NK cell development and function by cytokines. Front Immunol. 2013;4:450. doi:10.3389/fimmu.2013.00450.
40. Lee YJ, Jameson SC, Hogquist KA. Alternative memory in the CD8 T cell lineage. Trends Immunol. 2011;32:50–6. doi:10.1016/j.it.2010.12.004.
41. Weinreich MA, Odumade OA, Jameson SC, Hogquist KA. T cells expressing the transcription factor PLZF regulate the development of memory-like CD8+ T cells. Nat Immunol. 2010;11:709–16. doi:10.1038/ni.1898.
42. Gordon SM, et al. Requirements for eomesodermin and promyelocytic leukemia zinc finger in the development of innate-like CD8+ T cells. J Immunol. 2011;186:4573–8. doi:10.4049/jimmunol.1100037.
43. Lai D, et al. KLF13 sustains thymic memory-like CD8(+) T cells in BALB/c mice by regulating IL-4-generating invariant natural killer T cells. J Exp Med. 2011;208:1093–103. doi:10.1084/jem.20101527.
44. Carty SA, Koretzky GA, Jordan MS. Interleukin-4 regulates eomesodermin in CD8+ T cell development and differentiation. PLoS One. 2014;9:e106659. doi:10.1371/journal.pone.0106659.
45. Pulendran B, Artis D. New paradigms in type 2 immunity. Science. 2012;337:431–5. doi:10.1126/science.1221064.
46. Urban Jr JF, Maliszewski CR, Madden KB, Katona IM, Finkelman FD. IL-4 treatment can cure established gastrointestinal nematode infections in immunocompetent and immunodeficient mice. J Immunol. 1995;154:4675–84.
47. Grunig G, et al. Requirement for IL-13 independently of IL-4 in experimental asthma. Science. 1998;282:2261–3.
48. Bancroft AJ, McKenzie AN, Grencis RK. A critical role for IL-13 in resistance to intestinal nematode infection. J Immunol. 1998;160:3453–61.
49. Urban Jr JF, Katona IM, Paul WE, Finkelman FD. Interleukin 4 is important in protective immunity to a gastrointestinal nematode infection in mice. Proc Natl Acad Sci U S A. 1991;88:5513–7.
50. Brusselle GG, et al. Attenuation of allergic airway inflammation in IL-4 deficient mice. Clin Exp Allergy J Br Soc Allergy Clin Immunol. 1994;24:73–80.

51. Corry DB, et al. Interleukin 4, but not interleukin 5 or eosinophils, is required in a murine model of acute airway hyperreactivity. J Exp Med. 1996;183:109–17.
52. Rankin JA, et al. Phenotypic and physiologic characterization of transgenic mice expressing interleukin 4 in the lung: lymphocytic and eosinophilic inflammation without airway hyperreactivity. Proc Natl Acad Sci U S A. 1996;93:7821–5.
53. Wills-Karp M, et al. Interleukin-13: central mediator of allergic asthma. Science. 1998;282:2258–61.
54. Urban Jr JF, et al. IL-13, IL-4Ralpha, and Stat6 are required for the expulsion of the gastrointestinal nematode parasite Nippostrongylus brasiliensis. Immunity. 1998;8:255–64.
55. Shin K, et al. TH2 cells and their cytokines regulate formation and function of lymphatic vessels. Nat Commun. 2015;6:6196. doi:10.1038/ncomms7196.
56. Savetsky IL, et al. Th2 cytokines inhibit lymphangiogenesis. PLoS One. 2015;10:e0126908. doi:10.1371/journal.pone.0126908.
57. Walsh JT, et al. MHCII-independent CD4+ T cells protect injured CNS neurons via IL-4. J Clin Invest. 2015;125:699–714. doi:10.1172/JCI76210.
58. Karp CL, Murray PJ. Non-canonical alternatives: what a macrophage is 4. J Exp Med. 2012;209:427–31. doi:10.1084/jem.20120295.
59. Nguyen KD, et al. Alternatively activated macrophages produce catecholamines to sustain adaptive thermogenesis. Nature. 2011;480:104–8. doi:10.1038/nature10653.
60. Tsao CH, Shiau MY, Chuang PH, Chang YH, Hwang J. Interleukin-4 regulates lipid metabolism by inhibiting adipogenesis and promoting lipolysis. J Lipid Res. 2014;55:385–97. doi:10.1194/jlr.M041392.
61. Wu D, et al. Eosinophils sustain adipose alternatively activated macrophages associated with glucose homeostasis. Science. 2011;332:243–7. doi:10.1126/science.1201475.
62. Qiu Y, et al. Eosinophils and type 2 cytokine signaling in macrophages orchestrate development of functional beige fat. Cell. 2014;157:1292–308. doi:10.1016/j.cell.2014.03.066.
63. Rao RR, et al. Meteorin-like is a hormone that regulates immune-adipose interactions to increase beige fat thermogenesis. Cell. 2014;157:1279–91. doi:10.1016/j.cell.2014.03.065.
64. Li-Weber M, Krammer PH. Regulation of IL4 gene expression by T cells and therapeutic perspectives. Nat Rev Immunol. 2003;3:534–43. doi:10.1038/nri1128.
65. Takemoto N, et al. Th2-specific DNase I-hypersensitive sites in the murine IL-13 and IL-4 intergenic region. Int Immunol. 1998;10:1981–5.
66. Loots GG, et al. Identification of a coordinate regulator of interleukins 4, 13, and 5 by cross-species sequence comparisons. Science. 2000;288:136–40.
67. Takemoto N, et al. Cutting edge: chromatin remodeling at the IL-4/IL-13 intergenic regulatory region for Th2-specific cytokine gene cluster. J Immunol. 2000;165:6687–91.
68. Mohrs M, et al. Deletion of a coordinate regulator of type 2 cytokine expression in mice. Nat Immunol. 2001;2:842–7. doi:10.1038/ni0901-842.
69. Tanaka S, et al. The enhancer HS2 critically regulates GATA-3-mediated Il4 transcription in T(H)2 cells. Nat Immunol. 2011;12:77–85. doi:10.1038/ni.1966.
70. Henkel G, et al. A DNase I-hypersensitive site in the second intron of the murine IL-4 gene defines a mast cell-specific enhancer. J Immunol. 1992;149:3239–46.
71. Agarwal S, Rao A. Modulation of chromatin structure regulates cytokine gene expression during T cell differentiation. Immunity. 1998;9:765–75.
72. Lee GR, Fields PE, Flavell RA. Regulation of IL-4 gene expression by distal regulatory elements and GATA-3 at the chromatin level. Immunity. 2001;14:447–59.
73. Zhu J, Cote-Sierra J, Guo L, Paul WE. Stat5 activation plays a critical role in Th2 differentiation. Immunity. 2003;19:739–48.
74. Agarwal S, Avni O, Rao A. Cell-type-restricted binding of the transcription factor NFAT to a distal IL-4 enhancer in vivo. Immunity. 2000;12:643–52.
75. Guo L, Hu-Li J, Paul WE. Probabilistic regulation of IL-4 production in Th2 cells: accessibility at the Il4 locus. Immunity. 2004;20:193–203.

76. Kock J, et al. Nuclear factor of activated T cells regulates the expression of interleukin-4 in Th2 cells in an all-or-none fashion. J Biol Chem. 2014;289:26752–61. doi:10.1074/jbc.M114.587865.
77. Tanaka S, et al. The interleukin-4 enhancer CNS-2 is regulated by Notch signals and controls initial expression in NKT cells and memory-type CD4 T cells. Immunity. 2006;24:689–701. doi:10.1016/j.immuni.2006.04.009.
78. Solymar DC, Agarwal S, Bassing CH, Alt FW, Rao A. A 3′ enhancer in the IL-4 gene regulates cytokine production by Th2 cells and mast cells. Immunity. 2002;17:41–50.
79. Harada Y, et al. The 3′ enhancer CNS2 is a critical regulator of interleukin-4-mediated humoral immunity in follicular helper T cells. Immunity. 2012;36:188–200. doi:10.1016/j.immuni.2012.02.002.
80. Vijayanand P, et al. Interleukin-4 production by follicular helper T cells requires the conserved Il4 enhancer hypersensitivity site V. Immunity. 2012;36:175–87. doi:10.1016/j.immuni.2011.12.014.
81. Yagi R, Tanaka S, Motomura Y, Kubo M. Regulation of the Il4 gene is independently controlled by proximal and distal 3′ enhancers in mast cells and basophils. Mol Cell Biol. 2007;27:8087–97. doi:10.1128/MCB.00631-07.
82. Djuretic IM, et al. Transcription factors T-bet and Runx3 cooperate to activate Ifng and silence Il4 in T helper type 1 cells. Nat Immunol. 2007;8:145–53. doi:10.1038/ni1424.
83. Lee GR, Fields PE, Griffin TJ, Flavell RA. Regulation of the Th2 cytokine locus by a locus control region. Immunity. 2003;19:145–53.
84. Koh BH, et al. Th2 LCR is essential for regulation of Th2 cytokine genes and for pathogenesis of allergic asthma. Proc Natl Acad Sci U S A. 2010;107:10614–9. doi:10.1073/pnas.1005383107.
85. Fields PE, Lee GR, Kim ST, Bartsevich VV, Flavell RA. Th2-specific chromatin remodeling and enhancer activity in the Th2 cytokine locus control region. Immunity. 2004;21:865–76. doi:10.1016/j.immuni.2004.10.015.
86. Lee DU, Rao A. Molecular analysis of a locus control region in the T helper 2 cytokine gene cluster: a target for STAT6 but not GATA3. Proc Natl Acad Sci U S A. 2004;101:16010–5. doi:10.1073/pnas.0407031101.
87. Spilianakis CG, Flavell RA. Long-range intrachromosomal interactions in the T helper type 2 cytokine locus. Nat Immunol. 2004;5:1017–27. doi:10.1038/ni1115.
88. Lee GR, Spilianakis CG, Flavell RA. Hypersensitive site 7 of the TH2 locus control region is essential for expressing TH2 cytokine genes and for long-range intrachromosomal interactions. Nat Immunol. 2005;6:42–8. doi:10.1038/ni1148.
89. Williams A, et al. Hypersensitive site 6 of the Th2 locus control region is essential for Th2 cytokine expression. Proc Natl Acad Sci U S A. 2013;110:6955–60. doi:10.1073/pnas.1304720110.
90. Kubo M, et al. T-cell subset-specific expression of the IL-4 gene is regulated by a silencer element and STAT6. EMBO J. 1997;16:4007–20. doi:10.1093/emboj/16.13.4007.
91. Ansel KM, et al. Deletion of a conserved Il4 silencer impairs T helper type 1-mediated immunity. Nat Immunol. 2004;5:1251–9. doi:10.1038/ni1135.
92. Blonska M, Joo D, Zweidler-McKay PA, Zhao Q, Lin X. CARMA1 controls Th2 cell-specific cytokine expression through regulating JunB and GATA3 transcription factors. J Immunol. 2012;188:3160–8. doi:10.4049/jimmunol.1102943.
93. Hosken NA, Shibuya K, Heath AW, Murphy KM, O'Garra A. The effect of antigen dose on CD4+ T helper cell phenotype development in a T cell receptor-alpha beta-transgenic model. J Exp Med. 1995;182:1579–84.
94. Pfeiffer C, et al. Altered peptide ligands can control CD4 T lymphocyte differentiation in vivo. J Exp Med. 1995;181:1569–74.
95. Constant S, Pfeiffer C, Woodard A, Pasqualini T, Bottomly K. Extent of T cell receptor ligation can determine the functional differentiation of naive CD4+ T cells. J Exp Med. 1995;182:1591–6.

96. Jorritsma PJ, Brogdon JL, Bottomly K. Role of TCR-induced extracellular signal-regulated kinase activation in the regulation of early IL-4 expression in naive CD4+ T cells. J Immunol. 2003;170:2427–34.
97. Elo LL, et al. Genome-wide profiling of interleukin-4 and STAT6 transcription factor regulation of human Th2 cell programming. Immunity. 2010;32:852–62. doi:10.1016/j.immuni.2010.06.011.
98. Kaplan MH, Wurster AL, Smiley ST, Grusby MJ. Stat6-dependent and -independent pathways for IL-4 production. J Immunol. 1999;163:6536–40.
99. van Panhuys N, et al. Basophils are the major producers of IL-4 during primary helminth infection. J Immunol. 2011;186:2719–28. doi:10.4049/jimmunol.1000940.
100. Stokes K, et al. Cutting edge: STAT6 signaling in eosinophils is necessary for development of allergic airway inflammation. J Immunol. 2015;194:2477–81. doi:10.4049/jimmunol.1402096.
101. Kaplan MH, Schindler U, Smiley ST, Grusby MJ. Stat6 is required for mediating responses to IL-4 and for development of Th2 cells. Immunity. 1996;4:313–9.
102. Shimoda K, et al. Lack of IL-4-induced Th2 response and IgE class switching in mice with disrupted Stat6 gene. Nature. 1996;380:630–3. doi:10.1038/380630a0.
103. Takeda K, et al. Essential role of Stat6 in IL-4 signalling. Nature. 1996;380:627–30. doi:10.1038/380627a0.
104. Kurata H, Lee HJ, O'Garra A, Arai N. Ectopic expression of activated Stat6 induces the expression of Th2-specific cytokines and transcription factors in developing Th1 cells. Immunity. 1999;11:677–88.
105. Hanson EM, Dickensheets H, Qu CK, Donnelly RP, Keegan AD. Regulation of the dephosphorylation of Stat6. Participation of Tyr-713 in the interleukin-4 receptor alpha, the tyrosine phosphatase SHP-1, and the proteasome. J Biol Chem. 2003;278:3903–11. doi:10.1074/jbc.M211747200.
106. Johnson DJ, et al. Shp1 regulates T cell homeostasis by limiting IL-4 signals. J Exp Med. 2013;210:1419–31. doi:10.1084/jem.20122239.
107. Sahoo A, Alekseev A, Obertas L, Nurieva R. Grail controls Th2 cell development by targeting STAT6 for degradation. Nat Commun. 2014;5:4732. doi:10.1038/ncomms5732.
108. Ouyang W, et al. Stat6-independent GATA-3 autoactivation directs IL-4-independent Th2 development and commitment. Immunity. 2000;12:27–37.
109. Asnagli H, Afkarian M, Murphy KM. Cutting edge: identification of an alternative GATA-3 promoter directing tissue-specific gene expression in mouse and human. J Immunol. 2002;168:4268–71.
110. Scheinman EJ, Avni O. Transcriptional regulation of GATA3 in T helper cells by the integrated activities of transcription factors downstream of the interleukin-4 receptor and T cell receptor. J Biol Chem. 2009;284:3037–48. doi:10.1074/jbc.M807302200.
111. Cai S, Lee CC, Kohwi-Shigematsu T. SATB1 packages densely looped, transcriptionally active chromatin for coordinated expression of cytokine genes. Nat Genet. 2006;38:1278–88. doi:10.1038/ng1913.
112. Kim ST, Fields PE, Flavell RA. Demethylation of a specific hypersensitive site in the Th2 locus control region. Proc Natl Acad Sci U S A. 2007;104:17052–7. doi:10.1073/pnas.0708293104.
113. Wei L, et al. Discrete roles of STAT4 and STAT6 transcription factors in tuning epigenetic modifications and transcription during T helper cell differentiation. Immunity. 2010;32:840–51. doi:10.1016/j.immuni.2010.06.003.
114. Finkelman FD, et al. Stat6 regulation of in vivo IL-4 responses. J Immunol. 2000;164:2303–10.
115. Voehringer D, Shinkai K, Locksley RM. Type 2 immunity reflects orchestrated recruitment of cells committed to IL-4 production. Immunity. 2004;20:267–77.
116. Liang HE, et al. Divergent expression patterns of IL-4 and IL-13 define unique functions in allergic immunity. Nat Immunol. 2012;13:58–66. doi:10.1038/ni.2182.

117. van Panhuys N, et al. In vivo studies fail to reveal a role for IL-4 or STAT6 signaling in Th2 lymphocyte differentiation. Proc Natl Acad Sci U S A. 2008;105:12423–8. doi:10.1073/pnas.0806372105.
118. Jankovic D, et al. Single cell analysis reveals that IL-4 receptor/Stat6 signaling is not required for the in vivo or in vitro development of CD4+ lymphocytes with a Th2 cytokine profile. J Immunol. 2000;164:3047–55.
119. Hwang ES, White IA, Ho IC. An IL-4-independent and CD25-mediated function of c-maf in promoting the production of Th2 cytokines. Proc Natl Acad Sci U S A. 2002;99:13026–30. doi:10.1073/pnas.202474499.
120. Ben-Sasson SZ, Le Gros G, Conrad DH, Finkelman FD, Paul WE. IL-4 production by T cells from naive donors. IL-2 is required for IL-4 production. J Immunol. 1990;145:1127–36.
121. Cote-Sierra J, et al. Interleukin 2 plays a central role in Th2 differentiation. Proc Natl Acad Sci U S A. 2004;101:3880–5. doi:10.1073/pnas.0400339101.
122. Kagami S, et al. Both stat5a and stat5b are required for antigen-induced eosinophil and T-cell recruitment into the tissue. Blood. 2000;95:1370–7.
123. Takatori H, et al. Indispensable role of Stat5a in Stat6-independent Th2 cell differentiation and allergic airway inflammation. J Immunol. 2005;174:3734–40.
124. Zhu J, et al. Conditional deletion of Gata3 shows its essential function in T(H)1-T(H)2 responses. Nat Immunol. 2004;5:1157–65. doi:10.1038/ni1128.
125. Rani A, et al. IL-2 regulates expression of C-MAF in human CD4 T cells. J Immunol. 2011;187:3721–9. doi:10.4049/jimmunol.1002354.
126. Liao W, et al. Priming for T helper type 2 differentiation by interleukin 2-mediated induction of interleukin 4 receptor alpha-chain expression. Nat Immunol. 2008;9:1288–96. doi:10.1038/ni.1656.
127. Mukherjee S, et al. STAT5-induced lunatic fringe during Th2 development alters delta-like 4-mediated Th2 cytokine production in respiratory syncytial virus-exacerbated airway allergic disease. J Immunol. 2014;192:996–1003. doi:10.4049/jimmunol.1301991.
128. Adler SH, et al. Notch signaling augments T cell responsiveness by enhancing CD25 expression. J Immunol. 2003;171:2896–903.
129. Takatori H, et al. Stat5a inhibits IL-12-induced Th1 cell differentiation through the induction of suppressor of cytokine signaling 3 expression. J Immunol. 2005;174:4105–12.
130. Bruchard M, et al. The receptor NLRP3 is a transcriptional regulator of T2 differentiation. Nat Immunol. 2015. doi:10.1038/ni.3202.
131. Radtke F, MacDonald HR, Tacchini-Cottier F. Regulation of innate and adaptive immunity by Notch. Nat Rev Immunol. 2013;13:427–37. doi:10.1038/nri3445.
132. Tanigaki K, et al. Regulation of alphabeta/gammadelta T cell lineage commitment and peripheral T cell responses by Notch/RBP-J signaling. Immunity. 2004;20:611–22.
133. Amsen D, et al. Direct regulation of Gata3 expression determines the T helper differentiation potential of Notch. Immunity. 2007;27:89–99. doi:10.1016/j.immuni.2007.05.021.
134. Amsen D, et al. Instruction of distinct CD4 T helper cell fates by different notch ligands on antigen-presenting cells. Cell. 2004;117:515–26.
135. Fang TC, et al. Notch directly regulates Gata3 expression during T helper 2 cell differentiation. Immunity. 2007;27:100–10. doi:10.1016/j.immuni.2007.04.018.
136. Tu L, et al. Notch signaling is an important regulator of type 2 immunity. J Exp Med. 2005;202:1037–42. doi:10.1084/jem.20050923.
137. Skokos D, Nussenzweig MC. CD8- DCs induce IL-12-independent Th1 differentiation through Delta 4 Notch-like ligand in response to bacterial LPS. J Exp Med. 2007;204:1525–31. doi:10.1084/jem.20062305.
138. Maekawa Y, et al. Delta1-Notch3 interactions bias the functional differentiation of activated CD4+ T cells. Immunity. 2003;19:549–59.
139. Sun J, Krawczyk CJ, Pearce EJ. Suppression of Th2 cell development by Notch ligands Delta1 and Delta4. J Immunol. 2008;180:1655–61.

140. Ong CT, Sedy JR, Murphy KM, Kopan R. Notch and presenilin regulate cellular expansion and cytokine secretion but cannot instruct Th1/Th2 fate acquisition. PLoS One. 2008;3:e2823. doi:10.1371/journal.pone.0002823.
141. Krawczyk CM, Sun J, Pearce EJ. Th2 differentiation is unaffected by Jagged2 expression on dendritic cells. J Immunol. 2008;180:7931–7.
142. Bailis W, et al. Notch simultaneously orchestrates multiple helper T cell programs independently of cytokine signals. Immunity. 2013;39:148–59. doi:10.1016/j.immuni.2013.07.006.
143. Staal FJ, Luis TC, Tiemessen MM. WNT signalling in the immune system: WNT is spreading its wings. Nat Rev Immunol. 2008;8:581–93. doi:10.1038/nri2360.
144. Notani D, et al. Global regulator SATB1 recruits beta-catenin and regulates T(H)2 differentiation in Wnt-dependent manner. PLoS Biol. 2010;8:e1000296. doi:10.1371/journal.pbio.1000296.
145. Yu Q, et al. T cell factor 1 initiates the T helper type 2 fate by inducing the transcription factor GATA-3 and repressing interferon-gamma. Nat Immunol. 2009;10:992–9. doi:10.1038/ni.1762.
146. Maier E, et al. Inhibition of suppressive T cell factor 1 (TCF-1) isoforms in naive CD4+ T cells is mediated by IL-4/STAT6 signaling. J Biol Chem. 2011;286:919–28. doi:10.1074/jbc.M110.144949.
147. Hossain MB, et al. Lymphoid enhancer factor interacts with GATA-3 and controls its function in T helper type 2 cells. Immunology. 2008;125:377–86. doi:10.1111/j.1365-2567.2008.02854.x.
148. Hebenstreit D, et al. LEF-1 negatively controls interleukin-4 expression through a proximal promoter regulatory element. J Biol Chem. 2008;283:22490–7. doi:10.1074/jbc.M804096200.
149. Albert V, Hall MN. mTOR signaling in cellular and organismal energetics. Curr Opin Cell Biol. 2015;33:55–66. doi:10.1016/j.ceb.2014.12.001.
150. Zinzalla V, Stracka D, Oppliger W, Hall MN. Activation of mTORC2 by association with the ribosome. Cell. 2011;144:757–68. doi:10.1016/j.cell.2011.02.014.
151. Delgoffe GM, et al. The mTOR kinase differentially regulates effector and regulatory T cell lineage commitment. Immunity. 2009;30:832–44. doi:10.1016/j.immuni.2009.04.014.
152. Delgoffe GM, et al. The kinase mTOR regulates the differentiation of helper T cells through the selective activation of signaling by mTORC1 and mTORC2. Nat Immunol. 2011;12:295–303. doi:10.1038/ni.2005.
153. Lee K, et al. Mammalian target of rapamycin protein complex 2 regulates differentiation of Th1 and Th2 cell subsets via distinct signaling pathways. Immunity. 2010;32:743–53. doi:10.1016/j.immuni.2010.06.002.
154. Yang K, et al. T cell exit from quiescence and differentiation into Th2 cells depend on Raptor-mTORC1-mediated metabolic reprogramming. Immunity. 2013;39:1043–56. doi:10.1016/j.immuni.2013.09.015.
155. Heikamp EB, et al. The AGC kinase SGK1 regulates TH1 and TH2 differentiation downstream of the mTORC2 complex. Nat Immunol. 2014;15:457–64. doi:10.1038/ni.2867.
156. Imanishi T, et al. Nucleic acid sensing by T cells initiates Th2 cell differentiation. Nat Commun. 2014;5:3566. doi:10.1038/ncomms4566.
157. Ho IC, et al. Human GATA-3: a lineage-restricted transcription factor that regulates the expression of the T cell receptor alpha gene. EMBO J. 1991;10:1187–92.
158. Ko LJ, et al. Murine and human T-lymphocyte GATA-3 factors mediate transcription through a cis-regulatory element within the human T-cell receptor delta gene enhancer. Mol Cell Biol. 1991;11:2778–84.
159. Ting CN, Olson MC, Barton KP, Leiden JM. Transcription factor GATA-3 is required for development of the T-cell lineage. Nature. 1996;384:474–8. doi:10.1038/384474a0.
160. Pai SY, et al. Critical roles for transcription factor GATA-3 in thymocyte development. Immunity. 2003;19:863–75.
161. Tindemans I, Serafini N, Di Santo JP, Hendriks RW. GATA-3 function in innate and adaptive immunity. Immunity. 2014;41:191–206. doi:10.1016/j.immuni.2014.06.006.

162. Kim PJ, et al. GATA-3 regulates the development and function of invariant NKT cells. J Immunol. 2006;177:6650–9.
163. Tai TS, Pai SY, Ho IC. GATA-3 regulates the homeostasis and activation of CD8+ T cells. J Immunol. 2013;190:428–37. doi:10.4049/jimmunol.1201361.
164. Wang Y, et al. GATA-3 controls the maintenance and proliferation of T cells downstream of TCR and cytokine signaling. Nat Immunol. 2013;14:714–22. doi:10.1038/ni.2623.
165. Zheng W, Flavell RA. The transcription factor GATA-3 is necessary and sufficient for Th2 cytokine gene expression in CD4 T cells. Cell. 1997;89:587–96.
166. Ouyang W, et al. Inhibition of Th1 development mediated by GATA-3 through an IL-4-independent mechanism. Immunity. 1998;9:745–55.
167. Pai SY, Truitt ML, Ho IC. GATA-3 deficiency abrogates the development and maintenance of T helper type 2 cells. Proc Natl Acad Sci U S A. 2004;101:1993–8. doi:10.1073/pnas.0308697100.
168. Lee HJ, et al. GATA-3 induces T helper cell type 2 (Th2) cytokine expression and chromatin remodeling in committed Th1 cells. J Exp Med. 2000;192:105–15.
169. Zhang DH, Cohn L, Ray P, Bottomly K, Ray A. Transcription factor GATA-3 is differentially expressed in murine Th1 and Th2 cells and controls Th2-specific expression of the interleukin-5 gene. J Biol Chem. 1997;272:21597–603.
170. Kishikawa H, Sun J, Choi A, Miaw SC, Ho IC. The cell type-specific expression of the murine IL-13 gene is regulated by GATA-3. J Immunol. 2001;167:4414–20.
171. Hosokawa H, et al. Methylation of gata3 protein at arg-261 regulates transactivation of the il5 gene in T helper 2 cells. J Biol Chem. 2015;290:13095–103. doi:10.1074/jbc.M114.621524.
172. Yagi R, Zhu J, Paul WE. An updated view on transcription factor GATA3-mediated regulation of Th1 and Th2 cell differentiation. Int Immunol. 2011;23:415–20. doi:10.1093/intimm/dxr029.
173. Wei G, et al. Genome-wide analyses of transcription factor GATA3-mediated gene regulation in distinct T cell types. Immunity. 2011;35:299–311. doi:10.1016/j.immuni.2011.08.007.
174. Horiuchi S, et al. Genome-wide analysis reveals unique regulation of transcription of Th2-specific genes by GATA3. J Immunol. 2011;186:6378–89. doi:10.4049/jimmunol.1100179.
175. Kanhere A, et al. T-bet and GATA3 orchestrate Th1 and Th2 differentiation through lineage-specific targeting of distal regulatory elements. Nat Commun. 2012;3:1268. doi:10.1038/ncomms2260.
176. Ho IC, Hodge MR, Rooney JW, Glimcher LH. The proto-oncogene c-maf is responsible for tissue-specific expression of interleukin-4. Cell. 1996;85:973–83.
177. Nurieva RI, et al. Transcriptional regulation of th2 differentiation by inducible costimulator. Immunity. 2003;18:801–11.
178. Bauquet AT, et al. The costimulatory molecule ICOS regulates the expression of c-Maf and IL-21 in the development of follicular T helper cells and TH-17 cells. Nat Immunol. 2009;10:167–75. doi:10.1038/ni.1690.
179. Hodge MR, et al. NF-AT-Driven interleukin-4 transcription potentiated by NIP45. Science. 1996;274:1903–5.
180. Fathman JW, et al. NIP45 controls the magnitude of the type 2 T helper cell response. Proc Natl Acad Sci U S A. 2010;107:3663–8. doi:10.1073/pnas.0914700107.
181. Ho IC, Lo D, Glimcher LH. c-maf promotes T helper cell type 2 (Th2) and attenuates Th1 differentiation by both interleukin 4-dependent and -independent mechanisms. J Exp Med. 1998;188:1859–66.
182. Kim JI, Ho IC, Grusby MJ, Glimcher LH. The transcription factor c-Maf controls the production of interleukin-4 but not other Th2 cytokines. Immunity. 1999;10:745–51.
183. Leavenworth JW, Ma X, Mo YY, Pauza ME. SUMO conjugation contributes to immune deviation in nonobese diabetic mice by suppressing c-Maf transactivation of IL-4. J Immunol. 2009;183:1110–9. doi:10.4049/jimmunol.0803671.
184. Lin BS, et al. SUMOylation attenuates c-Maf-dependent IL-4 expression. Eur J Immunol. 2010;40:1174–84. doi:10.1002/eji.200939788.

185. Lai CY, et al. Tyrosine phosphorylation of c-Maf enhances the expression of IL-4 gene. J Immunol. 2012;189:1545–50. doi:10.4049/jimmunol.1200405.
186. Liu CC, et al. Reciprocal regulation of C-Maf tyrosine phosphorylation by Tec and Ptpn22. PLoS One. 2015;10:e0127617. doi:10.1371/journal.pone.0127617.
187. Rincon M, Derijard B, Chow CW, Davis RJ, Flavell RA. Reprogramming the signalling requirement for AP-1 (activator protein-1) activation during differentiation of precursor CD4+ T-cells into effector Th1 and Th2 cells. Genes Funct. 1997;1:51–68.
188. Li B, Tournier C, Davis RJ, Flavell RA. Regulation of IL-4 expression by the transcription factor JunB during T helper cell differentiation. EMBO J. 1999;18:420–32. doi:10.1093/emboj/18.2.420.
189. Hartenstein B, et al. Th2 cell-specific cytokine expression and allergen-induced airway inflammation depend on JunB. EMBO J. 2002;21:6321–9.
190. Goplen N, et al. ERK1 is important for Th2 differentiation and development of experimental asthma. FASEB J: Off Publ Fed Am Soc Exp Biol. 2012;26:1934–45. doi:10.1096/fj.11-196477.
191. Garaude J, et al. SUMOylation regulates the transcriptional activity of JunB in T lymphocytes. J Immunol. 2008;180:5983–90.
192. Fang D, et al. Dysregulation of T lymphocyte function in itchy mice: a role for Itch in TH2 differentiation. Nat Immunol. 2002;3:281–7. doi:10.1038/ni763.
193. Oliver PM, et al. Ndfip1 protein promotes the function of itch ubiquitin ligase to prevent T cell activation and T helper 2 cell-mediated inflammation. Immunity. 2006;25:929–40. doi:10.1016/j.immuni.2006.10.012.
194. Yang XO, et al. Requirement for the basic helix-loop-helix transcription factor Dec2 in initial TH2 lineage commitment. Nat Immunol. 2009;10:1260–6. doi:10.1038/ni.1821.
195. Liu Z, et al. Dec2 promotes Th2 cell differentiation by enhancing IL-2R signaling. J Immunol. 2009;183:6320–9. doi:10.4049/jimmunol.0900975.
196. Kwon SJ, et al. KLF13 cooperates with c-Maf to regulate IL-4 expression in CD4+ T cells. J Immunol. 2014;192:5703–9. doi:10.4049/jimmunol.1302830.
197. Guo J, et al. Yin-Yang 1 activates interleukin-4 gene expression in T cells. J Biol Chem. 2001;276:48871–8. doi:10.1074/jbc.M101592200.
198. Hwang SS, et al. Transcription factor YY1 is essential for regulation of the Th2 cytokine locus and for Th2 cell differentiation. Proc Natl Acad Sci U S A. 2013;110:276–81. doi:10.1073/pnas.1214682110.
199. Guo J, Lin X, Williams MA, Hamid Q, Georas SN. Yin-Yang 1 regulates effector cytokine gene expression and T(H)2 immune responses. J Allergy Clin Immunol.2008;122:195–201, 201 e191-195. doi:10.1016/j.jaci.2008.03.012.
200. Dickinson LA, Joh T, Kohwi Y, Kohwi-Shigematsu T. A tissue-specific MAR/SAR DNA-binding protein with unusual binding site recognition. Cell. 1992;70:631–45.
201. Yasui D, Miyano M, Cai S, Varga-Weisz P, Kohwi-Shigematsu T. SATB1 targets chromatin remodelling to regulate genes over long distances. Nature. 2002;419:641–5. doi:10.1038/nature01084.
202. Ahlfors H, et al. SATB1 dictates expression of multiple genes including IL-5 involved in human T helper cell differentiation. Blood. 2010;116:1443–53. doi:10.1182/blood-2009-11-252205.
203. Son JS, Chae CS, Hwang JS, Park ZY, Im SH. Enhanced chromatin accessibility and recruitment of JUNB mediate the sustained IL-4 expression in NFAT1 deficient T helper 2 cells. PLoS One. 2011;6:e22042. doi:10.1371/journal.pone.0022042.
204. Elser B, et al. IFN-gamma represses IL-4 expression via IRF-1 and IRF-2. Immunity. 2002;17:703–12.
205. Maruyama S, et al. A novel function of interferon regulatory factor-1: inhibition of Th2 cells by down-regulating the Il4 gene during Listeria infection. Int Immunol. 2015;27:143–52. doi:10.1093/intimm/dxu092.

206. Naoe Y, et al. Repression of interleukin-4 in T helper type 1 cells by Runx/Cbf beta binding to the Il4 silencer. J Exp Med. 2007;204:1749–55. doi:10.1084/jem.20062456.
207. Kohu K, et al. The Runx3 transcription factor augments Th1 and down-modulates Th2 phenotypes by interacting with and attenuating GATA3. J Immunol. 2009;183:7817–24. doi:10.4049/jimmunol.0802527.
208. Chang HC, et al. PU.1 expression delineates heterogeneity in primary Th2 cells. Immunity. 2005;22:693–703. doi:10.1016/j.immuni.2005.03.016.
209. Dent AL, Shaffer AL, Yu X, Allman D, Staudt LM. Control of inflammation, cytokine expression, and germinal center formation by BCL-6. Science. 1997;276:589–92.
210. Kusam S, Toney LM, Sato H, Dent AL. Inhibition of Th2 differentiation and GATA-3 expression by BCL-6. J Immunol. 2003;170:2435–41.
211. Kuwahara M, et al. The transcription factor Sox4 is a downstream target of signaling by the cytokine TGF-beta and suppresses T(H)2 differentiation. Nat Immunol. 2012;13:778–86. doi:10.1038/ni.2362.
212. Lee DU, Agarwal S, Rao A. Th2 lineage commitment and efficient IL-4 production involves extended demethylation of the IL-4 gene. Immunity. 2002;16:649–60.
213. Makar KW, et al. Active recruitment of DNA methyltransferases regulates interleukin 4 in thymocytes and T cells. Nat Immunol. 2003;4:1183–90. doi:10.1038/ni1004.
214. Santangelo S, Cousins DJ, Winkelmann NE, Staynov DZ. DNA methylation changes at human Th2 cytokine genes coincide with DNase I hypersensitive site formation during CD4(+) T cell differentiation. J Immunol. 2002;169:1893–903.
215. Hutchins AS, et al. Gene silencing quantitatively controls the function of a developmental trans-activator. Mol Cell. 2002;10:81–91.
216. Kaneko T, et al. Chromatin remodeling at the Th2 cytokine gene loci in human type 2 helper T cells. Mol Immunol. 2007;44:2249–56. doi:10.1016/j.molimm.2006.11.004.
217. Wei G, et al. Global mapping of H3K4me3 and H3K27me3 reveals specificity and plasticity in lineage fate determination of differentiating CD4+ T cells. Immunity. 2009;30:155–67. doi:10.1016/j.immuni.2008.12.009.
218. Seumois G, et al. Epigenomic analysis of primary human T cells reveals enhancers associated with TH2 memory cell differentiation and asthma susceptibility. Nat Immunol. 2014;15:777–88. doi:10.1038/ni.2937.
219. Gamper CJ, Agoston AT, Nelson WG, Powell JD. Identification of DNA methyltransferase 3a as a T cell receptor-induced regulator of Th1 and Th2 differentiation. J Immunol. 2009;183:2267–76. doi:10.4049/jimmunol.0802960.
220. Yu Q, et al. DNA methyltransferase 3a limits the expression of interleukin-13 in T helper 2 cells and allergic airway inflammation. Proc Natl Acad Sci U S A. 2012;109:541–6. doi:10.1073/pnas.1103803109.
221. Thomas RM, Gamper CJ, Ladle BH, Powell JD, Wells AD. De novo DNA methylation is required to restrict T helper lineage plasticity. J Biol Chem. 2012;287:22900–9. doi:10.1074/jbc.M111.312785.
222. Makar KW, Wilson CB. DNA methylation is a nonredundant repressor of the Th2 effector program. J Immunol. 2004;173:4402–6.
223. Koyanagi M, et al. EZH2 and histone 3 trimethyl lysine 27 associated with Il4 and Il13 gene silencing in Th1 cells. J Biol Chem. 2005;280:31470–7. doi:10.1074/jbc.M504766200.
224. Tumes DJ, et al. The polycomb protein Ezh2 regulates differentiation and plasticity of CD4(+) T helper type 1 and type 2 cells. Immunity. 2013;39:819–32. doi:10.1016/j.immuni.2013.09.012.
225. Grausenburger R, et al. Conditional deletion of histone deacetylase 1 in T cells leads to enhanced airway inflammation and increased Th2 cytokine production. J Immunol. 2010;185:3489–97. doi:10.4049/jimmunol.0903610.
226. Li Q, et al. Critical role of histone demethylase Jmjd3 in the regulation of CD4+ T-cell differentiation. Nat Commun. 2014;5:5780. doi:10.1038/ncomms6780.

227. Yamashita M, et al. Crucial role of MLL for the maintenance of memory T helper type 2 cell responses. Immunity. 2006;24:611–22. doi:10.1016/j.immuni.2006.03.017.
228. Nakata Y, et al. c-Myb, Menin, GATA-3, and MLL form a dynamic transcription complex that plays a pivotal role in human T helper type 2 cell development. Blood. 2010;116:1280–90. doi:10.1182/blood-2009-05-223255.
229. Kimura M, et al. Regulation of Th2 cell differentiation by mel-18, a mammalian polycomb group gene. Immunity. 2001;15:275–87.
230. Lehnertz B, et al. Activating and inhibitory functions for the histone lysine methyltransferase G9a in T helper cell differentiation and function. J Exp Med. 2010;207:915–22. doi:10.1084/jem.20100363.
231. Turner M, Galloway A, Vigorito E. Noncoding RNA and its associated proteins as regulatory elements of the immune system. Nat Immunol. 2014;15:484–91. doi:10.1038/ni.2887.
232. Lu TX, Munitz A, Rothenberg ME. MicroRNA-21 is up-regulated in allergic airway inflammation and regulates IL-12p35 expression. J Immunol. 2009;182:4994–5002. doi:10.4049/jimmunol.0803560.
233. Lu TX, et al. MicroRNA signature in patients with eosinophilic esophagitis, reversibility with glucocorticoids, and assessment as disease biomarkers. J Allergy Clin Immunol. 2012;129:1064–1075 e1069. doi:10.1016/j.jaci.2012.01.060.
234. Vennegaard MT, et al. Allergic contact dermatitis induces upregulation of identical microRNAs in humans and mice. Contact Dermatitis. 2012;67:298–305. doi:10.1111/j.1600-0536.2012.02083.x.
235. Sawant DV, Wu H, Kaplan MH, Dent AL. The Bcl6 target gene microRNA-21 promotes Th2 differentiation by a T cell intrinsic pathway. Mol Immunol. 2013;54:435–42. doi:10.1016/j.molimm.2013.01.006.
236. Chen RF, et al. MicroRNA-21 expression in neonatal blood associated with antenatal immunoglobulin E production and development of allergic rhinitis. Clin Exp Allergy J Br Soc Allergy Clin Immunol. 2010;40:1482–90. doi:10.1111/j.1365-2222.2010.03592.x.
237. Thai TH, et al. Regulation of the germinal center response by microRNA-155. Science. 2007;316:604–8. doi:10.1126/science.1141229.
238. Rodriguez A, et al. Requirement of bic/microRNA-155 for normal immune function. Science. 2007;316:608–11. doi:10.1126/science.1139253.
239. Sonkoly E, et al. MiR-155 is overexpressed in patients with atopic dermatitis and modulates T-cell proliferative responses by targeting cytotoxic T lymphocyte-associated antigen 4. J Allergy Clin Immunol. 2010;126(581–589):e581–520. doi:10.1016/j.jaci.2010.05.045.
240. Malmhall C, et al. MicroRNA-155 is essential for T(H)2-mediated allergen-induced eosinophilic inflammation in the lung. J Allergy Clin Immunol. 2014;133:1429–1438, 1438 e1421-1427. doi:10.1016/j.jaci.2013.11.008.
241. Simpson LJ, et al. A microRNA upregulated in asthma airway T cells promotes TH2 cytokine production. Nat Immunol. 2014;15:1162–70. doi:10.1038/ni.3026.
242. Ranzani V, et al. The long intergenic noncoding RNA landscape of human lymphocytes highlights the regulation of T cell differentiation by linc-MAF-4. Nat Immunol. 2015;16:318–25. doi:10.1038/ni.3093.
243. Spurlock 3rd CF, et al. Expression and functions of long noncoding RNAs during human T helper cell differentiation. Nat Commun. 2015;6:6932. doi:10.1038/ncomms7932.
244. Yusuf I, et al. Germinal center T follicular helper cell IL-4 production is dependent on signaling lymphocytic activation molecule receptor (CD150). J Immunol. 2010;185:190–202. doi:10.4049/jimmunol.0903505.
245. Johnston RJ, Choi YS, Diamond JA, Yang JA, Crotty S. STAT5 is a potent negative regulator of TFH cell differentiation. J Exp Med. 2012;209:243–50. doi:10.1084/jem.20111174.
246. Nurieva RI, et al. STAT5 protein negatively regulates T follicular helper (Tfh) cell generation and function. J Biol Chem. 2012;287:11234–9. doi:10.1074/jbc.M111.324046.
247. Kroenke MA, et al. Bcl6 and Maf cooperate to instruct human follicular helper CD4 T cell differentiation. J Immunol. 2012;188:3734–44. doi:10.4049/jimmunol.1103246.

248. Sherman MA, Nachman TY, Brown MA. Cutting edge: IL-4 production by mast cells does not require c-maf. J Immunol. 1999;163:1733–6.
249. Sherman MA, Secor VH, Lee SK, Lopez RD, Brown MA. STAT6-independent production of IL-4 by mast cells. Eur J Immunol. 1999;29:1235–42. doi:10.1002/(SICI)1521-4141(199904)29:04<1235::AID-IMMU1235>3.0.CO;2-0.
250. Gregory GD, Raju SS, Winandy S, Brown MA. Mast cell IL-4 expression is regulated by Ikaros and influences encephalitogenic Th1 responses in EAE. J Clin Invest. 2006;116:1327–36. doi:10.1172/JCI27227.
251. Li Y, Qi X, Liu B, Huang H. The STAT5-GATA2 pathway is critical in basophil and mast cell differentiation and maintenance. J Immunol. 2015;194:4328–38. doi:10.4049/jimmunol.1500018.
252. Nei Y, et al. GATA-1 regulates the generation and function of basophils. Proc Natl Acad Sci U S A. 2013;110:18620–5. doi:10.1073/pnas.1311668110.
253. Li Y, Qi X, Liu B, Huang H. The STAT5-GATA2 pathway is critical in basophil and mast cell differentiation and maintenance. J Immunol. 2015. doi:10.4049/jimmunol.1500018.
254. Hural JA, Kwan M, Henkel G, Hock MB, Brown MA. An intron transcriptional enhancer element regulates IL-4 gene locus accessibility in mast cells. J Immunol. 2000;165:3239–49.
255. Weiss DL, et al. Nuclear factor of activated T cells is associated with a mast cell interleukin 4 transcription complex. Mol Cell Biol. 1996;16:228–35.
256. Qi X, Nishida J, Chaves L, Ohmori K, Huang H. CCAAT/enhancer-binding protein alpha (C/EBPalpha) is critical for interleukin-4 expression in response to FcepsilonRI receptor cross-linking. J Biol Chem. 2011;286:16063–73. doi:10.1074/jbc.M110.213389.
257. Ugajin T, et al. Zinc-binding metallothioneins are key modulators of IL-4 production by basophils. Mol Immunol. 2015;66:180–8. doi:10.1016/j.molimm.2015.03.002.
258. Weidinger S, Baurecht H, Naumann A, Novak N. Genome-wide association studies on IgE regulation: are genetics of IgE also genetics of atopic disease? Curr Opin Allergy Clin Immunol. 2010;10:408–17. doi:10.1097/ACI.0b013e32833d7d2d.
259. Martinez FD, Vercelli D. Asthma. Lancet. 2013;382:1360–72. doi:10.1016/S0140-6736(13)61536-6.
260. Moffatt MF, et al. A large-scale, consortium-based genomewide association study of asthma. N Engl J Med. 2010;363:1211–21. doi:10.1056/NEJMoa0906312.
261. Li X, et al. Genome-wide association study of asthma identifies RAD50-IL13 and HLA-DR/DQ regions. J Allergy Clin Immunol. 2010;125:328–335 e311. doi:10.1016/j.jaci.2009.11.018.
262. Walker C, et al. Allergic and nonallergic asthmatics have distinct patterns of T-cell activation and cytokine production in peripheral blood and bronchoalveolar lavage. Am Rev Respir Dis. 1992;146:109–15. doi:10.1164/ajrccm/146.1.109.
263. Robinson DS, et al. Predominant TH2-like bronchoalveolar T-lymphocyte population in atopic asthma. N Engl J Med. 1992;326:298–304. doi:10.1056/NEJM199201303260504.
264. Schieck M, et al. A polymorphism in the TH 2 locus control region is associated with changes in DNA methylation and gene expression. Allergy. 2014;69:1171–80. doi:10.1111/all.12450.
265. Stefanowicz D, et al. DNA methylation profiles of airway epithelial cells and PBMCs from healthy, atopic and asthmatic children. PLoS One. 2012;7:e44213. doi:10.1371/journal.pone.0044213.
266. Begin P, Nadeau KC. Epigenetic regulation of asthma and allergic disease. Allergy Asthma Clin Immunol: Off J Can Soc Allergy Clin Immunol. 2014;10:27. doi:10.1186/1710-1492-10-27.
267. Liang L, et al. An epigenome-wide association study of total serum immunoglobulin E concentration. Nature. 2015;520:670–4. doi:10.1038/nature14125.
268. Yang IV, et al. DNA methylation and childhood asthma in the inner city. J Allergy Clin Immunol. 2015;136:69–80. doi:10.1016/j.jaci.2015.01.025.
269. Hirota T, et al. Genome-wide association study identifies three new susceptibility loci for adult asthma in the Japanese population. Nat Genet. 2011;43:893–6. doi:10.1038/ng.887.

270. Torgerson DG, et al. Meta-analysis of genome-wide association studies of asthma in ethnically diverse North American populations. Nat Genet. 2011;43:887–92. doi:10.1038/ng.888.
271. Gudbjartsson DF, et al. Sequence variants affecting eosinophil numbers associate with asthma and myocardial infarction. Nat Genet. 2009;41:342–7. doi:10.1038/ng.323.
272. Ying S, et al. Thymic stromal lymphopoietin expression is increased in asthmatic airways and correlates with expression of Th2-attracting chemokines and disease severity. J Immunol. 2005;174:8183–90.
273. Ying S, et al. Expression and cellular provenance of thymic stromal lymphopoietin and chemokines in patients with severe asthma and chronic obstructive pulmonary disease. J Immunol. 2008;181:2790–8.
274. Prefontaine D, et al. Increased expression of IL-33 in severe asthma: evidence of expression by airway smooth muscle cells. J Immunol. 2009;183:5094–103. doi:10.4049/jimmunol.0802387.
275. Kurowska-Stolarska M, et al. IL-33 amplifies the polarization of alternatively activated macrophages that contribute to airway inflammation. J Immunol. 2009;183:6469–77. doi:10.4049/jimmunol.0901575.
276. Corrigan CJ, et al. Allergen-induced expression of IL-25 and IL-25 receptor in atopic asthmatic airways and late-phase cutaneous responses. J Allergy Clin Immunol. 2011;128:116–24. doi:10.1016/j.jaci.2011.03.043.
277. Cheng D, et al. Epithelial interleukin-25 is a key mediator in Th2-high, corticosteroid-responsive asthma. Am J Respir Crit Care Med. 2014;190:639–48. doi:10.1164/rccm.201403-0505OC.
278. Beale J, et al. Rhinovirus-induced IL-25 in asthma exacerbation drives type 2 immunity and allergic pulmonary inflammation. Sci Transl Med. 2014;6:256ra134. doi:10.1126/scitranslmed.3009124.
279. Sonnenberg GF, Artis D. Innate lymphoid cells in the initiation, regulation and resolution of inflammation. Nat Med. 2015;21:698–708. doi:10.1038/nm.3892.
280. Kabata H, Moro K, Koyasu S, Asano K. Group 2 innate lymphoid cells and asthma. Allergol Int: Off J Jpn Soc Allergol. 2015;64:227–34. doi:10.1016/j.alit.2015.03.004.
281. Allakhverdi Z, et al. CD34+ hematopoietic progenitor cells are potent effectors of allergic inflammation. J Allergy Clin Immunol. 2009;123:472–8. doi:10.1016/j.jaci.2008.10.022.
282. Christianson CA, et al. Persistence of asthma requires multiple feedback circuits involving type 2 innate lymphoid cells and IL-33. J Allergy Clin Immunol. 2015. doi:10.1016/j.jaci.2014.11.037.
283. Xue L, et al. Prostaglandin D2 activates group 2 innate lymphoid cells through chemoattractant receptor-homologous molecule expressed on TH2 cells. J Allergy Clin Immunol. 2014;133:1184–94. doi:10.1016/j.jaci.2013.10.056.
284. Mjosberg J, et al. The transcription factor GATA3 is essential for the function of human type 2 innate lymphoid cells. Immunity. 2012;37:649–59. doi:10.1016/j.immuni.2012.08.015.
285. Doherty TA, et al. Lung type 2 innate lymphoid cells express cysteinyl leukotriene receptor 1, which regulates TH2 cytokine production. J Allergy Clin Immunol. 2013;132:205–13. doi:10.1016/j.jaci.2013.03.048.
286. Endo Y, et al. The interleukin-33-p38 kinase axis confers memory T helper 2 cell pathogenicity in the airway. Immunity. 2015;42:294–308. doi:10.1016/j.immuni.2015.01.016.
287. Guo L, et al. IL-1 family members and STAT activators induce cytokine production by Th2, Th17, and Th1 cells. Proc Natl Acad Sci U S A. 2009;106:13463–8. doi:10.1073/pnas.0906988106.
288. Omori M, Ziegler S. Induction of IL-4 expression in CD4(+) T cells by thymic stromal lymphopoietin. J Immunol. 2007;178:1396–404.
289. Angkasekwinai P, et al. Interleukin 25 promotes the initiation of proallergic type 2 responses. J Exp Med. 2007;204:1509–17. doi:10.1084/jem.20061675.
290. Rochman I, Watanabe N, Arima K, Liu YJ, Leonard WJ. Cutting edge: direct action of thymic stromal lymphopoietin on activated human CD4+ T cells. J Immunol. 2007;178:6720–4.

291. Kitajima M, Lee HC, Nakayama T, Ziegler SF. TSLP enhances the function of helper type 2 cells. Eur J Immunol. 2011;41:1862–71. doi:10.1002/eji.201041195.
292. Min B, et al. Basophils produce IL-4 and accumulate in tissues after infection with a Th2-inducing parasite. J Exp Med. 2004;200:507–17. doi:10.1084/jem.20040590.
293. Oeser K, Schwartz C, Voehringer D. Conditional IL-4/IL-13-deficient mice reveal a critical role of innate immune cells for protective immunity against gastrointestinal helminths. Mucosal Immunol. 2015;8:672–82. doi:10.1038/mi.2014.101.
294. Sullivan BM, et al. Genetic analysis of basophil function in vivo. Nat Immunol. 2011;12:527–35. doi:10.1038/ni.2036.
295. Schwartz C, et al. Basophil-mediated protection against gastrointestinal helminths requires IgE-induced cytokine secretion. Proc Natl Acad Sci U S A. 2014;111:E5169–77. doi:10.1073/pnas.1412663111.
296. Sokol CL, Barton GM, Farr AG, Medzhitov R. A mechanism for the initiation of allergen-induced T helper type 2 responses. Nat Immunol. 2008;9:310–8. doi:10.1038/ni1558.
297. Rosenstein RK, Bezbradica JS, Yu S, Medzhitov R. Signaling pathways activated by a protease allergen in basophils. Proc Natl Acad Sci U S A. 2014;111:E4963–71. doi:10.1073/pnas.1418959111.
298. Phillips C, Coward WR, Pritchard DI, Hewitt CR. Basophils express a type 2 cytokine profile on exposure to proteases from helminths and house dust mites. J Leukoc Biol. 2003;73:165–71.
299. Egawa M, et al. Inflammatory monocytes recruited to allergic skin acquire an anti-inflammatory M2 phenotype via basophil-derived interleukin-4. Immunity. 2013;38:570–80. doi:10.1016/j.immuni.2012.11.014.
300. Siracusa MC, et al. TSLP promotes interleukin-3-independent basophil haematopoiesis and type 2 inflammation. Nature. 2011;477:229–33. doi:10.1038/nature10329.
301. Glatman Zaretsky A, et al. T follicular helper cells differentiate from Th2 cells in response to helminth antigens. J Exp Med. 2009;206:991–9. doi:10.1084/jem.20090303.
302. King IL, Mohrs M. IL-4-producing CD4+ T cells in reactive lymph nodes during helminth infection are T follicular helper cells. J Exp Med. 2009;206:1001–7. doi:10.1084/jem.20090313.
303. Thomas SY, Chyung YH, Luster AD. Natural killer T cells are not the predominant T cell in asthma and likely modulate, not cause, asthma. J Allergy Clin Immunol. 2010;125:980–4. doi:10.1016/j.jaci.2010.01.032.
304. Koyama K. NK1.1+ cell depletion in vivo fails to prevent protection against infection with the murine nematode parasite Trichuris muris. Parasite Immunol. 2002;24:527–33.
305. Mallevaey T, et al. Invariant and noninvariant natural killer T cells exert opposite regulatory functions on the immune response during murine schistosomiasis. Infect Immun. 2007;75:2171–80. doi:10.1128/IAI.01178-06.
306. Albacker LA, et al. Invariant natural killer T cells recognize a fungal glycosphingolipid that can induce airway hyperreactivity. Nat Med. 2013;19:1297–304. doi:10.1038/nm.3321.
307. Travers J, Rothenberg ME. Eosinophils in mucosal immune responses. Mucosal Immunol. 2015;8:464–75. doi:10.1038/mi.2015.2.
308. Chu DK, et al. Indigenous enteric eosinophils control DCs to initiate a primary Th2 immune response in vivo. J Exp Med. 2014;211:1657–72. doi:10.1084/jem.20131800.
309. Jacobsen EA, et al. Differential activation of airway eosinophils induces IL-13-mediated allergic Th2 pulmonary responses in mice. Allergy. 2015. doi:10.1111/all.12655.
310. Borish LC, et al. Efficacy of soluble IL-4 receptor for the treatment of adults with asthma. J Allergy Clin Immunol. 2001;107:963–70. doi:10.1067/mai.2001.115624.
311. Hart TK, et al. Preclinical efficacy and safety of pascolizumab (SB 240683): a humanized anti-interleukin-4 antibody with therapeutic potential in asthma. Clin Exp Immunol. 2002;130:93–100.

312. Wenzel S, et al. Dupilumab in persistent asthma with elevated eosinophil levels. N Engl J Med. 2013;368:2455–66. doi:10.1056/NEJMoa1304048.
313. Beck LA, et al. Dupilumab treatment in adults with moderate-to-severe atopic dermatitis. N Engl J Med. 2014;371:130–9. doi:10.1056/NEJMoa1314768.
314. Krug N, et al. Allergen-induced asthmatic responses modified by a GATA3-specific DNAzyme. N Engl J Med. 2015;372:1987–95. doi:10.1056/NEJMoa1411776.

Chapter 4
Regulation of IL-6 in Immunity and Diseases

Toshio Tanaka, Masashi Narazaki, Kazuya Masuda, and Tadamitsu Kishimoto

Abstract Interleukin-6 (IL-6) is a prototypical cytokine with functional pleiotropy and plays an important role in host defense. When infections or tissue injuries occur, IL-6 is promptly produced by monocytes and macrophages and contributes to removal of infectious agents and restoration of damaged tissues through activation of immune, hematological, and acute-phase responses. Once stress is removed from the host, IL-6 synthesis ends, but uncontrolled excessive or persistent IL-6 production plays a pathological role in the development of various inflammatory diseases and cancers, indicating that IL-6 is a double-edged sword for the host. Thus, the proper IL-6 expression is very important for host defense and is strictly controlled by chromatin structure, transcriptional regulation, and posttranscriptional modification. Differentiation status of cells, various transcription factors, RNA-binding proteins, and microRNAs are involved in this process. Since it is assumed that dysregulation of any of these regulatory molecules may cause abnormal IL-6 expression in a particular disease, further elucidation of the factors and processes involved in IL-6 expression can be expected to facilitate to clarification of pathogenesis and to identification of novel target molecule(s) for specific diseases.

Keywords Arid5a • IL-6 • MicroRNAs • Regnase-1

T. Tanaka
Department of Clinical Application of Biologics, Osaka University Graduate School of Medicine, Osaka University, Osaka 565-0871, Japan

M. Narazaki
Department of Respiratory Medicine, Allergy and Rheumatic Diseases, Osaka University Graduate School of Medicine, Osaka University, Osaka 565-0871, Japan

K. Masuda • T. Kishimoto (✉)
Laboratory of Immune Regulation, World Premier International Immunology Frontier Research Center, Osaka University, Osaka 565-0871, Japan
e-mail: kishimoto@ifrec.osaka-u.ac.jp

X. Ma (ed.), *Regulation of Cytokine Gene Expression in Immunity and Diseases*, Advances in Experimental Medicine and Biology 941,
DOI 10.1007/978-94-024-0921-5_4

4.1 Introduction

Interleukin-6 (IL-6) is a prototypical cytokine featuring functional redundancy and pleiotropy [3, 16, 17]. Under healthy conditions the serum level of IL-6 is less than 4 pg/ml, but when stresses such as infections or tissue injuries occur, IL-6 is immediately produced by monocytes and macrophages by the stimulation of pattern recognition receptors with pathogen-associated molecular patterns or damage-associated molecular patterns and serum IL-6 level increases to several tens to hundreds of pg/ml, depending on the infection or injury [15, 32, 35]. IL-6 binds to its specific IL-6 receptor (IL-6R), and the IL-6/IL-6R complex associates with and activates signal transducer gp130 that is shared with the IL-6 family cytokines. IL-6 then activates hepatocytes, immune-competent cells, and hematological cells, to produce acute-phase, immune, and hematological responses. These responses are essential for elimination of infectious agents and healing of tissues, so that IL-6 is a major cytokine contributing to host defense. Once such a stress has been completely removed from the host, IL-6 synthesis ceases and serum IL-6 concentration returns to normal. However, excessive expression of IL-6 during the period of infection or tissue damage results in acute, severe fatal complications or persistent IL-6 production in chronic inflammatory diseases [15, 32]. In the case of chronic inflammatory diseases, including rheumatoid arthritis (RA), serum concentration of IL-6 is usually elevated, ranging from ten to several hundred pg/ml, while in the case of acute severe inflammatory diseases such as fatal complications resulting from septic shock, its level can increase from several ng/ml to μg/ml [38], at which level IL-6 can stimulate almost all cells expressing gp130 [9], since IL-6 associates with serum soluble IL-6R and could activate gp130, even though target cells lack transmembrane IL-6R. This hyper-IL-6 state leads to multiple organ failure. Therefore, strict control of IL-6 expression is very important for the host. In this review we focus on the physiological regulatory mechanisms of IL-6 expression and the involvement of its collapse in disease development.

4.2 Regulation of IL-6 Gene Expression

IL-6 is produced by not only immune-mediated cells but also mesenchymal cells, endothelial cells, fibroblasts, and many other cells in response to various stimuli and even spontaneously by cancer cells [3, 16]. Expression of IL-6 in certain cells is regulated at various stages, including chromatin remodeling, transcription, mRNA export, posttranscriptional, and translational levels. In addition, a gene polymorphism in the IL-6 gene reportedly modifies its expression. A G/C polymorphism located at position -174 in the flanking region of the IL-6 gene affects IL-6 transcription and is associated with systemic onset juvenile idiopathic arthritis [8] or other chronic inflammatory diseases including coronary arterial disease [11]. Epigenetic variability also regulates IL-6 production. Two major epigenetic effects

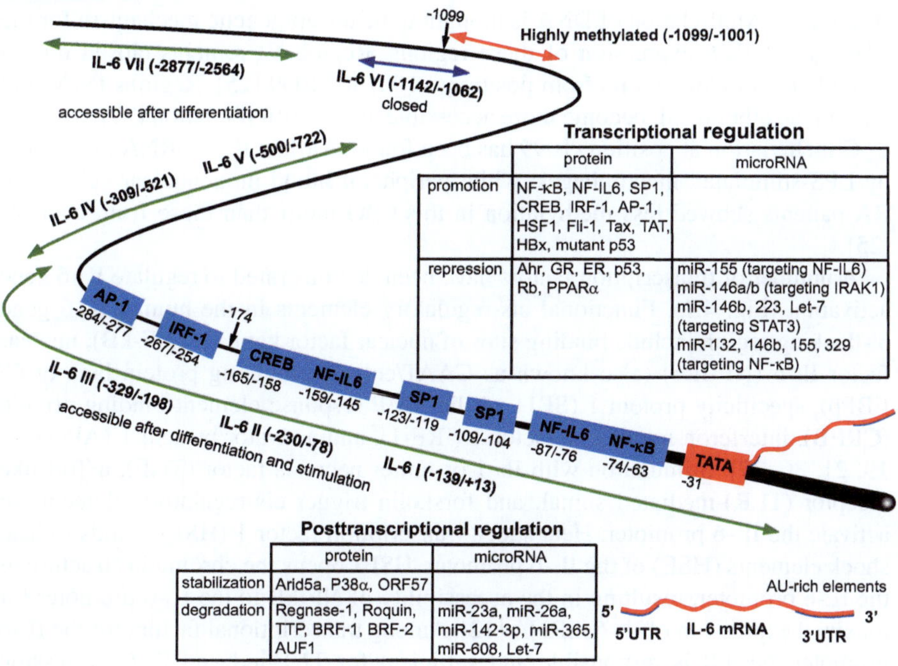

Fig. 4.1 Schematic overview of regulatory mechanisms of human IL-6 expression. IL-6 gene expression is regulated by chromatin remodeling as well as transcriptional and posttranscriptional mechanisms in which several proteins and microRNAs are involved. Activation of these proteins and microRNAs determines the fate of IL-6 mRNA. A G/C polymorphism at position -174 is associated with inflammatory disease and at position -1099 CpG is less methylated in RA patients. NF-κB, nuclear factor kappa B; NF-IL6, nuclear factor of IL-6; SP1, specificity protein 1; CREB, cyclic AMP response element-binding protein; IRF-1, interferon regulatory factor 1; AP-1, activation protein 1; HSF1, heat shock transcription factor 1; Fli-1, Friend leukemia virus integration 1; Tax, transactivator protein; TAT, transactivator of the transcription; HBVX, hepatitis B virus X protein; Ahr, aryl hydrocarbon receptor; GR, glucocorticoid receptor; ER, estrogen receptor; Rb, retinoblastoma; PPAR-α, peroxisome proliferator-activated receptor-α; miR, microRNA; IRAK1, IL-1 receptor-associated kinase 1; STAT3, signal transducer and activator of transcription 3; ORF, open reading frame; TTP, tristetraprolin; BRF-1, butyrate response factor-1; AUF1, AU-binding factor 1; Let-7, lethal-7

are the modification of histone proteins in chromatin and the methylation of the genomic DNA. Cells undergo programmed changes in their gene expression patterns during differentiation, and their gene expression is regulated by structural remodeling of chromatin [1]. The main IL-6 producing cells are monocytes and macrophages. Promyeloid HL-60 cells initially do not produce IL-6, but during differentiation from promyeloid cells to monocytes, HL-60 cells gain the ability to produce IL-6 after stimulation of lipopolysaccharide (LPS). A chromatin study identified seven IL-6 promoter regions that undergo changes in accessibility during cell differentiation (Fig. 4.1). The I, II, and III regions of IL-6 are inaccessible before, but become accessible after differentiation and even more so after LPS stimulation. Methylation of DNA is thought to be the epigenetic mechanism for the silencing of IL-6 expression of these regions. Region VI remains closed due to methylation of CpG motifs from positions -1001 to -1099 [25]. Regions IV, V, and VII, on the other hand, become more accessible during differentiation [27]. A single CpG methylation at position -1099 has been found to affect IL-6 mRNA expression by LPS-stimulated macrophages, while peripheral blood mononuclear cells from RA patients showed less methylation in this CpG motif than those from controls [25].

A number of transcription factors have been demonstrated to regulate IL-6 gene activation (Fig. 4.1). Functional cis-regulatory elements in the human IL-6 gene 5′-flanking region include binding sites of nuclear factor kappa B (NF-kB), nuclear factor IL-6 (NF-IL6) (also known as CAAT/enhancer-binding protein beta or C/EBPβ), specificity protein 1 (SP1), cyclic AMP response element-binding protein (CREB), interferon regulatory factor 1 (IRF-1), and activator protein 1 (AP-1) [2, 19, 21, 36, 37]. Stimulation with IL-1 or tumor necrosis factor (TNF), a Toll-like receptor (TLR)-mediated signal, and forskolin trigger cis-regulatory elements to activate the IL-6 promoter. Heat shock transcription factor 1 (HSF1) binds to heat shock elements (HSE) of the IL-6 promoter. HSF1 opens the chromatin structure of the IL-6 promoter, resulting in the accessibility of NF-kB to the IL-6 promoter for maximal expression of IL-6 mRNA [12]. Among transcriptional factors for the IL-6 promoter, NF-kB is responsible and sufficient for TNF-induced IL-6 expression since it interacts with CREB-binding protein (CBP)/p300 for maximal transcriptional stimulation [36]. Local hyperacetylation of histones by CBP/p300 then leads to nucleosomal relaxation and opens chromatin-embedded promoter regions. On the other hand, some transcription factors suppress IL-6 expression [33]. Peroxisome proliferator-activated receptors (PPARs) are ligand-activated transcription factors consisting of three subtypes, α, β, and γ. Fibrate-activated PPAR-α interacts with c-Jun and the p65 NF-kB subunit, which negatively regulates IL-6 transcription. Hormone receptors reportedly also suppress IL-6 expression. For instance, the activation of glucocorticoid receptors can inhibit IL-6 expression, which is partly responsible for the anti-inflammatory effects of corticosteroids. IL-6 expression is suppressed by the activation of estrogen receptors, while a decrease in estrogen secretion as a result of menopause or ovariectomy may lead to an increase in serum IL-6. Retinoblastoma protein and p53 have been shown to repress IL-6 gene promoter activity, whereas mutated p53 loses this inhibitory function. It was also found that the aryl hydrocarbon receptor, a ligand-activated transcriptional regulator that binds to dioxin and other exogenous contaminants, forms a complex with NF-kB to inhibit the promoter activity of IL-6. Thus, various proteins affect IL-6 transcription. In addition, microRNAs (miRs) directly or indirectly suppress the activity of transcription factors regulating the IL-6 gene [33]. miR-155 interacts with the 3′ untranslated region (UTR) of NF-IL6 and suppresses transcription activity of NF-IL6 by inhibiting its expression, while miR-146a/b or miR-146b and miR-223 suppress transcription of IL-6 indirectly by targeting interleukin-1 receptor-associated kinase 1 (IRAK1) or signal transducers and activator of transcription 3

(STAT3), respectively. Moreover, miRs such as miR-132, miR-146b, miR-155, and miR-329 have been found to inhibit IL-6 expression by targeting NF-kB.

Another regulatory mechanism is posttranscription through which cytokine mRNAs are regulated through both the 5′ and 3′UTRs. The 5′UTR dictates initiation of mRNA translation, while AU-rich elements (AREs) and stem-loop structures located in the 3′UTR region determine the stability of mRNA [5, 31]. A number of RNA-binding proteins and miRs bind to the 3′UTRs and regulate the stability of IL-6 mRNA (Fig. 4.1). For example, mitogen-activated protein kinase (MAPK) p38-α promotes IL-6 mRNA stabilization via 3′UTR of IL-6. Conversely, RNA-binding proteins such as tristetraprolin (TTP), butyrate response factor-1 (BRF-1), BRF-2, and AU-binding factor 1 (AUF1) promote IL-6 mRNA degradation, while several miRs such as miR-23a, miR-26a, miR-142-3p, miR-365, and miR-608 are reported to reduce IL-6 mRNA levels through direct interaction with IL-6 3′UTR.

It was found that a regulatory RNase-1 (Regnase-1, also known as Zc3h12a) is an endonuclease involved in a set of target mRNAs including IL-6 via binding to a stem-loop structure [13, 22]. The knockout mice exhibited spontaneous autoimmune diseases accompanied by splenomegaly and lymphadenopathy. Recently, it was found that another RNA-binding protein, Roquin, recognizes target mRNAs overlapping with Regnase-1 via the same stem-loop structures in their 3′UTRs [23]. However, their localizations and states of the target mRNAs are different. Regnase-1 degrades transcriptionally active mRNA in the cytoplasm, endoplasmic reticulum, and ribosome, whereas Roquin controls transcriptionally inactive mRNA in stress granules and processing bodies, so that Regnase-1 and Roquin degrade mRNAs spatiotemporally (Fig. 4.2). In contrast, we recently identified a novel RNA-binding protein, AT-rich interactive domain-containing protein 5a (Arid5a), which selectively stabilizes IL-6 but not TNF-α or IL-12 mRNA through binding to the 3′UTR of IL-6 mRNA [20]. Arid5a was enhanced in macrophages in response to LPS, IL-1β, and IL-6 and also was induced under Th17-polarizing conditions. Arid5a gene deficiency impairs elevation of IL-6 level in LPS-injected mice and preferential Th17 cell development in naïve T cells. Moreover, Arid5a counteracts the destabilizing function of Regnase-1 (Fig. 4.2), indicating that the balance between Arid5a and Regnase-1 plays an important role in IL-6 mRNA stability, while predominance of Arid5a over Regnase-1 can promote inflammatory processes and may induce the development of autoimmune inflammatory diseases [34].

4.3 Dysregulated IL-6 Gene Expression in Diseases

As explained earlier, dysregulated excessive or persistent production of IL-6 plays a role in the development of acute severe inflammatory diseases or chronic diseases and cancers [7, 15, 35]. However, the primary cells, which pathologically produce IL-6, differ for different IL-6-related diseases, and the pathological mechanisms through which IL-6 is synthesized excessively or persistently are not fully clarified. It is conceivable that abnormality of a specific molecule involved in IL-6 expression

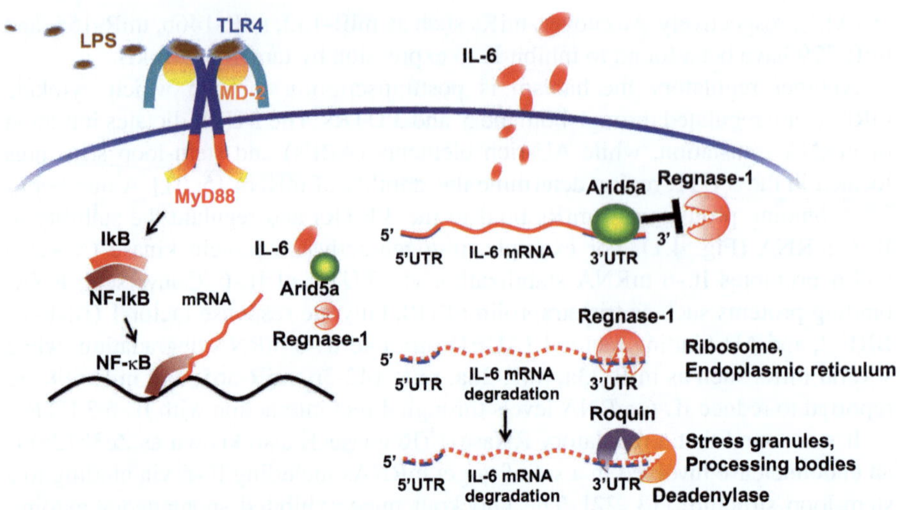

Fig. 4.2 IL-6 synthesis and posttranscriptional regulation of IL-6 mRNA by Arid5a, Regnase-1, and Roquin. Pathogen-associated molecular patterns are recognized by pathogen recognition receptors to induce proinflammatory cytokines. In this figure, the Toll-like receptor (TLR) 4 and MD-2 complex recognizes lipopolysaccharide (LPS) and induces IL-6, Arid5a, or Regnase-1 mRNA via MyD88, followed by activation of the nuclear factor kappa B (NF-kB) signaling pathway. Regnase-1 degrades transcriptionally active mRNA in the ribosome and endoplasmic reticulum, while Roquin degrades inactive mRNA in stress granules and processing bodies, for the initiation of which Roquin requires deadenylase. Arid5a inhibits the destabilizing effect of Regnase-1, so that the balance between Arid5a and Regnase-1 is important for the regulation of IL-6 mRNA. MD-2, myeloid differentiation protein 2; MyD88, myeloid differentiation primary response 88; IkB, inhibitor of NF-kB; UTR, untranslated region

may contribute to dysregulated IL-6 production for the development of a specific disease. IL-6 acts as a paracrine or autocrine growth factor for many cancer cells and induces cancer-related inflammation. It is associated with worsening of the prognosis for cancer by promoting angiogenesis, metastasis, and modification of immunity. In highly metastatic breast cancer cells, hyperactivated NF-kB and AP-1 promote chromatin accessibility of the IL-6 promoter region and enhance transcription of the IL-6 gene [24]. It has been reported that some cancer-related virus products can intensify IL-6 expression. For example, human T-lymphotropic virus 1 (HTLV-1) activates IL-6 production via interaction of the virus-derived transactivator protein (Tax) with NF-kB [18]. Human immunodeficiency virus 1 (HIV-1) transactivator of the transcription (TAT) protein can also enhance NF-kB and NF-IL6 DNA-binding activity [4, 30]. Moreover, the human hepatitis B virus X protein (HBx) enhances the DNA binding of NF-IL6 [26]. Open reading frame 57 derived from the Kaposi sarcoma-associated herpesvirus (also known as human herpesvirus 8) has been shown to promote the stabilization of both viral and human IL-6 mRNA by competing with the binding of miR-1293 to the viral or of miR-608 to the human form of IL-6 mRNA [14]. These findings suggest that virus products can aggravate inflammation by augmenting IL-6 mRNA expression.

Fli-1 (Friend leukemia virus integration 1), a member of the ETS (E-26) transcription factor family, has been implicated in the pathogenesis of systemic lupus erythematosus (SLE), since peripheral blood lymphocytes from SLE patients show increased expression of Fli-1 and the level of this expression is associated with the disease activity [10]. There are three binding regions (-820 to -613, -419 to -281, and -344 to -195) of Fli-1 in the mouse IL-6 gene promoter, and overexpression of Fli-1 causes IL-6 production [29]. In an Fli-1-heterozygous MRL/*lpr* mouse, which is an SLE model mouse, IL-6 expression in sera and kidneys is reduced, and the disease development is markedly inhibited in comparison with the findings for wild MRL/*lpr* mouse [42].

The gene encoding the tumor suppressor miR-146b is a direct STAT3 target gene. miR-146b also inhibits NF-kB-dependent IL-6 production and subsequently STAT3 activation, while its expression is reduced in breast cancer cells, and this reduction in miR-146b expression is associated with IL-6/STAT3-derived migration and invasion in breast cancer cells [40]. miR-142-3p, with its expression inhibited by IL-6, downregulates IL-6 expression by targeting the 3′UTR of IL-6 mRNA. In patients with glioblastoma, who show upregulated IL-6 expression, the hypermethylated miR-142-3p promoter demonstrates poor survival outcome [6]. Moreover, miR-26a targets the IL-6 gene, and downregulation of miR-26a has been found to be associated with poor prognosis, recurrence, and metastasis of hepatocellular carcinoma [41]. Lethal-7 (let-7) miR reportedly targets both IL-6 and STAT3 mRNAs. As mentioned earlier, the HBx protein augments IL-6 transcription by enhancing the DNA binding NF-IL6, while it inhibits the expression of let-7, leading to stabilization of IL-6 and STAT3 mRNAs. These effects of HBx are at least partially related to tumorigenesis of hepatocellular carcinoma [39]. These findings indicate that the diminished expression of several miRs can lead to IL-6 expression and have an effect on tumorigenesis.

Finally, as already mentioned, Arid5a is a stabilizing RNA-binding protein for IL-6 mRNA, and we hypothesize that the balance between Arid5a and Regnase-1 determines IL-6 mRNA stability and that the predominance of Arid5a over Regnase-1 prolongs IL-6 mRNA half-life and induces the development of autoimmune inflammatory diseases [15, 33, 34]. Indeed, Arid5a expression was found to have increased in the peripheral blood CD4-positive T cells from patients with RA, and treatment of patients with an IL-6 inhibitor resulted in the reduction of Arid5a expression in association with the suppression of the disease activity [28]. Accrued but so far limited evidence indicates that various different molecules, such as virus products, transcription factors, miRs, and RNA-binding proteins, are implicated in dysregulated IL-6 production and associated disease development.

4.4 Concluding Remarks

IL-6 is undeniably a key cytokine in the development of various diseases, since it has been verified that the IL-6 inhibitor, tocilizumab, is markedly effective for the treatment of various autoimmune inflammatory diseases [15, 35]. While IL-6 is produced excessively or continuously in these diseases by a distinct cell population, the pathological mechanism(s) responsible for the dysregulated IL-6 production remains unknown. Its clarification will undoubtedly lead to the identification of more specific target molecules and investigations into the pathogenesis of specific diseases.

References

1. Agarwal S, Rao A. Modulation of chromatin structure regulates cytokine gene expression during T cell differentiation. Immunity. 1998;9:765–75.
2. Akira S, Kishimoto T. IL-6 and NF-IL6 in acute-phase response and viral infection. Immunol Rev. 1992;127:25–50.
3. Akira S, Taga T, Kishimoto T. Interleukin-6 in biology and medicine. Adv Immunol. 1993;54:1–78.
4. Ambrosino C, Ruocco MR, Chen X, Mallardo M, Baudi F, Trematerra S, Quinto I, Venuta S, Scala G. HIV-1 Tat induces the expression of the interleukin-6 (IL6) gene by binding to the IL6 leader RNA and by interacting with CAAT enhancer-binding protein beta (NF-IL6) transcription factors. J Biol Chem. 1997;272:14883–92.
5. Anderson P. Post-transcriptional control of cytokine production. Nat Immunol. 2008;9:353–9.
6. Chiou GY, Chien CS, Wang ML, Chen MT, Yang YP, Yu YL, Chien Y, Chang YC, Shen CC, Chio CC, et al. Epigenetic regulation of the miR142-3p/interleukin-6 circuit in glioblastoma. Mol Cell. 2013;52:693–706.
7. Coussens LM, Werb Z. Inflammation and cancer. Nature. 2002;420:860–7.
8. Fishman D, Faulds G, Jeffery R, Mohamed-Ali V, Yudkin JS, Humphries S, Woo P. The effect of novel polymorphisms in the interleukin-6 (IL-6) gene on IL-6 transcription and plasma IL-6 levels, and an association with systemic-onset juvenile chronic arthritis. J Clin Invest. 1998;102:1369–76.
9. Garbers C, Aparicio-Siegmund S, Rose-John S. The IL-6/gp130/STAT3 signaling axis: recent advances towards specific inhibition. Curr Opin Immunol. 2015;34:75–82.
10. Georgiou P, Maroulakou I, Green J, Dantis P, Romanospica V, Kottaridis S, Lautenberger J, Watson D, Papas T, Fischinger P, et al. Expression of ets family of genes in systemic lupus erythematosus and Sjogren's syndrome. Int J Oncol. 1996;9:9–18.
11. Hou H, Wang C, Sun F, Zhao L, Dun A, Sun Z. Association of interleukin-6 gene polymorphism with coronary artery disease: an updated systematic review and cumulative meta-analysis. Inflamm Res. 2015;64:707–20.
12. Inouye S, Fujimoto M, Nakamura T, Takaki E, Hayashida N, Hai T, Nakai A. Heat shock transcription factor 1 opens chromatin structure of interleukin-6 promoter to facilitate binding of an activator or a repressor. J Biol Chem. 2007;282:33210–7.
13. Iwasaki H, Takeuchi O, Teraguchi S, Matsushita K, Uehata T, Kuniyoshi K, Satoh T, Saitoh T, Matsushita M, Standley DM, et al. The IkB kinase complex regulates the stability of cytokine-encoding mRNA induced by TLR-IL-1R by controlling degradation of regnase-1. Nat Immunol. 2011;12:1167–75.
14. Kang JG, Pripuzova N, Majerciak V, Kruhlak M, Le SY, Zheng ZM. Kaposi's sarcoma-associated herpesvirus ORF57 promotes escape of viral and human interleukin-6 from microRNA-mediated suppression. J Virol. 2011;85:2620–30.
15. Kang S, Tanaka T, Kishimoto T. Therapeutic uses of anti-interleukin-6 receptor antibody. Int Immunol. 2015;27:21–9.
16. Kishimoto T. The biology of interleukin-6. Blood. 1989;74:1–10.

17. Kishimoto T. Interleukin-6: from basic science to medicine-40 years in immunology. Annu Rev Immunol. 2005;23:1–21.
18. Leung K, Nabel GJ. HTLV-1 transactivator induces interleukin-2 receptor expression through an NF-kappa B-like factor. Nature. 1988;333:776–8.
19. Liebermann TA, Baltimore D. Activation of interleukin-6 gene expression through the NF-kappa B transcription factor. Mol Cell Biol. 1990;10:2327–34.
20. Masuda K, Ripley B, Nishimura R, Mino T, Takeuchi O, Shioi G, Kiyonari H, Kishimoto T. Arid5a controls IL-6 mRNA stability, which contributes to elevation of IL-6 level in vivo. Proc Natl Acad Sci U S A. 2013;110:9409–14.
21. Matsusaka T, Fujikawa K, Nishio Y, Mukaida N, Matsushima K, Kishimoto T, Akira S. Transcriptional factor NF-IL-6 and NF-kappa B synergistically activate transcription of the inflammatory cytokines, interleukin 6 and interleukin 8. Proc Natl Acad Sci U S A. 1993;90:10193–7.
22. Matsushita K, Takeuchi O, Standley DM, Kumagai Y, Kawagoe T, Miyake T, Satoh T, Kato H, Tsujimura T, Nakamura H, et al. Zc3h12a is an RNase essential for controlling immune responses by regulating mRNA decay. Nature. 2009;458:1185–90.
23. Mino T, Murakawa Y, Fukao A, Vandenbon A, Wessels HH, Ori D, Uehata T, Tartey S, Akira S, Suzuki Y, et al. Regnase-1 and roquin regulate a common element in inflammatory mRNAs by spatiotemporally distinct mechanisms. Cell. 2015;161:1058–73.
24. Ndlovu MN, Van Lint C, Van Wesemael K, Callebert P, Chalbos D, Haegeman G, Vanden Berghe W. Hyperactivated NF-{kappa}B and AP-1 transcription factors promote highly accessible chromatin and constitutive transcription across the interleukin-6 gene promoter in metastatic breast cancer cells. Mol Cell Biol. 2009;29:5488–504.
25. Nile CJ, Read RC, Akil M, Duff GW, Wilson AG. Methylation status of a single CpG site in the *IL6* promoter is related to *IL6* messenger RNA levels and rheumatoid arthritis. Arthritis Rheum. 2008;58:2686–93.
26. Ohno H, Kaneko S, Lin Y, Kobayashi K, Murakami S. Human hepatitis B virus X protein augments the DNA binding of nuclear factor for IL-6 through its basic-leucine zipper domain. J Med Virol. 1999;58:11–8.
27. Poplutz MK, Wessels I, Rink L, Uciechowski P. Regulation of the Interleukin-6 gene expression during monocytic differentiation of HL-60 cells by chromatin remodeling and methylation. Immunobiology. 2014;219:619–26.
28. Saito Y, Kagami S, Sanayama Y, Ikeda K, Suto A, Kashiwakuma D, Furuta S, Iwamoto I, Nonaka K, Ohara O, et al. AT-rich-interactive domain-containing protein 5a functions as a negative regulator of retinoic acid receptor-related orphan nuclear receptor γt-induced Th17 cell differentiation. Arthritis Rheum. 2014;66:1185–94.
29. Sato S, Lennard Richard M, Brandon D, Jones Buie JN, Oates JC, Gilkeson GS, Zhang XK. A critical role of the transcriptional factor Fli-1 in murine lupus development by regulation of interleukin-6 expression. Arthritis Rheumatol. 2014;66:3436–44.
30. Scala G, Ruocco MR, Ambrosino C, Mallardo M, Giordano V, Baldassarre F, Dragonetti E, Quinto I, Venuta S. The expression of the interleukin 6 gene is induced by the human immunodeficiency virus 1 TAT protein. J Exp Med. 1994;179:961–71.
31. Stoecklin G, Gross B, Ming XF, Moroni C. A novel mechanism of tumor suppression by destabilizing AU-rich growth factor mRNA. Oncogene. 2003;22:3554–61.
32. Tanaka T, Narazaki M, Kishimoto T. Therapeutic targeting of the interleukin-6 receptor. Annu Rev Pharmacol Toxicol. 2012;52:199–219.
33. Tanaka T, Narazaki M, Masuda K, Kishimoto T. Interleukin-6; pathogenesis and treatment of autoimmune inflammatory diseases. Inflamm Regen. 2013;33:54–65.
34. Tanaka T, Kishimoto T. The biology and medical implications of interleukin-6. Cancer Immunol Res. 2014;2:288–94.
35. Tanaka T, Narazaki M, Kishimoto T. IL-6 in inflammation, immunity, and disease. Cold Spring Harb Perspect Biol. 2014;6:a016295.

36. Vanden Berghe W, De Bosscher K, Boone E, Plaisance S, Haegeman G. Nuclear factor-kB engages CBP/p300 and histone acetyltransferase activity for transcriptional activation of the interleukin-6 gene promoter. J Biol Chem. 1999;274:32091–8.
37. Vanden Berghe W, Vermeulen L, De Wilde G, De Bosscher K, Boone E, Haegeman G. Signal transduction by tumor necrosis factor and gene regulation of the inflammatory cytokine interleukin-6. Biochem Pharmacol. 2000;60:1185–95.
38. Waage A, Brandtzaeg P, Halstensen A, Kierulf P, Espevik T. The complex pattern of cytokines in serum from patients with meningococcal septic shock. Association between interleukin 6, interleukin 1, and fatal outcome. J Exp Med. 1989;169:333–8.
39. Wang Y, Lu Y, Toh ST, Sung WK, Tan P, Chow P, Chung AY, Jooi LL, Lee CG. Lethal-7 is down-regulated by the hepatitis B virus x protein and targets signal transducer and activator of transcription 3. J Hepatol. 2010;53:57–66.
40. Xiang, M., Birkbak, N.J., Vafaizadeh, V., Walker, S.R., Yeh, J.E., Liu, S., Kroll, Y., Boldin, M., Taganov, K., Groner, B., et al. STAT3 induction of miR-146b forms a feedback loop to inhibit the NF-kB to IL-6 signaling axis and STAT3-driven cancer phenotypes. Sci Signal. 2014;7:ra11.
41. Yang X, Liang L, Zhang XF, Jia HL, Qin Y, Zhu XC, Gao XM, Qiao P, Zheng Y, Sheng YY, et al. MicroRNA-26a suppresses tumor growth and metastasis of human hepatocellular carcinoma by targeting interleukin-6-Stat3 pathway. Hepatology. 2013;58:158–70.
42. Zhang XK, Gallant S, Molano I, Moussa OM, Ruiz P, Spyropoulos DD, Watson DK, Gilkeson G. Decreased expression of the Ets family transcription factor Fli-1 markedly prolongs survival and significantly reduces renal disease in MRL/lpr mice. J Immunol. 2004;173:6481–9.

Chapter 5
Regulation of Interleukin-10 Expression

Sascha Rutz and Wenjun Ouyang

Abstract Interleukin (IL)-10 is an essential anti-inflammatory cytokine that plays important roles as a negative regulator of immune responses to microbial antigens. Loss of IL-10 results in the spontaneous development of inflammatory bowel disease as a consequence of an excessive immune response to the gut microbiota. IL-10 also functions to prevent excessive inflammation during the course of infection. IL-10 can be produced in response to pro-inflammatory signals by virtually all immune cells, including T cells, B cells, macrophages, and dendritic cells. Given its function in maintaining the delicate balance between effective immunity and tissue protection, it is evident that IL-10 expression is highly dynamic and needs to be tightly regulated. The transcriptional regulation of IL-10 production in myeloid cells and T cells is the topic of this review. Drivers of IL-10 expression as well as their downstream signaling pathways and transcription factors will be discussed. We will examine in more detail how various signals in CD4⁺ T cells converge on common transcriptional circuits, which fine-tune IL-10 expression in a context-dependent manner.

Keywords Interleukin-10 • Transcriptional regulation • Immune suppression • Inflammation • Tolerance • c-Maf • Blimp-1 • T cell • Myeloid cell

---

S. Rutz (✉)
Department of Cancer Immunology, Genentech,
1 DNA Way, South San Francisco, CA 94080, USA
e-mail: saschar@gene.com

W. Ouyang
Department of Immunology, Genentech, South San Francisco, CA, USA

Present Affiliation: Department of Inflammation and Oncology, Amgen,
1120 Veterans Blvd, South San Francisco, CA 94080, USA
e-mail: wouyang@amgen.com

X. Ma (ed.), *Regulation of Cytokine Gene Expression in Immunity and Diseases*,
Advances in Experimental Medicine and Biology 941,
DOI 10.1007/978-94-024-0921-5_5

5.1 Introduction

Multicellular organisms have developed ever more sophisticated and effective immune systems to defend themselves against a wide variety of pathogens. Equally importantly, the immune system has the capacity to limit its potentially deleterious adverse effects on the host itself by utilizing various strategies, such as self-tolerance and anti-inflammatory pathways. Defects in these strategies can result in the development of autoimmune and inflammatory diseases. The anti-inflammatory cytokine IL-10, identified by Mosmann and colleagues in 1989 [52], is a critical negative regulator of immune responses. Loss of IL-10 leads to inflammatory diseases, most notably the development of IBD [131].

IL-10 is the founding member of the IL-10 family of cytokines, which also includes IL-19, IL-20, IL-22, IL-24, IL-26, and the more distantly related IL-28A, IL-28B, and IL-29 [131, 149]. IL-10 was initially described as a secreted cytokine synthesis inhibitory factor (CSIF) produced by Th2 T cell clones, which inhibits the production of several cytokines from Th1 cells [52]. Since this first characterization it has become clear that IL-10 is in fact expressed by a wide variety of cells of the innate and adaptive arms of the immune system, including macrophages, monocytes, dendritic cells (DCs), mast cells, eosinophils, neutrophils, natural killer (NK) cells, CD4$^+$ and CD8$^+$ T cells, and B cells [119, 131].

IL-10 forms non-covalently linked homodimers, which bind to two receptor chains, IL-10R1 and IL-10R2 [92, 176]. IL-10R1 binds IL-10 with high affinity and is unique to the IL-10 receptor, whereas IL-10R2 is a common component of the receptors for IL-22, IL-26, IL-28A, IL-28B, and IL-29 [131, 149]. While the IL-10R2 chain is ubiquitously expressed, IL-10R1 is mainly present on leukocytes. In fact, IL-10 is the only member of the IL-10 cytokine family, which primarily targets leukocytes [131, 149].

IL-10 signals through the Janus kinase (Jak)/signal transducer and activator of transcription (STAT) signaling pathway (Fig. 5.1). Jak1 and Tyk2 are associated with IL-10R1 and IL-10R2, respectively. Binding of IL-10 to the receptor leads to Jak-dependent phosphorylation of the receptor. This allows for the recruitment of STAT3 and to a lesser extent STAT1. Jak1 and Tyk2 phosphorylate STAT3 on tyrosine 705, leading to its dissociation from the receptor and the formation of an active homodimer [50, 70, 117, 162, 175].

IL-10 has a broad spectrum of anti-inflammatory functions and can suppress immune responses to foreign or self-antigens. It mainly targets antigen-presenting cells, such as monocytes and macrophages, by inhibiting the release of pro-inflammatory mediators, including TNF-α, IL-1β, IL-6, IL-8, G-CSF, and GM-CSF, from these cells [40, 51]. IL-10 also inhibits antigen presentation by reducing the expression of MHC II and co-stimulating (e.g. CD86) and adhesion (e.g. CD54) molecules [34, 41, 194]. Moreover, IL-10 inhibits the production and/or secretion of cytokines required for CD4$^+$ T cell differentiation, such as IL-12 and IL-23 [38, 158]. Apart from these indirect ways of inhibiting T cell responses through the downregulation of APC functions, IL-10 can also directly inhibit both proliferation and cytokine production of CD4$^+$ T cells [65].

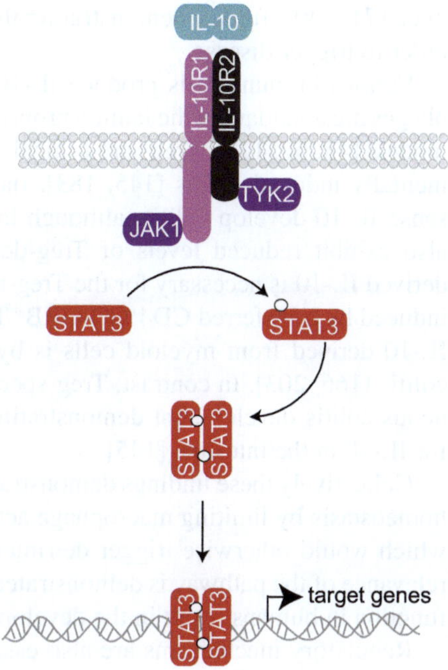

Fig. 5.1 IL-10 signaling. The IL-10 receptor consists of IL-10R1 and IL-10R2 chains. IL-10 binds to the IL-10R1 chain with high affinity. Binding of IL-10 to the receptor leads to phosphorylation and activation of tyrosine kinases Jak1 and Tyk2, which in turn phosphorylate Tyr705 in the C-terminal region of STAT3. This allows STAT3 dimerization and nuclear translocation, and the transcription of IL-10-dependent genes

5.2 IL-10 Function During Homeostasis and Infection

IL-10, like transforming growth factor beta (TGF-β), is a regulatory cytokine with pleiotropic roles in the immune system. However, the prominent function of TGF-β is to maintain T cell tolerance to self or innocuous environmental antigens via its direct effects on differentiation and homeostasis of effector and regulatory T cells (Tregs). Deficiencies in the TGF-β pathway result in hyperactivation and uncontrolled expansion of T cells leading to a lethal multi-organ autoimmune disorder [101]. In contrast, IL-10 functions primarily as a feedback inhibitor of excessive T cell responses to microbial antigens. Most prominently, IL-10-deficient mice spontaneously develop colitis demonstrating that IL-10 has an essential role in maintaining peripheral immune tolerance [95]. Colitis in these mice is mediated by activation and differentiation of effector T cells and is inhibited in IL-10-deficient mice housed under germ-free conditions [95, 160]. This suggests that IL-10 functions to maintain homeostatic T cell tolerance to commensal bacteria in the intestine. IL-10 acts on myeloid cells in order to maintain colonic homeostasis. Mice bearing a deletion of the IL-10 receptor within the myeloid compartment develop colitis indistinguishable from mice with global IL-10 receptor deficiency [165, 203]. The absence of IL-10 signaling results in excessive production of pro-inflammatory cytokines, such as IL-1β, IL-23, and IL-6 by myeloid cells [71]. Deficiency in the signaling adaptor myeloid differentiation primary response gene 88 (MyD88), which abrogates most of Toll-like receptor (TLR) signaling – one of the major pathways by which myeloid cells sense microbial products – blocks the development of colitis in IL-10-deficient mice [71, 140], further demonstrating that indeed microbial stimuli are required in order to trigger disease.

Various immune cells produce IL-10 in the intestine. IL-10 producing macrophages are abundant in the lamina propria and can induce or expand Tregs in various models [43, 67, 122]. Tregs are critical in the prevention of spontaneous or experimentally induced colitis [145, 183]. Indeed, mice with Tregs that are not able to sense IL-10 develop colitis, although later than IL-10-deficient mice. These mice also exhibit reduced levels of Treg-derived IL-10 [29]. Similarly, macrophage-derived IL-10 is necessary for the Treg-mediated prevention of experimental colitis induced by transferred $CD4^{+}CD45RB^{+}$ T cells [122]. However, the selective loss of IL-10 derived from myeloid cells is by itself not sufficient to drive spontaneous colitis [166, 203]. In contrast, Treg-specific IL-10 deficiency does result in spontaneous colitis development demonstrating that indeed Tregs are the critical source for IL-10 in the intestine [145].

Collectively these findings demonstrate that IL-10 is essential to maintain colonic homeostasis by limiting macrophage activation in response to commensal bacteria, which would otherwise trigger detrimental effector T cell responses. The clinical relevance of the pathway is demonstrated by the fact that mutations that block IL-10 function in humans result in the development of severe early onset colitis [64].

Regulatory mechanisms are also essential to properly control and limit inflammatory responses during ongoing immune responses against pathogens. IL-10 is produced as a negative feedback mechanism [33, 129]. Depending on the type of immune response, loss of IL-10 may lead to an enhanced immune reaction and therefore faster pathogen clearance; adversely it can give rise to excessive inflammation resulting in tissue damage or mortality. IL-10 derived from macrophages is critical in order to limit innate responses to certain microbial stimuli, such as the TLR4 ligand lipopolysaccharide (LPS) [166]. Mice with global IL-10 deficiency as well as those with IL-10 deficient macrophages, but not wild-type mice, succumb to septic shock following administration of low doses of LPS [16, 166].

IL-10 production from effector T cells, and to a lesser extent from regulatory T cells, is critical in many infections that trigger an adaptive immune response. During *Toxoplasma gondii* infection, for example, IL-10 produced by Th1 cells is essential to limit an otherwise excessive Th1 cell response [48, 77]. Lack of IL-10 production from T cells in this model is associated with enhanced T cell activation and differentiation. Mice with a T cell-specific deficiency in IL-10 succumb to severe immunopathology upon infection with *T. gondii* akin to mice with global IL-10 deficiency [77, 142]. Similarly, infection with *Plasmodium chabaudi* leads to IL-10 secretion from $CD4^{+}$ T cells. Deletion of IL-10 from T cells results in decreased survival, greater weight loss, and increased levels of effector cytokines, such as IFN-γ and TNF-α [55]. The contribution of IL-10 derived from B cells during infection is less well understood. However, one report found that B cell-derived IL-10 non-redundantly decreases virus-specific $CD8^{+}$ T cell responses and plasma cell expansion during murine cytomegalovirus infection [108].

In some cases the IL-10-mediated negative feedback promotes chronic infection as it dampens the immune response enough to prevent efficient pathogen clearance

[33]. This is the case, for example, during infection with the *Leishmania major* strain NIH/S, which induces nonhealing lesions in infected wild-type mice due to IL-10 production from Th1 cells [8]. Certain pathogens, such as the Epstein-Barr virus, even encode IL-10 homologs in their genome as a means to evade the host's immune response [120].

5.3 Drivers of IL-10 Expression in Myeloid Cells

IL-10 production in myeloid cells is triggered by microbial products (Fig. 5.2), which are recognized through pattern recognition receptors, including TLRs, C-type lectin receptors, retinoic acid-inducible gene 1 (RIG-I)-like receptors, and nucleotide-binding oligomerization domain (NOD)-like receptors. Several TLRs, including TLR2, TLR4, TLR5, TLR7, and TLR9, have been shown to induce IL-10 production in human and murine macrophages and DCs [2, 18, 27, 30, 44, 47, 51, 56, 59, 68, 69, 80, 143]. TLR2 ligands, such as Pam3cys, appear to be particularly potent inducers of IL-10 production by myeloid cells [44]. Accordingly, TLR2-deficient mice exhibit reduced IL-10 production from macrophages during *Candida albicans* infection [127]. Despite their common expression of TLRs, myeloid cells differ in their ability to produce IL-10 in response to TLR ligation. TLR9 activation, for example, induces IL-10 production more readily from murine macrophages than from DCs [18, 80].

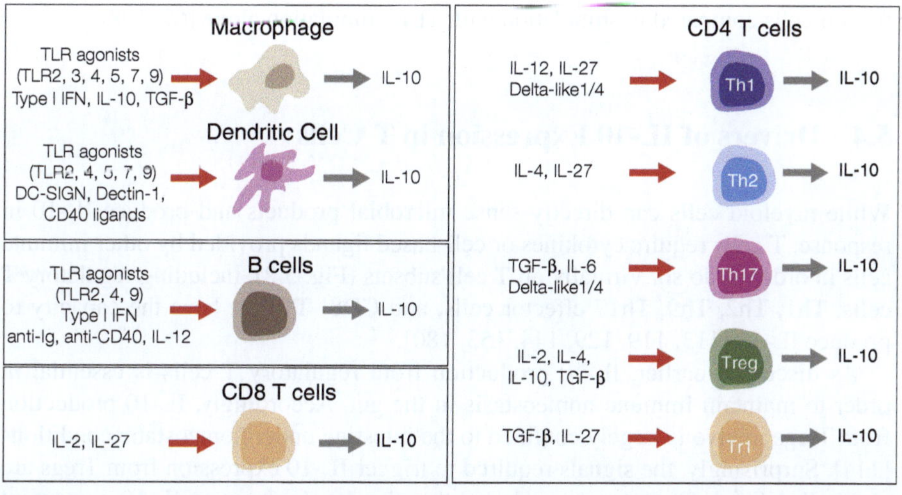

Fig. 5.2 IL-10 expression in different immune cells. Many cell types of the innate and adaptive immune system express IL-10 in response to microbial stimuli or cytokines. Macrophages and dendritic cells produce IL-10 mostly in response to microbial stimuli, such as TLR ligands. Different T cell subsets secrete IL-10 when stimulated by a variety of cytokines. In particular IL-27 and TGF-β broadly induce IL-10 production from various T cell subsets

Myeloid cells also have the capacity to integrate signals received through TLRs with other pathways in order to modulate their IL-10 production. TLR-independent triggers of IL-10 production include the C-type lectin receptors DC-SIGN [59] and Dectin-1 [42, 143]. Co-activation of TLR2 and Dectin-1 enhances IL-10 production in DCs relative to TLR2 or Dectin-1 stimulation alone [42]. Similarly, CD40 ligation together with TLR activation further enhances IL-10 production in DCs [47].

In addition to microbial stimuli, myeloid cells modulate their IL-10 production in response to cytokines secreted by other immune cells. Type I interferon enhances IL-10 production from TLR4-stimulated murine macrophages and human monocytes. In fact, a sustained IL-10 production by macrophages in response to LPS requires an IFN-β-mediated autocrine feedback loop in order to maintain IL-10 transcription [5, 27, 132]. Type I IFN also promotes IL-10 production in *Mycobacterium tuberculosis*-infected macrophages [112, 116]. In contrast, IFN-γ reduces the production of IL-10 in TLR2-activated human macrophages [72].

Phagocytosis of apoptotic cells results in an anti-inflammatory response characterized by IL-10, PGE2, and TGF-β [190]. LPS-activated peripheral blood mononuclear cells in the presence of apoptotic peripheral blood lymphocytes are able to produce higher levels of IL-10 compared to LPS alone [190]. IL-10 may also act to drive its own transcription in macrophages and/or monocytes. Stimulation of monocyte-derived macrophages with IL-10 leads to an increase in IL-10 mRNA [170].

B cells are a source for IL-10 in vitro and in vivo (Fig. 5.2). A combination of anti-Ig and anti-CD40 stimulation induces IL-10 expression, a process that is further enhanced by IL-12 [167]. B cells express a number of TLRs. Agonists of TLR2, TLR4, or TLR9 have all been shown to promote IL-10 production [1, 97, 156]. Similar to macrophages, IFN-α in combination with TLR agonists increases IL-10 production from B cells compared to stimulation with TLR stimulation alone [62, 198].

5.4 Drivers of IL-10 Expression in T Cells

While myeloid cells can directly sense microbial products and produce IL-10 in response, T cells require cytokines or cell-based ligands provided by other immune cells in order to do so. Virtually all T cell subsets (Fig. 5.2), including regulatory T cells, Th1, Th2, Th9, Th17 effector cells, and $CD8^+$ T cells, have the capacity to produce IL-10 [113, 119, 129, 144, 153, 180].

As discussed earlier, IL-10 production from regulatory T cells is essential in order to maintain immune homeostasis in the gut. Accordingly, IL-10 production from Tregs in vivo is largely confined to the intestine under homeostatic conditions [114]. Surprisingly, the signals required to trigger IL-10 expression from Tregs are not well defined. One report concluded that the development of IL-10-competent Tregs does not require IL-10 itself, but is instead dependent on TGF-β [114]. However, another study found that Tregs in the intestine indeed sense IL-10 and that IL-10R-deficient Tregs have reduced levels of IL-10 expression [29]. IL-2 and IL-4 induce IL-10 production from regulatory T cells in vitro and in vivo [13, 35, 96].

In accordance with its function as a negative feedback regulator, effector T cells transiently express IL-10 linked to a state of full activation and effector cytokine production [153, 155]. Although this conclusion is largely based on in vitro studies, in many cases, the induction of IL-10 appears to be an intricate part of the T cell differentiation program, triggered by the same stimuli. Accordingly, strong TCR stimulation together with the Th1-polarizing cytokine IL-12 is required for IL-10 production from Th1 cells, while the Th2-polarizing cytokine IL-4 is essential for IL-10 production from Th2 cells [28, 60, 118, 155, 164]. TGF-β and IL-6 drive the production of IL-10 as part of the Th17 cell differentiation program [173, 195]. The more recently defined subset of IL-9-producing Th9 cells requires IL-4 and TGF-β for its differentiation and IL-10 production [39, 187].

Similar to their role in myeloid cells, type I IFNs promote IL-10 production from $CD4^+$ T cells [5, 32, 100].

5.4.1 *IL-10 Induction by IL-27 and TGF-β*

IL-27 and TGF-β are potent inducers of IL-10 production in various T cell subsets in vivo (Fig. 5.2). IL-27 is a member of the IL-12 cytokine family and is produced mostly by innate immune cells during infections [73, 84]. IL-27, which is normally not part of T cell differentiation protocols, indeed promotes IL-10 production in $CD4^+$ T cells under neutral and Th1-, Th2-, or Tr1-polarizing conditions in vitro [11, 14, 53, 124, 134, 173]. More importantly, IL-27 limits Th1, Th2, and Th17 cell responses in various infection and autoimmune disease models [9, 53, 73, 84, 172]. For example, $IL\text{-}10^+IFN\text{-}\gamma^+$ Th1 cells are absent in IL-27 receptor-deficient mice, and these mice develop lethal $CD4^+$ T cell-mediated inflammation upon *T. gondii* infection [172, 189]. Acting together with IL-2, IL-27 also induces IL-10 expression in cytotoxic $CD8^+$ T cells [174].

TGF-β, on the other hand, is required for the differentiation and IL-10 production of regulatory T cells [114] and Th17 cells in the intestine. These T cells are generated mainly in response to the commensal microbiota, in the absence of strong pro-inflammatory stimuli, which precludes high levels of IL-27. However, TGF-β does further augment IL-27-induced IL-10 expression in $CD4^+$ T cells in vitro, conditions most commonly used to generate Tr1 cells [11, 173]. In contrast, IL-10 produced by Th17 cells in response to IL-6/TGF-β is not further enhanced by IL-27 [173]. Mechanistically IL-27 and TGF-β rely on distinct but partially overlapping transcriptional programs to induce IL-10 expression [128].

5.4.2 *IL-10 Induction by Notch*

In addition to secreting cytokines that shape the differentiation of $CD4^+$ T cells, myeloid cells also upregulate Notch ligands in response to microbial stimuli that have been shown to drive T cell polarization and IL-10 production [6, 139].

We found that the Notch pathway is a critical regulator of IL-10 production under pro-inflammatory conditions in Th1 cells. Notch synergizes with IL-12 or IL-27 to strongly enhance IL-10 production in IFN-γ-producing Th1 cells [147]. Activation of the Notch pathway via its ligands Delta-like (Dll) 1 or Dll4 promotes the co-expression of IL-10 and IFN-γ in Th1 cells in vitro [83, 147]. The Notch pathway also promotes IL-10 production from Th17 cells in vitro [131]. Notch-mediated IL-10 induction in vivo critically depends on Dll4, the expression of which is strongly induced on dendritic cells upon TLR stimulation, but is largely restricted to plasmacytoid DCs in the steady state [83].

5.5 The *Il10* Gene Locus and Its Epigenetic Regulation

The *Il10* gene locus in mice exhibits a high degree of homology to the one in human. In both cases *Il10* is located within the *Il10* gene family cluster on chromosome 1 [88], immediately downstream of *Il19*, *Il20*, and *Il24* (Fig. 5.3a). *Il10* exhibits the prototypical genomic organization of all IL-10 family cytokines being comprised of five exons and four introns. Genomic alignments of the mouse and human *Il10* loci have identified a number of highly conserved noncoding sequences (CNS), indicating important roles in the regulation of *Il10* expression (Fig. 5.3b). The highest degree of conservation is observed for CNS-29.8, CNS-26, CNS-20, CNS-9, CNS-4.5, and CNS-0.12 (in kb relative to *Il10* transcriptional start site (TSS)) upstream as well as CNS+1.65, CNS+3, and CNS+6.5 downstream of the TSS [79, 98, 191].

Regions within the *Il10* locus that are hypersensitive to DNase I digestion (HSS) localize mostly to the CNS regions, indicating accessible chromatin in these areas. For instance, bone marrow-derived macrophages stimulated with TLR ligands (LPS, CpG, or zymosan) exhibit five HSS regions located −4.5, −2, and −0.12 kb upstream and +1.65 and +2.98 kb downstream of the TSS [154]. Non-stimulated macrophages also show some degree of sensitivity to DNase I digestion at these sites. This is consistent with the immediate IL-10 expression observed in these cells upon exposure to microbial stimuli. Only HSS −4.5 kb was detected in BM-derived DCs when stimulated with TLR ligands [154]. Again, the reduced accessibility in DCs as compared to macrophages is in line with the weaker IL-10 expression in DCs in response to the same stimuli.

In contrast, chromatin accessibility in T cells is more dynamic. The *Il10* gene locus in naïve T cells is in a closed, transcriptionally inactive conformation with only one HSS at −8.8 kb [79]. Extensive chromatin remodeling during T cell differentiation leads to the formation of additional, mostly common HSS in Th1 and Th2 cells [74, 79, 154, 191]. Some of these sites are more prominent in Th2 than in Th1 cells (HSS −30.4, −29.8, −21, −17.5, and −0.12 kb and +6.45 kb) [79]. Most described HSS in the *Il10* locus are common between T cells and myeloid cells, apart from HSS −4.5 kb which appears to be specific to macrophage and DCs [154].

The chromatin structure at the *Il10* locus is critical in the regulation of IL-10 expression. Among other mechanisms, chromatin accessibility for the transcriptional

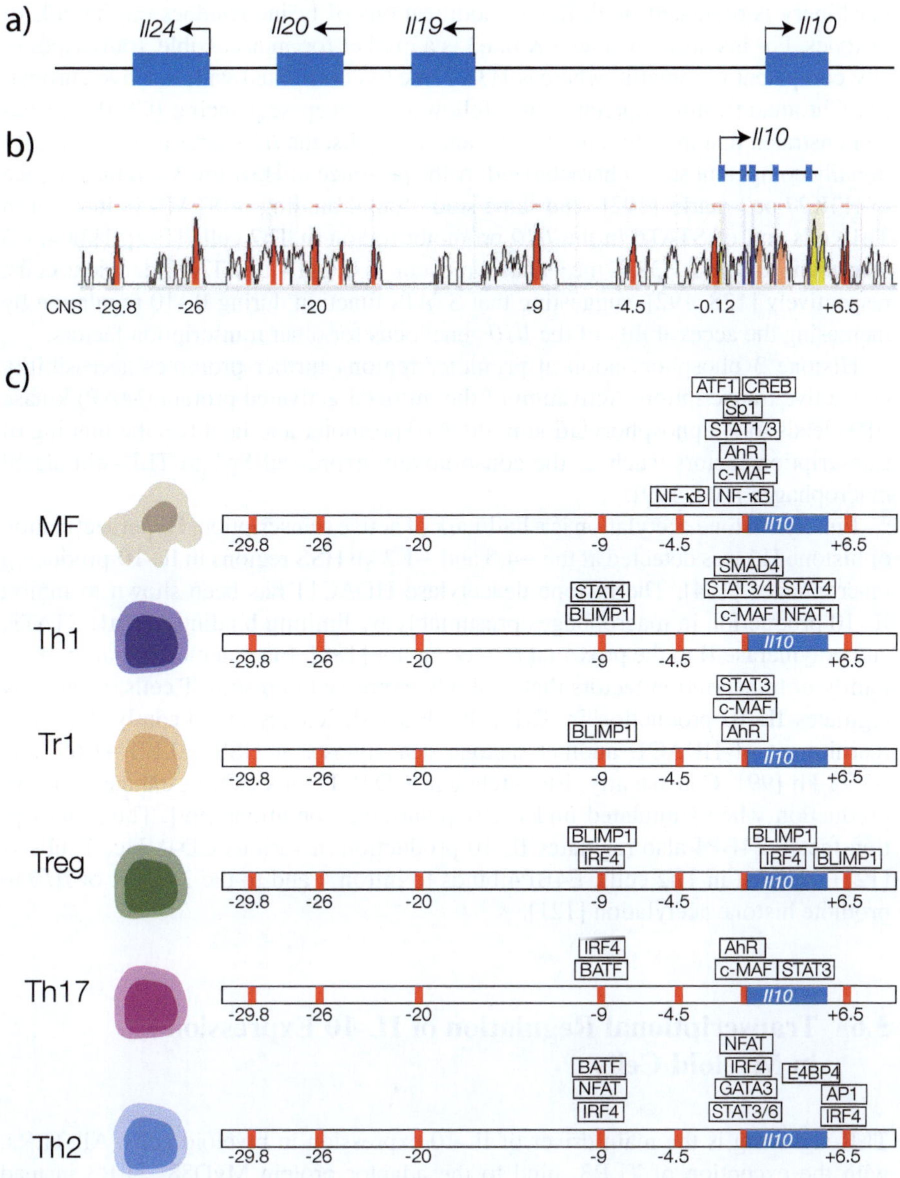

Fig. 5.3 The *Il10* locus and transcriptional regulation of IL-10 expression in various cell types. **(a)** *Il10* is part of the IL-10 cluster on chromosome 1 that also contains *Il19*, *Il20*, and *Il24*. **(b)** Sequence conservation between mouse *Il10* and human *IL10* genes. Highly conserved noncoding sequence (CNS) regions in the *Il10* locus are colored in *red*, and positions are relative to the transcriptional start site. **(c)** Transcription factors implicated in IL-10 regulation in various cell types and their binding within the *Il10* locus

machinery is regulated by different modifications of lysine residues on the tails of histones. For instance, histone 3 K4me3 is a marker for an accessible, transcriptionally competent chromatin, whereas H3K27me3 is associated with inactive chromatin. Chromatin immunoprecipitation followed by deep sequencing (ChIP-seq) has demonstrated that in differentiated Th1 and Th2 cells, the *Il10* locus is in a transcriptionally competent state, characterized by the presence of H3K4me3 and the absence of H3K27me3 marks [192]. The same study found binding of STAT4 in intron 4 in Th1 cells and of STAT6 in the *Il10* promoter region in Th2 cells [192]. H3K4me3 marks are lost and H3K27me3 marks reoccur in STAT4 or STAT6 deficient cells, respectively [128, 192], suggesting that STATs function during IL-10 regulation by increasing the accessibility of the *Il10* gene locus for other transcription factors.

Histone 3 phosphorylation at promoter regions further promotes accessibility and active transcription. Activation of the mitogen-activated protein (MAP) kinase ERK leads to H3 phosphorylation in the *Il10* promoter and facilities the binding of transcription factors, such as the constitutively expressed Sp1 in TLR-stimulated macrophages [104, 199].

Finally, histone acetylation is a hallmark of active transcription. Hyperacetylation of histone H4 was detected at the −4.5 and −1.2 kb HSS regions in IL-10-producing macrophages [154]. The histone deacetylase HDAC11 has been shown to inhibit IL-10 production in macrophages presumably by limiting binding of Sp1, STAT3, and polymerase II to the proximal *Il10* promoter [188]. Ets-1, a member of the ETS family of transcription factors that is highly expressed in resting T cells, negatively regulates IL-10 production in Th1 cells. Ets-1 deficiency significantly decreases recruitment of HDAC1, another histone deacetylase, at HSS −0.12, +1.65, and +2.98 kb [99]. Consistently, Ets1-deficient $CD4^+$ T cells exhibit enhanced IL-10 production when stimulated under Th1-polarizing conditions [99]. The transcription factor E4BP4 also regulates IL-10 production in various $CD4^+$ T cell subsets [121]. At least in Th2 cells, E4BP4 binds to intron 4 and to the 3′ UTR of *Il10* to promote histone acetylation [121].

5.6 Transcriptional Regulation of IL-10 Expression in Myeloid Cells

TLR signaling is the main driver of IL-10 expression in myeloid cells. All TLRs, with the exception of TLR3, bind to the adaptor protein MyD88. TLR3 instead binds the adaptor TRIF (TIR domain-containing adaptor protein inducing IFN-β), while TLR4 recruits both MyD88 and TRIF [4]. Both MyD88- and TRIF-dependent TLR signaling are able to induce IL-10 production [18]. TLR ligation leads to the activation of several downstream pathways, including the MAP kinase pathway, the PI3K/AKT pathway, and the NF-κB pathway [85].

Activation of the MAP3 kinase TPL-2 and downstream ERK1 and ERK2 is critical for IL-10 production in macrophages and DCs in response to TLR2, TLR4, and TLR9 activation [12, 44, 80, 196]. Accordingly, inhibitors of the ERK1/2 pathway reduce IL-10 secretion from macrophages [103]. Similarly, a knockout of TPL-2 reduces TLR-induced IL-10 production [12]. ERK activation is also downstream of Dectin-1-induced IL-10 production in DCs [45, 168]. The degree of ERK activation correlates with the levels of IL-10 production observed in macrophages, mDCs, and pDCs [80] demonstrating the central importance of this pathway. The MAP kinase pathway eventually results in activation of members of the AP1 and ATF transcription factor families. The AP1 factor c-Fos, for instance, has been shown to be involved in TLR-induced IL-10 expression in macrophages and DCs [2, 44, 72, 80].

In addition to ERK, p38 [30, 54, 72, 78, 89, 196] and JNK [26, 30, 72] also contribute to IL-10 production in TLR-stimulated macrophages, monocytes, and DCs. For example, knockout of p38α or treatment with p38α/β inhibitors reduce IL-10 secretion in macrophages [86]. Macrophages deficient for DUSP1, a dual-specificity phosphatase involved in deactivating p38 signaling in response to LPS, display prolonged p38α activation relative to wild-type cells, and this correlates with an increase in IL-10 production [151]. p38γ and -δ also seem to be required for normal LPS-induced IL-10 production, presumably through their regulation of TPL-2 expression. p38γ/δ knockout macrophages have very low TPL-2 levels and fail to activate ERK1/2 downstream of TLR activation [141]. Both ERK and p38 may function cooperatively in their regulation of IL-10 production, through their joined activation of MSK1 and MSK2, mitogen- and stress-activated protein kinases, which promote IL-10 production in TLR4-stimulated macrophages. Downstream of MSK1 and MSK2 the transcription factors CREB and ATF1 bind and trans-activate the *Il10* promoter [7, 133]. The MAPK–MSK1/2–CREB pathway is also downstream of Dectin-1 induced IL-10 expression [49]. Both ERK1/2 and p38α directly phosphorylate Sp1 [37, 177]. Numerous studies using *Il10* promoter reporter genes have shown that Sp1 binding sites are required for the induction of IL-10 transcription [20, 106, 178, 199].

The PI3K/AKT pathway also contributes to IL-10 expression in myeloid cells [109, 130, 193], either by antagonizing GSK3-β, a constitutively active kinase which inhibits the production of IL-10 [125, 130], or through ERK [109] and mTOR [130, 193] activation. mTOR may further regulate IL-10 production through the activation of STAT3 [193].

NF-κB is activated in response to TLR stimulation and contributes to IL-10 regulation. Accordingly, IKK2-deficient macrophages, which have impaired NF-κB activation, show reduced production of IL-10 [82]. In TLR4-stimulated macrophages, the NF-κB subunit p65 is recruited to an NF-κB binding site at HSS −4.5 kb [154]. Its binding to a different site mediates IL-10 production in response to dsRNA [26]. p105 (NF-κB1) is a binding partner for TPL-2. Accordingly, knockout of p105 blocks TLR-induced ERK1/2 activation and downstream IL-10 expression [12]. In addition to binding to TPL-2, p105 can be cleaved to generate the p50 NF-κB subunit. The p50 homodimer is recruited to a site proximal to the *Il10* transcriptional start site and promotes IL-10 production in macrophages in response to TLR4 activation [22].

Several transcription factors, including c-Maf, the aryl hydrocarbon receptor (AhR), and STATs, mediate IL-10 production in macrophages as well as in T cells (see below). c-Maf belongs to the Maf family of basic region and leucine zipper transcription factors and has been discussed as somewhat of a master transcription factor for IL-10 in T cells and macrophages [23, 146, 153]. c-Maf binds to the *Il10* promoter and enhances IL-10 production in LPS-stimulated macrophages [23]. Conversely, c-Maf-deficient macrophages exhibit strongly impaired IL-10 production upon LPS stimulation. c-Maf is expressed constitutively in resting monocytes and macrophages and further upregulated by IL-4. The ability of IL-4 to act in combination with LPS to induce IL-10 is blunted in c-Maf knockout cells [23]. The ligand-activated transcription factor AhR is required for optimal IL-10 production in TLR4- but not TLR9-stimulated macrophages, where it forms a complex with STAT1 [90]. LPS-induced IL-10 production is inhibited in AhR-deficient or STAT1-deficient peritoneal macrophages compared to WT cells [90]. The IL-10 enhancing function of type I IFN relies in part on the recruitment of STAT1 or STAT3 to the *Il10* promoter [66, 202]. Inactivation of the STAT-binding motif completely ablates trans-activation by type I IFN.

Type I IFN also activates the transcription factor interferon regulatory factor 1 (IRF1), which enhances IL-10 production [202]. In contrast IRF5, the expression of which is induced downstream of TLR ligation, binds to the *Il10* promoter but functions as a repressor of IL-10 in human GM-CSF differentiated monocytes [94].

5.7 Transcriptional Regulation of IL-10 in T Cells

T cells need to undergo activation and differentiation into effector T cells in order to express IL-10. Accordingly, they integrate signals through their T cell receptor (TCR) and cytokine receptors in regulating IL-10 expression.

5.7.1 TCR Signaling in IL-10 Regulation in T Cells

TCR ligation, which is the minimal requirement for any kind of T cell activation but not sufficient for subsequent effector differentiation, activates several downstream signaling pathways culminating in the activation of transcription factors such as AP1, NFAT, and NF-κB. Similar to its role in macrophages and DCs, the MAP kinase ERK is a common positive regulator of IL-10 expression in different T helper cell subsets [155]. Based on the use of small-molecule kinase inhibitors, the ERK1/2 and p38 MAP kinase signaling pathways also seem to regulate IL-10 production in $CD8^+$ T cells [179]. ERK is activated via Ras downstream of the TCR [46] and eventually leads to the activation of transcription factors of the AP1 family. Several members of the AP1 family have been shown to promote IL-10 expression in various T cell subsets by binding to AP1 motifs in the *Il10* locus. Early studies found that in Th2, but not in Th1 cells, JunB and to a lesser extent c-Jun bind at HSS +6.45 kb and promote IL-10 expression [79, 191]. Ectopic expression of c-Jun and JunB enhances IL-10 production in activated naïve T cells, whereas a dominant negative c-Jun reduces IL-10 production [191]. The transcription factor IRF4 positively regulates IL-10 expression in Th2 cells and also binds to the *Il10* promoter as well as the CNS+6.45 region [3].

Upon TCR activation, NFAT1 translocates from the cytoplasm into the nucleus where it is known to interact with AP1 and other transcriptional partners to promote cytokine gene transcription [107]. Consistently, NFAT1 binds to the *Il10* promoter in Th2 and to intron 4 in Th1 cell lines [74]. NFAT1/IRF4 co-binding to CNS-9 synergistically enhances IL-10 expression in Th2 cells [98].

More recently the basic leucine zipper transcription factor ATF-like (BATF), another AP1 family member, together with JunB and IRF4 were found to function as "pioneer factors" in T cell differentiation by binding to a multitude of loci, including CNS-9 in the *Il10* locus, early after T cell activation [31, 63, 102, 182]. Mutating either the IRF or AP1 motif within CNS-9 results in a diminished luciferase reporter activity consistent with functional cooperation between these factors in *Il10* gene regulation [102]. Deficiencies in BATF or IRF4 significantly impair Th2 or Th17 differentiation and effector cytokine production demonstrating a much broader function beyond IL-10 regulation [3, 31, 123]. This further demonstrates the close interrelation of IL-10 expression and T cell differentiation. However, the broad function of BATF/IRF4 across different T cell subsets is more consistent with a scenario in which T cell activation provides a framework for both IL-10 expression and T cell differentiation, both of which are then further specified by additional polarizing factors, such as STATs.

5.7.2 STAT/SMAD Pathways in IL-10 Regulation in T Cells

STATs are cytoplasmic transcription factors that translocate to the nucleus to regulate gene expression in response to cytokines and growth factors. In this regard STATs are essential in mediating T cell polarization. STAT4 is downstream of IL-12 in driving Th1 differentiation, and STAT6 downstream of IL-4 mediates Th2 polarization, whereas STAT3 is activated by IL-6 and IL-23 and is critical for Th17 differentiation. All of these STATs have been implicated as critical drivers of IL-10 expression in the respective T cell subsets [28, 147, 155, 195]. STAT3 and to some extent STAT1 downstream of IL-27 and IL-21 have roles in inducing IL-10 beyond a particular T cell subset [14, 53, 115, 169, 170, 173, 195]. Two STAT binding sites have been identified in the murine and human *Il10* promoter [184, 202]. Accordingly, STAT4 and STAT6 bind to the *Il10* promoter in Th1 and Th2 cells, respectively [128, 192]. STAT4 binding has also been detecting in the CNS-9 [128] and intron 4 [192] region in Th1 cells. STAT3 binds to the same intron 4 region in Th17 cells

[102]. However, STAT3 deficiency does not block the ability of TGF-β/IL-6 treatment to promote IL-10 production, although it blocks the effects of IL-27 [195]. In conclusion, STATs downstream of a variety of cytokines critically contribute to IL-10 expression, arguably in some cases as an integral part of the respective differentiation program, but also more broadly in a subset-nonspecific manner.

Similarly, SMADs downstream of TGF-β regulate IL-10 production in T helper cells, although TGF-β in many cases interfers with or alters T helper cell differentiation. For example, Th17 cells generated with IL-6 and TGF-β produce large amounts of IL-10 in addition to the signature cytokine IL-17. In contrast, Th17 cells differentiated in the absence of TGF-β lack IL-10 expression, whereas IL-17 levels are comparable [61]. Even more strikingly, TGF-β potently induces IL-10 production from Th1 cells [91, 128]. Downstream SMAD4 binds and trans-activates the *Il10* promoter in Th1 cells [91]. At the same time, however, TGF-β extinguishes IFN-γ expression in these cells [91, 128]. Similarly, TGF-β, in the presence of IL-4, stirs T cell differentiation toward a Th9 phenotype, away from Th2 [39, 187]. Yet, SMAD3 together with GATA3 positively regulate IL-10 production in response to TGF-β in Th2 cells [17]. TGF-β signaling seems to promote IL-10 expression broadly and mostly independently of the T cell differentiation program.

5.7.3 T Cell Lineage Transcription Factors in Regulation of IL-10 Expression

As discussed previously, IL-10 is not produced by naïve T cells, but instead requires T cell activation following antigen recognition. This activation usually occurs in the context of certain cytokines that favor T cell differentiation. Therefore both processes usually coincide.

One way of dissecting this further is to assess the direct contribution of master transcription factors that are induced during T cell differentiation, and that in fact drive and define this process. T-bet is induced by STAT4 downstream of IL-12 and drives Th1 differentiation. Th2 differentiation requires IL-4 signaling through STAT6 and the expression of GATA binding protein 3 (GATA3). Th17 differentiation is dependent on TGF-β and IL-6, which together induce the expression of RAR-related orphan receptor gamma (RORγt) [201].

The requirement for these master transcription factors for IL-10 expression is probably best studied in the case of GATA3 in Th2 cells. GATA3 is recruited to two locations in the *Il10* locus, but it does not trans-activate the *Il10* promoter. Instead, GATA3 binding facilities chromatin remodeling and leads to histone acetylation at the *Il10* locus [164]. This leads to "epigenetic imprinting," which allows Th2 cells to establish a stable memory for IL-10 expression [28, 87, 164]. However, GATA3 is not required for IL-10 production in differentiated Th2 cells [200]. The role for T-bet in IL-10 expression in Th1 cells has been studied in less detail. However, one of the main functions of T-bet during Th1 polarization is the induction of the

IL-12RβII chain to enable IL-12 signaling [159], which in turn drives IL-10 expression. Similarly, we find that the induction of IL-10 downstream for Notch signaling is not impaired in T-bet-deficient T cells cultured under Th1-polarizing conditions. It is, however, strictly dependent on STAT4 [147]. A detailed recent analysis of the transcriptional network in Th17 cells suggests that RORγt indeed acts as a repressor for *Il10*, whereas it positively regulates the expression of the Th17 signature cytokine *Il17a* [31]. Overall these findings suggest that although IL-10 expression coincides with effector cytokine expression in various T helper cell subsets, it is, at least to some degree, decoupled from the respective differentiation program. The term "master transcription factors" is further called into question by recent findings showing that T-bet, GATA3, and RORγt only regulate the expression of a relatively small subset of genes directly – which includes the signature cytokine genes – whereas STATs affect gene expression much more broadly [31, 185]. Additional factors, first and foremost STATs and SMADs downstream of pro-inflammatory cytokines and TGF-β, determine the expression of IL-10 in response to environmental factors.

5.7.4 The Role of c-Maf in IL-10 Regulation in T Cells

c-Maf expression is induced by several important drivers of IL-10 expression, including IL-27, TGF-β, and ICOS ligand [10, 15, 134, 195]. Indeed, c-Maf has been shown to be critical for IL-10 expression in Th17 and Tr1 cells [10, 134, 195]. But its expression also correlates with IL-10 production in Th1 and Th2 cells [155]. Although its role in Th9 cells has not been studied, it is likely that c-Maf also controls IL-10 production in this T cell subset, given the fact that Th9 cells differentiate in the presence of IL-4 and TGF-β [39, 187], both inducers of c-Maf.

Although c-Maf can trans-activate *Il10* by itself to some extent [10, 195], robust IL-10 expression requires interaction with additional transcriptional regulators. In the case of Tr1 cells generated in the presence of IL-27 or IL-27/TGF-β [134], c-Maf cooperates with AhR in the regulation of IL-10 production in mouse as well as human [10, 58]. As such, knocking down c-Maf or AhR in Tr1 cells decreases *Il10* mRNA expression. Both factors bind and synergistically trans-activate the *Il10* gene promoter [10]. The induction of c-Maf in Tr1 cells is thought to be further regulated by IL-21 and possibly by ICOS, since both IL-21- and ICOS-deficient T cells show reduced IL-10 production and c-Maf expression [135]. In Th17 cells, c-Maf is induced downstream of TGF-β and acts mainly as a suppressor of pro-inflammatory gene expression most likely by antagonizing BATF [31, 148]. However, IL-10 is among the few genes induced by c-Maf in Th17 cells [31, 195]. Although c-Maf expression is low in Th1 cells compared to other T cell subsets, it is driven by IL-12 and can be potently induced by activation of the Notch pathway in order to drive IL-10 expression [128].

5.7.5 The Role of Blimp-1 in IL-10 Regulation in T Cells

Blimp-1 encoded by the *Prdm1* gene is a transcriptional repressor that positively regulates IL-10 production in both $CD4^+$ and $CD8^+$ T cells [76, 111, 174]. Blimp-1 expression is limited to highly polarized effector $CD4^+$ T cells and therefore associated with effector cytokine secretion [110, 111].

We recently reported that Blimp-1 is critical for IL-10 production in Th1 cells, where it binds mainly to CNS-9 [128]. Blimp-1 expression in Th1 cells in vitro is dependent on STAT4, which explains its late induction during Th1 polarization. The early phase of Th1 differentiation is IL-12 independent and instead relies on IFN-γ-mediated induction of T-bet [159]. Consistent with the transient nature of IL-10 expression in Th1 cells, Blimp-1 activity is restricted to an effector state and is likely to coincide with high availability of pro-inflammatory cytokines, such as IL-12. Blimp-1-deficient Th1 cells lack IL-10 production in vitro and in vivo [128].

T cell-specific Blimp-1 deficiency results in enhanced inflammation and immunopathology during *T. gondii* infection. IL-27-induced IL-10 production in $CD4^+$ T cells is completely dependent on Blimp-1 [128], further suggesting a broad function of Blimp-1 in IL-10 regulation in T cells. In IL-27-induced Tr1 cells, the transcription factor egr2 is upstream of Blimp-1 [76]. Accordingly, egr2-deficient Tr1 cells have reduced IL-10 production but enhanced secretion of IL-17 or IFN-γ [76]. Although IL-12 and IL-27 seem to use different pathways to induce Blimp-1, IL-12 by signaling through STAT4 and IL-27 by signaling through STAT1/STAT3 [173], their co-expression in many Th1-driven immune responses makes it likely that both cytokines synergize in promoting Blimp-1-dependent IL-10 expression in Th1 cells in vivo.

Similarly, Blimp-1 is critical for IL-10 production in $CD8^+$ T cells, where its expression requires $CD4^+$ T cell help and is limited to effector and memory $CD8^+$ T cells [81, 111, 174]. Blimp-1 is also involved in IL-10 regulation in regulatory T cells and is expressed in an effector regulatory T cell population that is found at the site of inflammation [35]. Blimp-1 deficiency does not prevent Treg development but strongly impairs the production of IL-10 by these cells in response to TCR stimulation. In Treg cells Blimp-1 acts synergistically with IRF4 [35]. Given its expression pattern, Blimp-1 is perfectly suited to restrict IL-10 production to an effector phase at the peak of an acute inflammatory response. However, Blimp-1 is not universally required for IL-10 expression in T cells, as differentiation of Blimp-1-deficient $CD4^+$ T cells into Th2 cells results in normal IL-10 expression [81].

5.7.6 Transcriptional Regulation of IL-10 Expression Through Inflammation

T cells integrate diverse environmental stimuli to appropriately express IL-10. Pro-inflammatory cytokines, such as IL-12 and IL-27, secreted by activated antigen-presenting cells signal ongoing inflammation and eventually the need for self-limitation of these responses. Blimp-1 is being recognized as the main driver of IL-10 production in pro-inflammatory effector T cells under these conditions [128, 186]. Not only is Blimp-1 essential for IL-10 expression in Th1, Tr1 cells, and $CD8^+$ T cells; it also drives IL-10 production in regulatory T cells that have entered an "effector stage" induced by cytokines such as IL-2, IL-4, and others [36, 128, 174]. All these cells are severely impaired in their capacity to produce IL-10 in the absence of Blimp-1. How does Blimp-1-mediated IL-10 expression relate to c-Maf, another widely expressed transcription factor regulating IL-10 [146]? c-Maf expression can be induced by pro-inflammatory stimuli including IL-27 [134] but also by Notch ligands expressed on APCs in response to TLR stimulation. This process greatly enhances IL-10 production from Th1 cells, but still requires Blimp-1 [128]. Therefore c-Maf acts cooperatively with Blimp-1 in inducing IL-10 under inflammatory conditions.

However, IL-10 production is also required to maintain immune homeostasis in the absence of acute inflammation, mainly in the intestine. TGF-β is an important driver of IL-10 production under those circumstances. While TGF-β is a potent inducer of c-Maf, it strongly antagonizes Blimp-1 expression [10, 128, 134, 150, 195]. In fact, Th17 cells generated in the presence of TGF-β/IL-6 do express high levels of c-Maf but no Blimp-1. The suppressive effect of TGF-β on Blimp-1 is dominant over IL-27, explaining why IL-27 does not further enhance IL-10 from Th17 cells [173]. Ectopic expression of Blimp-1 in Th17 cells indeed strongly increases IL-10 production [128]. Tr1 cells can be generated in vitro with IL-27 in the presence or absence of TGF-β. The addition of increasing concentrations of TGF-β to these cultures shifts IL-10 expression from a Blimp-dependent to a Blimp-independent pathway [128]. In the presence of high levels of TGF-β, in Tr1 cells and Th17 cells, IL-10 expression fully relies on c-Maf interacting with AhR [10, 128]. Collectively, these observations across several T helper cell subsets suggest that IL-10 expression, rather than being regulated in a subset-specific manner as part of the differentiation program, is driven more universally and fine-tuned by the inflammation state.

5.8 Posttranscriptional Regulation of IL-10 Expression

Posttranscriptional control of cytokine production represents an additional layer of regulation that enables the cell to rapidly release cytokines in response to extracellular stimuli but also to shut down this response in a timely manner. Prominent mechanisms of posttranscriptional regulation of cytokines are RNA-binding proteins and microRNAs that control the stability and translational activity of the cytokine RNA [75].

A growing number of studies, mostly conducted in macrophages, has found pathways that regulate IL-10 expression on the posttranscriptional level. The 3′UTR of the IL-10 mRNA contains AU-rich elements (AREs) [136], which are characteristic of short-lived RNAs, that mediate mRNA decay through their recruitment of

ARE-binding proteins [75]. Two studies identified tristetraprolin (TTP) as the ARE-binding protein regulating the stability of IL-10 mRNA in TLR4-stimulated macrophages [57, 171]. As a consequence, TTP-deficient macrophages produce higher levels of IL-10 [57]. A more recent study found higher serum IL-10 levels in LPS-treated mice with a myeloid-specific deficiency in TTP [93]. TTP also targets other cytokine mRNAs for rapid degradation, including IL-6, TNF-α, and GM-CSF [24, 25, 126].

Interestingly, the mRNA-destabilizing function of TTP is inversely regulated by p38 MAPK activity [152, 181], such that after receiving an inflammatory stimulus the TTP-dependent decay is initially limited ensuring IL-10 expression at the height of the inflammatory response [93]. IL-10 signaling itself regulates IL-10 mRNA [21] in a TTP-dependent manner. IL-10 increases TTP expression and activity by antagonizing p38 activity [57, 157].

Another posttranscriptional regulatory circuit involves the RNA-binding protein ARE/poly(U) binding/degradation factor 1 (AUF1), which also binds to the 3′UTR of the IL-10 mRNA and reduces its half-life [19], and the upstream MKP-1, which regulates the translocation of AUF1 from the nucleus to the cytosol [197]. In the absence of MKP-1, both IL-10 mRNA stability and IL-10 secretion are increased [197].

MicroRNAs (miRNAs) target IL-10 and in doing so play important roles in autoimmune and inflammatory diseases such as SLE, reperfusion injury, and asthma [138]. IL-10 can be regulated by several microRNAs, including miR-106a, miR-4661, miR-98, miR-27, let7, and miR-142-3p/5p [138]. While miR-106a binds to the 3′UTR of the IL-10 mRNA and negatively regulates its expression [161], binding of miR-4661 to the 3′UTR results in a net increase in the half-life of IL-10 mRNA, by preventing TTP binding [105]. Another miRNA, miR-21, has been shown to indirectly regulate IL-10 via downregulation of the IL-10 inhibitor PDCD4 [163].

A number of viruses exploit the cell-intrinsic control of gene expression by miRNAs for immune evasion by encoding their own viral homologues of miRNAs. For example, Kaposi's sarcoma-associated herpesvirus encodes 17 mature microRNAs, two of which, miR-K12-3 and miR-K12-7, activate transcription and secretion of IL-10 [137].

5.9 Concluding Remarks

Given its crucial roles in maintaining an effective immune response against pathogens while reducing the risk of potentially deleterious excessive inflammation, IL-10 expression has to be finely tuned. As a negative feedback mechanism, it has to be responsive to the degree of inflammation and has to have the ability to be turned on or off quickly in any immune cell. Accordingly, IL-10 expression is regulated at multiple levels, by extracellular stimuli, transcriptionally, through epigenetics, and posttranscriptionally. A broad understanding of the signals that regulate IL-10 expression has evolved, according to which myeloid cells directly respond to microbial stimuli, most prominently TLR ligands, whereas IL-10 production from T cells requires the release of pro-inflammatory cytokines, such as IL-27 from antigen-presenting cells. These extracellular stimuli then translate into the activation of various signaling pathways and the downstream activation of transcription factors. In T cells this occurs in the context of transcriptional programs of T cell activation and differentiation. Over the past years, we have identified a number of transcription factors, such as c-Maf or Blimp-1, with prominent roles in IL-10 regulation. While we are making progress in elucidating how these factors function individually to enable IL-10 expression, a comprehensive understanding of the transcriptional regulation of IL-10 is only just beginning to emerge. Extensive research will be required to better understand the epigenetic regulation of the *Il10* locus, as it is crucial to a timely expression of IL-10. Currently we only have a rudimentary understanding of how chromatin accessibility at the *Il10* locus is controlled. In particular in T cells, this process appears to be highly dynamic, very different from "effector cytokines" and distinct in various T cell subsets.

While most of our knowledge of the transcriptional regulation of IL-10, in particular in T cells, is derived from the studies of cultured cells, moving forward it will be essential to apply and further develop this knowledge toward an integrated view of IL-10 expression during immune responses in vivo. Only by doing so we will be able to harness the great potential for therapeutic intervention inherent in this fundamental immune-regulatory pathway.

References

1. Agrawal S, Gupta S. TLR1/2, TLR7, and TLR9 signals directly activate human peripheral blood naive and memory B cell subsets to produce cytokines, chemokines, and hematopoietic growth factors. J Clin Immunol. 2011;31(1):89–98.
2. Agrawal S, et al. Cutting edge: different Toll-like receptor agonists instruct dendritic cells to induce distinct Th responses via differential modulation of extracellular signal-regulated kinase-mitogen-activated protein kinase and c-Fos. J Immunol. 2003;171(10):4984–9.
3. Ahyi A-NN, et al. IFN regulatory factor 4 regulates the expression of a subset of Th2 cytokines. J Immunol. 2009;183(3):1598–606.
4. Akira S, Takeda K. Toll-like receptor signalling. Nat Rev Immunol. 2004;4(7):499–511.
5. Aman MJ, et al. Interferon-alpha stimulates production of interleukin-10 in activated CD4+ T cells and monocytes. Blood. 1996;87(11):4731–6.
6. Amsen D, Antov A, Flavell RA. The different faces of Notch in T-helper-cell differentiation. Nat Rev Immunol. 2009;9(2):116–24.
7. Ananieva O, et al. The kinases MSK1 and MSK2 act as negative regulators of Toll-like receptor signaling. Nat Immunol. 2008;9(9):1028–36.
8. Anderson CF, et al. CD4(+)CD25(-)Foxp3(-) Th1 cells are the source of IL-10-mediated immune suppression in chronic cutaneous leishmaniasis. J Exp Med. 2007;204(2):285–97.
9. Anderson CF, et al. IL-27 regulates IL-10 and IL-17 from CD4+ cells in nonhealing Leishmania major infection. J Immunol. 2009;183(7):4619–27.

10. Apetoh L, et al. The aryl hydrocarbon receptor interacts with c-Maf to promote the differentiation of type 1 regulatory T cells induced by IL-27. Nat Immunol. 2010;11(9):854–61.
11. Awasthi A, et al. A dominant function for interleukin 27 in generating interleukin 10-producing anti-inflammatory T cells. Nat Immunol. 2007;8(12):1380–9.
12. Banerjee A, et al. Diverse Toll-like receptors utilize Tpl2 to activate extracellular signal-regulated kinase (ERK) in hemopoietic cells. Proc Natl Acad Sci U S A. 2006;103(9):3274–9.
13. Barthlott T, et al. CD25+ CD4+ T cells compete with naive CD4+ T cells for IL-2 and exploit it for the induction of IL-10 production. Int Immunol. 2005;17(3):279–88.
14. Batten M, et al. Cutting edge: IL-27 is a potent inducer of IL-10 but not FoxP3 in murine T cells. J Immunol. 2008;180(5):2752–6.
15. Bauquet AT, et al. The costimulatory molecule ICOS regulates the expression of c-Maf and IL-21 in the development of follicular T helper cells and TH-17 cells. Nat Immunol. 2009;10(2):167–75.
16. Berg DJ, et al. Interleukin-10 is a central regulator of the response to LPS in murine models of endotoxic shock and the Shwartzman reaction but not endotoxin tolerance. J Clin Invest. 1995;96(5):2339–47.
17. Blokzijl A, Dijke Ten P, Ibáñez CF. Physical and functional interaction between GATA-3 and Smad3 allows TGF-beta regulation of GATA target genes. Curr Biol. 2002;12(1):35–45.
18. Boonstra A, et al. Macrophages and myeloid dendritic cells, but not plasmacytoid dendritic cells, produce IL-10 in response to MyD88- and TRIF-dependent TLR signals, and TLR-independent signals. J Immunol. 2006;177(11):7551–8.
19. Brewer G, et al. Increased interleukin-10 mRNA stability in melanoma cells is associated with decreased levels of A + U-rich element binding factor AUF1. J Interf Cytokine Res. 2003;23(10):553–64.
20. Brightbill HD, et al. A prominent role for Sp1 during lipopolysaccharide-mediated induction of the IL-10 promoter in macrophages. J Immunol. 2000;164(4):1940–51.
21. Brown CY, et al. Differential regulation of the stability of cytokine mRNAs in lipopolysaccharide-activated blood monocytes in response to interleukin-10. J Biol Chem. 1996;271(33):20108–12.
22. Cao S, et al. NF-kappaB1 (p50) homodimers differentially regulate pro- and anti-inflammatory cytokines in macrophages. J Biol Chem. 2006;281(36):26041–50.
23. Cao S, et al. The protooncogene c-Maf is an essential transcription factor for IL-10 gene expression in macrophages. J Immunol. 2005;174(6):3484–92.
24. Carballo E, Lai WS, Blackshear PJ. Evidence that tristetraprolin is a physiological regulator of granulocyte-macrophage colony-stimulating factor messenger RNA deadenylation and stability. Blood. 2000;95(6):1891–9.
25. Carballo E, Lai WS, Blackshear PJ. Feedback inhibition of macrophage tumor necrosis factor-alpha production by tristetraprolin. Science. 1998;281(5379):1001–5.
26. Chakrabarti A, et al. Protein kinase R-dependent regulation of interleukin-10 in response to double-stranded RNA. J Biol Chem. 2008;283(37):25132–9.
27. Chang EY, et al. Cutting edge: involvement of the type I IFN production and signaling pathway in lipopolysaccharide-induced IL-10 production. J Immunol. 2007;178(11):6705–9.
28. Chang H-D, et al. Expression of IL-10 in Th memory lymphocytes is conditional on IL-12 or IL-4, unless the IL-10 gene is imprinted by GATA-3. Eur J Immunol. 2007;37(3):807–17.
29. Chaudhry A, et al. Interleukin-10 signaling in regulatory T cells is required for suppression of Th17 cell-mediated inflammation. Immunity. 2011;34(4):566–78.
30. Chi H, et al. Dynamic regulation of pro- and anti-inflammatory cytokines by MAPK phosphatase 1 (MKP-1) in innate immune responses. Proc Natl Acad Sci U S A. 2006;103(7):2274–9.
31. Ciofani M, et al. A validated regulatory network for Th17 cell specification. Cell. 2012;151(2):289–303.

32. Corre B, et al. Type I interferon potentiates T-cell receptor mediated induction of IL-10-producing CD4+ T cells. Eur J Immunol. 2013;43(10):2730–40.
33. Couper KN, Blount DG, Riley EM. IL-10: the master regulator of immunity to infection. J Immunol. 2008;180(9):5771–7.
34. Creery WD, et al. Differential modulation of B7-1 and B7-2 isoform expression on human monocytes by cytokines which influence the development of T helper cell phenotype. Eur J Immunol. 1996;26(6):1273–7.
35. Cretney, E. et al. The transcription factors Blimp-1 and IRF4 jointly control the differentiation and function of effector regulatory T cells. Nat Immunol. 2011;12(4):304–11.
36. Cretney E, Kallies A, Nutt SL. Differentiation and function of Foxp3(+) effector regulatory T cells. Trends Immunol. 2012;34(2):74–80.
37. D'Addario M, Arora PD, McCulloch CA. Role of p38 in stress activation of Sp1. Gene. 2006;379:51–61.
38. D'Andrea A, et al. Interleukin 10 (IL-10) inhibits human lymphocyte interferon gamma-production by suppressing natural killer cell stimulatory factor/IL-12 synthesis in accessory cells. J Exp Med. 1993;178(3):1041–8.
39. Dardalhon V, et al. IL-4 inhibits TGF-beta-induced Foxp3+ T cells, and together with TGF-beta, generates IL-9+ IL-10+ Foxp3(-) effector T cells. Nat Immunol. 2008;9(12):1347–55.
40. de Waal-Malefyt R, Abrams J, et al. Interleukin 10(IL-10) inhibits cytokine synthesis by human monocytes: an autoregulatory role of IL-10 produced by monocytes. J Exp Med. 1991;174(5):1209–20.
41. de Waal-Malefyt R, Haanen J, et al. Interleukin 10 (IL-10) and viral IL-10 strongly reduce antigen-specific human T cell proliferation by diminishing the antigen-presenting capacity of monocytes via downregulation of class II major histocompatibility complex expression. J Exp Med. 1991;174(4):915–24.
42. Dennehy KM, et al. Reciprocal regulation of IL-23 and IL-12 following co-activation of Dectin-1 and TLR signaling pathways. Eur J Immunol. 2009;39(5):1379–86.
43. Denning TL, et al. Lamina propria macrophages and dendritic cells differentially induce regulatory and interleukin 17-producing T cell responses. Nat Immunol. 2007;8(10):1086–94.
44. Dillon S, et al. A Toll-like receptor 2 ligand stimulates Th2 responses in vivo, via induction of extracellular signal-regulated kinase mitogen-activated protein kinase and c-Fos in dendritic cells. J Immunol. 2004;172(8):4733–43.
45. Dillon S, et al. Yeast zymosan, a stimulus for TLR2 and dectin-1, induces regulatory antigen-presenting cells and immunological tolerance. J Clin Invest. 2006;116(4):916–28.
46. Dong C, Davis R, Flavell R. Map kinases in the immune response. Annu Rev Immunol. 2002;20:55–72.
47. Edwards AD, et al. Microbial recognition via Toll-like receptor-dependent and -independent pathways determines the cytokine response of murine dendritic cell subsets to CD40 triggering. J Immunol. 2002;169(7):3652–60.
48. Eidenschenk C, et al. Role of IL-22 in microbial host defense. Curr Top Microbiol Immunol. 2014;380(Chapter 10):213–36.
49. Elcombe SE, et al. Dectin-1 regulates IL-10 production via a MSK1/2 and CREB dependent pathway and promotes the induction of regulatory macrophage markers. R. Lang, ed. PLoS One. 2013;8(3):e60086.
50. Finbloom DS, Winestock KD. IL-10 induces the tyrosine phosphorylation of tyk2 and Jak1 and the differential assembly of STAT1 alpha and STAT3 complexes in human T cells and monocytes. J Immunol. 1995;155(3):1079–90.
51. Fiorentino DF, et al. IL-10 inhibits cytokine production by activated macrophages. J Immunol. 1991;147(11):3815–22.
52. Fiorentino DF, Bond MW, Mosmann TR. Two types of mouse T helper cell. IV. Th2 clones secrete a factor that inhibits cytokine production by Th1 clones. J Exp Med. 1989;170(6):2081–95.
53. Fitzgerald DC, et al. Suppression of autoimmune inflammation of the central nervous system by interleukin 10 secreted by interleukin 27-stimulated T cells. Nat Immunol. 2007;8(12):1372–9.

54. Foey AD, et al. Regulation of monocyte IL-10 synthesis by endogenous IL-1 and TNF-alpha: role of the p38 and p42/44 mitogen-activated protein kinases. J Immunol. 1998;160(2):920–8.
55. Freitas do Rosário AP, et al. IL-27 promotes IL-10 production by effector Th1 CD4+ T cells: a critical mechanism for protection from severe immunopathology during malaria infection. J Immunol. 2012;188(3):1178–90.
56. Fujita S, et al. Regulatory dendritic cells act as regulators of acute lethal systemic inflammatory response. Blood. 2006;107(9):3656–64.
57. Gaba A, et al. Cutting edge: IL-10-mediated tristetraprolin induction is part of a feedback loop that controls macrophage STAT3 activation and cytokine production. J Immunol. 2012;189(5):2089–93.
58. Gandhi R, et al. Activation of the aryl hydrocarbon receptor induces human type 1 regulatory T cell-like and Foxp3(+) regulatory T cells. Nat Immunol. 2010;11(9):846–53.
59. Geijtenbeek TBH, et al. Mycobacteria target DC-SIGN to suppress dendritic cell function. J Exp Med. 2003;197(1):7–17.
60. Gerosa F, et al. Interleukin-12 primes human CD4 and CD8 T cell clones for high production of both interferon-gamma and interleukin-10. J Exp Med. 1996;183(6):2559–69.
61. Ghoreschi K, et al. Generation of pathogenic T(H) 17 cells in the absence of TGF-β signalling. Nature. 2010;467(7318):967–71.
62. Giordani L, et al. IFN-alpha amplifies human naive B cell TLR-9-mediated activation and Ig production. J Leukoc Biol. 2009;86(2):261–71.
63. Glasmacher E, et al. A genomic regulatory element that directs assembly and function of immune-specific AP-1-IRF complexes. Science. 2012;338(6109):975–80.
64. Glocker E-O, et al. IL-10 and IL-10 receptor defects in humans. Ann N Y Acad Sci. 2011;1246:102–7.
65. Groux H, et al. Interleukin-10 induces a long-term antigen-specific anergic state in human CD4+ T cells. J Exp Med. 1996;184(1):19–29.
66. Guarda G, et al. Type I interferon inhibits interleukin-1 production and inflammasome activation. Immunity. 2011;34(2):213–23.
67. Hadis U, et al. Intestinal tolerance requires gut homing and expansion of FoxP3+ regulatory T cells in the lamina propria. Immunity. 2011;34(2):237–46.
68. Häcker H, et al. Specificity in Toll-like receptor signalling through distinct effector functions of TRAF3 and TRAF6. Nature. 2006;439(7073):204–7.
69. Higgins SC, et al. Toll-like receptor 4-mediated innate IL-10 activates antigen-specific regulatory T cells and confers resistance to Bordetella pertussis by inhibiting inflammatory pathology. J Immunol. 2003;171(6):3119–27.
70. Ho AS, et al. Functional regions of the mouse interleukin-10 receptor cytoplasmic domain. Mol Cell Biol. 1995;15(9):5043–53.
71. Hoshi N, et al. MyD88 signalling in colonic mononuclear phagocytes drives colitis in IL-10-deficient mice. Nat Commun. 2012;3:1120.
72. Hu X, et al. IFN-gamma suppresses IL-10 production and synergizes with TLR2 by regulating GSK3 and CREB/AP-1 proteins. Immunity. 2006;24(5):563–74.
73. Hunter CA, Kastelein R. Interleukin-27: balancing protective and pathological immunity. Immunity. 2012;37(6):960–9.
74. Im S-H, et al. Chromatin-level regulation of the IL10 gene in T cells. J Biol Chem. 2004;279(45):46818–25.
75. Ivanov P, Anderson P. Post-transcriptional regulatory networks in immunity. Immunol Rev. 2013;253(1):253–72.
76. Iwasaki Y, et al. Egr-2 transcription factor is required for Blimp-1 mediated IL-10 production in IL-27 stimulated CD4(+) T cells. Eur J Immunol. 2013;43(4):1063–73.
77. Jankovic D, et al. Conventional T-bet(+)Foxp3(−) Th1 cells are the major source of host-protective regulatory IL-10 during intracellular protozoan infection. J Exp Med. 2007;204(2):273–83.

78. Jarnicki AG, et al. Attenuating regulatory T cell induction by TLR agonists through inhibition of p38 MAPK signaling in dendritic cells enhances their efficacy as vaccine adjuvants and cancer immunotherapeutics. J Immunol. 2008;180(6):3797–806.
79. Jones EA, Flavell RA. Distal enhancer elements transcribe intergenic RNA in the IL-10 family gene cluster. J Immunol. 2005;175(11):7437–46.
80. Kaiser F, et al. TPL-2 negatively regulates interferon-beta production in macrophages and myeloid dendritic cells. J Exp Med. 2009;206(9):1863–71.
81. Kallies A, Hawkins E, Belz G, Metcalf D, Hommel M, Corcoran L, Hodgkin P, Nutt S. Transcriptional repressor Blimp-1 is essential for T cell homeostasis and self-tolerance. Nat Immunol. 2006;7(5):466–74.
82. Kanters E, et al. Inhibition of NF-kappaB activation in macrophages increases atherosclerosis in LDL receptor-deficient mice. J Clin Invest. 2003;112(8):1176–85.
83. Kassner N, et al. Cutting edge: plasmacytoid dendritic cells induce IL-10 production in T cells via the delta-like-4/Notch axis. J Immunol. 2010;184(2):550–4.
84. Kastelein RA, Hunter CA, Cua DJ. Discovery and biology of IL-23 and IL-27: related but functionally distinct regulators of inflammation. Annu Rev Immunol. 2007;25:221–42.
85. Kawai T, Akira S. The role of pattern-recognition receptors in innate immunity: update on Toll-like receptors. Nat Immunol. 2010;11(5):373–84.
86. Kim C, et al. The kinase p38 alpha serves cell type-specific inflammatory functions in skin injury and coordinates pro- and anti-inflammatory gene expression. Nat Immunol. 2008;9(9):1019–27.
87. Kim JI, et al. The transcription factor c-Maf controls the production of interleukin-4 but not other Th2 cytokines. Immunity. 1999;10(6):745–51.
88. Kim JM, et al. Structure of the mouse IL-10 gene and chromosomal localization of the mouse and human genes. J Immunol. 1992;148(11):3618–23.
89. Kim L, et al. p38 MAPK autophosphorylation drives macrophage IL-12 production during intracellular infection. J Immunol. 2005;174(7):4178–84.
90. Kimura A, et al. Aryl hydrocarbon receptor in combination with Stat1 regulates LPS-induced inflammatory responses. J Exp Med. 2009;206(9):2027–35.
91. Kitani A, et al. Transforming growth factor (TGF)-beta1-producing regulatory T cells induce Smad-mediated interleukin 10 secretion that facilitates coordinated immunoregulatory activity and amelioration of TGF-beta1-mediated fibrosis. J Exp Med. 2003;198(8):1179–88.
92. Kotenko SV, et al. Identification and functional characterization of a second chain of the interleukin-10 receptor complex. EMBO J. 1997;16(19):5894–903.
93. Kratochvill F, et al. Tristetraprolin-driven regulatory circuit controls quality and timing of mRNA decay in inflammation. Mol Syst Biol. 2011;7(1):560.
94. Krausgruber T, et al. IRF5 promotes inflammatory macrophage polarization and TH1-TH17 responses. Nat Immunol. 2011;12(3):231–8.
95. Kühn R, et al. Interleukin-10-deficient mice develop chronic enterocolitis. Cell. 1993;75(2):263–74.
96. de La Rosa M, et al. Interleukin-2 is essential for CD4+CD25+ regulatory T cell function. European J Immunol. 2004;34(9):2480–8.
97. Lampropoulou V, et al. TLR-activated B cells suppress T cell-mediated autoimmunity. J Immunol. 2008;180(7):4763–73.
98. Lee C-G, et al. A distal cis-regulatory element, CNS-9, controls NFAT1 and IRF4-mediated IL-10 gene activation in T helper cells. Mol Immunol. 2009;46(4):613–21.
99. Lee C-G, et al. Interaction of Ets-1 with HDAC1 represses IL-10 expression in Th1 cells. J Immunol. 2012;188(5):2244–53.
100. Levings MK, et al. IFN-alpha and IL-10 induce the differentiation of human type 1 T regulatory cells. J Immunol. 2001;166(9):5530–9.
101. Li MO, Flavell RA. Contextual regulation of inflammation: a duet by transforming growth factor-beta and interleukin-10. Immunity. 2008;28(4):468–76.

102. Li P, et al. BATF-JUN is critical for IRF4-mediated transcription in T cells. Nature. 2012;490(7421):543–6.
103. Liu Y-W, et al. Lipopolysaccharide-induced transcriptional activation of interleukin-10 is mediated by MAPK- and NF-kappaB-induced CCAAT/enhancer-binding protein delta in mouse macrophages. Cell Signal. 2006;18(9):1492–500.
104. Lucas M, et al. ERK activation following macrophage FcgammaR ligation leads to chromatin modifications at the IL-10 locus. J Immunol. 2005;175(1):469–77.
105. Ma F, et al. MicroRNA-4661 upregulates IL-10 expression in TLR-triggered macrophages by antagonizing RNA-binding protein tristetraprolin-mediated IL-10 mRNA degradation. J Immunol. 2010;184(11):6053–9.
106. Ma W, et al. The p38 mitogen-activated kinase pathway regulates the human interleukin-10 promoter via the activation of Spl transcription factor in lipopolysaccharide-stimulated human macrophages. J Biol Chem. 2001;276(17):13664–74.
107. Macian F. NFAT proteins: key regulators of T-cell development and function. Nat Rev Immunol. 2005;5(6):472–84.
108. Madan R, et al. Nonredundant roles for B cell-derived IL-10 in immune counter-regulation. J Immunol. 2009;183(4):2312–20.
109. Martin M, et al. Role of the phosphatidylinositol 3 kinase-Akt pathway in the regulation of IL-10 and IL-12 by Porphyromonas gingivalis lipopolysaccharide. J Immunol. 2003;171(2):717–25.
110. Martins G, Calame K. Regulation and functions of Blimp-1 in T and B lymphocytes. How TCRS bind MHCS, peptides, and coreceptors. Annu Rev Immunol. 2008;26:133–169.
111. Martins GA, et al. Transcriptional repressor Blimp-1 regulates T cell homeostasis and function. Nat Immunol. 2006;7(5):457–65.
112. Mayer-Barber KD, et al. Innate and adaptive interferons suppress IL-1α and IL-1β production by distinct pulmonary myeloid subsets during Mycobacterium tuberculosis infection. Immunity. 2011;35(6):1023–34.
113. Maynard C, Weaver C. Diversity in the contribution of interleukin-10 to T-cell-mediated immune regulation. Immunol Rev. 2008;226:219–33.
114. Maynard CL, et al. Regulatory T cells expressing interleukin 10 develop from Foxp3+ and Foxp3- precursor cells in the absence of interleukin 10. Nat Immunol. 2007;8(9):931–41.
115. McGeachy MJ, et al. TGF-beta and IL-6 drive the production of IL-17 and IL-10 by T cells and restrain T(H)-17 cell-mediated pathology. Nat Immunol. 2007;8(12):1390–7.
116. McNab FW, et al. TPL-2-ERK1/2 signaling promotes host resistance against intracellular bacterial infection by negative regulation of type I IFN production. J Immunol. 2013;191(4):1732–43.
117. Meraz MA, et al. Targeted disruption of the Statl gene in mice reveals unexpected physiologic specificity in the JAK-STAT signaling pathway. Cell. 1996;84(3):431–42.
118. Meyaard L, et al. IL-12-induced IL-10 production by human T cells as a negative feedback for IL-12-induced immune responses. J Immunol. 1996;156(8):2776–82.
119. Moore K et al. Interleukin-10 and the interleukin-10 receptor. Annu Rev Immunol. 2001;19:683–765.
120. Moore KW, et al. Homology of cytokine synthesis inhibitory factor (IL-10) to the Epstein-Barr virus gene BCRFI. Science. 1990;248(4960):1230–4.
121. Motomura Y, et al. The transcription factor E4BP4 regulates the production of IL-10 and IL-13 in CD4+ T cells. Nat Immunol. 2011;12(5):450–9.
122. Murai M, et al. Interleukin 10 acts on regulatory T cells to maintain expression of the transcription factor Foxp3 and suppressive function in mice with colitis. Nat Immunol. 2009;10(11):1178–84.
123. Murphy TL, Tussiwand R, Murphy KM. Specificity through cooperation: BATF-IRF interactions control immune-regulatory networks. Nat Rev Immunol. 2013;13(7):499–509.
124. Murugaiyan G, et al. IL-27 is a key regulator of IL-10 and IL-17 production by human CD4+ T cells. J Immunol. 2009;183(4):2435–43.

125. Nandan D, et al. Myeloid cell IL-10 production in response to leishmania involves inactivation of glycogen synthase kinase-3β downstream of phosphatidylinositol-3 kinase. J Immunol. 2012;188(1):367–78.
126. Neininger A, et al. MK2 targets AU-rich elements and regulates biosynthesis of tumor necrosis factor and interleukin-6 independently at different post-transcriptional levels. J Biol Chem. 2002;277(5):3065–8.
127. Netea MG, et al. Toll-like receptor 2 suppresses immunity against Candida albicans through induction of IL-10 and regulatory T cells. J Immunol. 2004;172(6):3712–8.
128. Neumann C, et al. Role of Blimp-1 in programing Th effector cells into IL-10 producers. J Exp Med. 2014;204(2):285.
129. O'Garra A, Vieira P. T(H)1 cells control themselves by producing interleukin-10. Nat Rev Immunol. 2007;7(6):425–8.
130. Ohtani M, et al. Mammalian target of rapamycin and glycogen synthase kinase 3 differentially regulate lipopolysaccharide-induced interleukin-12 production in dendritic cells. Blood. 2008;112(3):635–43.
131. Ouyang W, et al. Regulation and functions of the IL-10 family of cytokines in inflammation and disease. Annu Rev Immunol. 2011;29(1):71–109.
132. Pattison MJ, MacKenzie KF, Arthur JSC. Inhibition of JAKs in macrophages increases lipopolysaccharide-induced cytokine production by blocking IL-10-mediated feedback. J Immunol. 2012;189(6):2784–92.
133. Platzer C, et al. Cyclic adenosine monophosphate-responsive elements are involved in the transcriptional activation of the human IL-10 gene in monocytic cells. Eur J Immunol. 1999;29(10):3098–104.
134. Pot C, et al. Cutting edge: IL-27 induces the transcription factor c-Maf, cytokine IL-21, and the costimulatory receptor ICOS that coordinately act together to promote differentiation of IL-10-producing Tr1 cells. J Immunol. 2009;183(2):797–801.
135. Pot C, et al. Induction of regulatory Tr1 cells and inhibition of T(H)17 cells by IL-27. Semin Immunol. 2011;23(6):438–45.
136. Powell MJ, et al. Posttranscriptional regulation of IL-10 gene expression through sequences in the 3′-untranslated region. J Immunol. 2000;165(1):292–6.
137. Qin Z, et al. Pivotal advance: Kaposi's sarcoma-associated herpesvirus (KSHV)-encoded microRNA specifically induce IL-6 and IL-10 secretion by macrophages and monocytes. J Leukoc Biol. 2010;87(1):25–34.
138. Quinn SR, O'Neill LA. The role of microRNAs in the control and mechanism of action of IL-10. Curr Top Microbiol Immunol. 2014;380(Chapter 7):145–55.
139. Radtke F, Macdonald HR, Tacchini-Cottier F. Regulation of innate and adaptive immunity by Notch. Nat Rev Immunol. 2013;13(6):427–37.
140. Rakoff-Nahoum S, Hao L, Medzhitov R. Role of toll-like receptors in spontaneous commensal-dependent colitis. Immunity. 2006;25(2):319–29.
141. Risco A, et al. p38γ and p38δ kinases regulate the Toll-like receptor 4 (TLR4)-induced cytokine production by controlling ERK1/2 protein kinase pathway activation. Proc Natl Acad Sci U S A. 2012;109(28):11200–5.
142. Roers A, et al. T cell-specific inactivation of the interleukin 10 gene in mice results in enhanced T cell responses but normal innate responses to lipopolysaccharide or skin irritation. J Exp Med. 2004;200(10):1289–97.
143. Rogers NC, et al. Syk-dependent cytokine induction by Dectin-1 reveals a novel pattern recognition pathway for C type lectins. Immunity. 2005;22(4):507–17.
144. Roncarolo MG, et al. Interleukin-10-secreting type 1 regulatory T cells in rodents and humans. Immunol Rev. 2006;212:28–50.
145. Rubtsov YP, et al. Regulatory T cell-derived interleukin-10 limits inflammation at environmental interfaces. Immunity. 2008;28(4):546–58.
146. Rutz S, Ouyang W. Regulation of interleukin-10 and interleukin-22 expression in T helper cells. Curr Opin Immunol. 2011;23(5):605–12.

147. Rutz S, et al. Notch regulates IL-10 production by T helper 1 cells. Proc Natl Acad Sci U S A. 2008;105(9):3497–502.
148. Rutz S, et al. Transcription factor c-Maf mediates the TGF-β-dependent suppression of IL-22 production in T(H)17 cells. Nat Immunol. 2011;12(12):1238–45.
149. Rutz S, Wang X, Ouyang W. The IL-20 subfamily of cytokines – from host defence to tissue homeostasis. Nat Rev Immunol. 2014;14(12):783–95.
150. Salehi S, et al. B lymphocyte-induced maturation protein-1 contributes to intestinal mucosa homeostasis by limiting the number of IL-17-producing CD4+ T cells. J Immunol. 2012;189(12):5682–93.
151. Salojin KV, et al. Essential role of MAPK phosphatase-1 in the negative control of innate immune responses. J Immunol. 2006;176(3):1899–907.
152. Sandler H, Stoecklin G. Control of mRNA decay by phosphorylation of tristetraprolin. Biochem Soc Trans. 2008;36(Pt 3):491–6.
153. Saraiva M, O'Garra A. The regulation of IL-10 production by immune cells. Nat Rev Immunol. 2010;10(3):170–81.
154. Saraiva M, et al. Identification of a macrophage-specific chromatin signature in the IL-10 locus. J Immunol. 2005;175(2):1041–6.
155. Saraiva M, et al. Interleukin-10 production by Th1 cells requires interleukin-12-induced STAT4 transcription factor and ERK MAP kinase activation by high antigen dose. Immunity. 2009;31(2):209–19.
156. Sayi A, et al. TLR-2-activated B cells suppress Helicobacter-induced preneoplastic gastric immunopathology by inducing T regulatory-1 cells. J Immunol. 2011;186(2):878–90.
157. Schaljo B, et al. Tristetraprolin is required for full anti-inflammatory response of murine macrophages to IL-10. J Immunol. 2009;183(2):1197–206.
158. Schuetze N, et al. IL-12 family members: differential kinetics of their TLR4-mediated induction by Salmonella enteritidis and the impact of IL-10 in bone marrow-derived macrophages. Int Immunol. 2005;17(5):649–59.
159. Schulz EG, et al. Sequential polarization and imprinting of type 1 T helper lymphocytes by interferon-gamma and interleukin-12. Immunity. 2009;30(5):673–83.
160. Sellon RK, et al. Resident enteric bacteria are necessary for development of spontaneous colitis and immune system activation in interleukin-10-deficient mice. Infect Immun. 1998;66(11):5224–31.
161. Sharma A, et al. Posttranscriptional regulation of interleukin-10 expression by hsa-miR-106a. Proc Natl Acad Sci U S A. 2009;106(14):5761–6.
162. Shaw MH, et al. Tyk2 negatively regulates adaptive Th1 immunity by mediating IL-10 signaling and promoting IFN-gamma-dependent IL-10 reactivation. J Immunol. 2006;176(12):7263–71.
163. Sheedy FJ, et al. Negative regulation of TLR4 via targeting of the proinflammatory tumor suppressor PDCD4 by the microRNA miR-21. Nat Immunol. 2010;11(2):141–7.
164. Shoemaker J, Saraiva M, O'Garra A. GATA-3 directly remodels the IL-10 locus independently of IL-4 in CD4+ T cells. J Immunol. 2006;176(6):3470–9.
165. Shouval DS, et al. Interleukin-10 receptor signaling in innate immune cells regulates mucosal immune tolerance and anti-inflammatory macrophage function. Immunity. 2014;40(5):706–19.
166. Siewe L, et al. Interleukin-10 derived from macrophages and/or neutrophils regulates the inflammatory response to LPS but not the response to CpG DNA. Eur J Immunol. 2006;36(12):3248–55.
167. Skok J, Poudrier J, Gray D. Dendritic cell-derived IL-12 promotes B cell induction of Th2 differentiation: a feedback regulation of Th1 development. J Immunol. 1999;163(8):4284–91.
168. Slack EC, et al. Syk-dependent ERK activation regulates IL-2 and IL-10 production by DC stimulated with zymosan. Eur J Immunol. 2007;37(6):1600–12.

169. Spolski R, et al. IL-21 mediates suppressive effects via its induction of IL-10. J Immunol. 2009;182(5):2859–67.
170. Staples KJ, et al. IL-10 induces IL-10 in primary human monocyte-derived macrophages via the transcription factor Stat3. J Immunol. 2007;178(8):4779–85.
171. Stoecklin G, et al. Genome-wide analysis identifies interleukin-10 mRNA as target of tristetraprolin. J Biol Chem. 2008;283(17):11689–99.
172. Stumhofer JS, et al. Interleukin 27 negatively regulates the development of interleukin 17-producing T helper cells during chronic inflammation of the central nervous system. Nat Immunol. 2006;7(9):937–45.
173. Stumhofer JS, et al. Interleukins 27 and 6 induce STAT3-mediated T cell production of interleukin 10. Nat Immunol. 2007;8(12):1363–71.
174. Sun J, et al. CD4+ T cell help and innate-derived IL-27 induce Blimp-1-dependent IL-10 production by antiviral CTLs. Nat Immunol. 2011;12(4):327–34.
175. Takeda K, et al. Enhanced Th1 activity and development of chronic enterocolitis in mice devoid of Stat3 in macrophages and neutrophils. Immunity. 1999;10(1):39–49.
176. Tan JC, et al. Characterization of interleukin-10 receptors on human and mouse cells. J Biol Chem. 1993;268(28):21053–9.
177. Tan NY, Khachigian LM. Sp1 phosphorylation and its regulation of gene transcription. Mol Cell Biol. 2009;29(10):2483–8.
178. Tone M, et al. IL-10 gene expression is controlled by the transcription factors Sp1 and Sp3. J Immunol. 2000;165(1):286–91.
179. Trandem K, et al. Highly activated cytotoxic CD8 T cells express protective IL-10 at the peak of coronavirus-induced encephalitis. J Immunol. 2011;186(6):3642–52.
180. Trinchieri G. Interleukin-10 production by effector T cells: Th1 cells show self control. J Exp Med. 2007;204(2):239–43.
181. Tudor C, et al. The p38 MAPK pathway inhibits tristetraprolin-directed decay of interleukin-10 and pro-inflammatory mediator mRNAs in murine macrophages. FEBS Lett. 2009;583(12):1933–8.
182. Tussiwand R, et al. Compensatory dendritic cell development mediated by BATF-IRF interactions. Nature. 2012;490(7421):502–7.
183. Uhlig HH, et al. Characterization of Foxp3+CD4+CD25+ and IL-10-secreting CD4+CD25+ T cells during cure of colitis. J Immunol. 2006;177(9):5852–60.
184. Unterberger C, et al. Role of STAT3 in glucocorticoid-induced expression of the human IL-10 gene. Mol Immunol. 2008;45(11):3230–7.
185. Vahedi G, et al. STATs shape the active enhancer landscape of T cell populations. Cell. 2012;151(5):981–93.
186. Vasanthakumar A, Kallies A. IL-27 paves different roads to Tr1. Eur J Immunol. 2013;43(4):882–5.
187. Veldhoen M, et al. Transforming growth factor-beta "reprograms" the differentiation of T helper 2 cells and promotes an interleukin 9-producing subset. Nat Immunol. 2008;9(12):1341–6.
188. Villagra A, et al. The histone deacetylase HDAC11 regulates the expression of interleukin 10 and immune tolerance. Nat Immunol. 2009;10(1):92–100.
189. Villarino A, et al. The IL-27R (WSX-1) is required to suppress T cell hyperactivity during infection. Immunity. 2003;19(5):645–55.
190. Voll RE, et al. Immunosuppressive effects of apoptotic cells. Nature. 1997;390(6658):350–1.
191. Wang Z-Y, et al. Regulation of IL-10 gene expression in Th2 cells by Jun proteins. J Immunol. 2005;174(4):2098–105.
192. Wei L, et al. Discrete roles of STAT4 and STAT6 transcription factors in tuning epigenetic modifications and transcription during T helper cell differentiation. Immunity. 2010;32(6):840–51.

193. Weichhart T, et al. The TSC-mTOR signaling pathway regulates the innate inflammatory response. Immunity. 2008;29(4):565–77.
194. Willems F, et al. Interleukin-10 inhibits B7 and intercellular adhesion molecule-1 expression on human monocytes. Eur J Immunol. 1994;24(4):1007–9.
195. Xu J, et al. c-Maf regulates IL-10 expression during Th17 polarization. J Immunol. 2009;182(10):6226–36.
196. Yi A-K, et al. Role of mitogen-activated protein kinases in CpG DNA-mediated IL-10 and IL-12 production: central role of extracellular signal-regulated kinase in the negative feedback loop of the CpG DNA-mediated Th1 response. J Immunol. 2002;168(9):4711–20.
197. Yu H, et al. MKP-1 regulates cytokine mRNA stability through selectively modulation subcellular translocation of AUF1. Cytokine. 2011;56(2):245–55.
198. Zhang X, et al. Type I interferons protect neonates from acute inflammation through interleukin 10-producing B cells. J Exp Med. 2007;204(5):1107–18.
199. Zhang X, Edwards J, Mosser D. Dynamic and transient remodeling of the macrophage IL-10 promoter during transcription. J Immunol. 2006;177(2):1282–8.
200. Zhu J et al. Conditional deletion of Gata3 shows its essential function in T(H)1-T(H)2 responses. Nat Immunol. 2004;5(11):1157–65.
201. Zhu J, Yamane H, Paul WE. Differentiation of effector CD4 T cell populations (*). Annu Rev Immunol. 2010;28:445–89.
202. Ziegler-Heitbrock L, et al. IFN-alpha induces the human IL-10 gene by recruiting both IFN regulatory factor 1 and Stat3. J Immunol. 2003;171(1):285–90.
203. Zigmond E, et al. Macrophage-restricted interleukin-10 receptor deficiency, but not IL-10 deficiency, causes severe spontaneous colitis. Immunity. 2014;40(5):720–33.

Chapter 6
Regulation of Interleukin-12 Production in Antigen-Presenting Cells

Hua Zheng, Yi Ban, Fang Wei, and Xiaojing Ma

Abstract Interleukin-12 is a heterodimeric cytokine produced primarily by pathogen-activated antigen-presenting cells, particularly macrophages and dendritic cells, during encountering with intracellular microbes. IL-12 plays a key role in the activation of natural killer cells and CD4$^+$ T helper cells in both innate and adaptive immune responses against infectious agents and immunosurveillance against endogenous malignancies. However, the potency of IL-12 makes it a target for stringent regulation. Indeed, the temporal, spatial, and quantitative expression of IL-12 during an immune response in a microenvironment contributes critically to the determination of the type, extent, and ultimate resolution of the reaction. Breaching of the delicate control and balance involving IL-12 frequently leads to autoimmune inflammatory disorders and pathogenesis. Thus, a better understanding of the regulatory mechanisms in the production and control of this cytokine is both scientifically significant and clinically beneficial. Here we provide an update on the research that has been conducted on this subject particularly in the last 10 years since the publication of a major thesis of this nature.

The authors declare no conflict of interest.

H. Zheng • F. Wei
State Key Laboratory of Microbial Metabolism, Sheng Yushou Center of Cell Biology and Immunology, School of Life Science and Biotechnology, Shanghai Jiaotong University, 800 Dongchuan Road, Shanghai 200240, China

Y. Ban
Department of Microbiology and Immunology, Weill Cornell Medical College, 1300 York Avenue, New York, NY 10065, USA

X. Ma (✉)
State Key Laboratory of Microbial Metabolism, Sheng Yushou Center of Cell Biology and Immunology, School of Life Science and Biotechnology, Shanghai Jiaotong University, 800 Dongchuan Road, Shanghai 200240, China

Department of Microbiology and Immunology, Weill Cornell Medical College, 1300 York Avenue, New York, NY 10065, USA
e-mail: xim2002@sjtu.edu.cn; xim2002@med.cornell.edu

X. Ma (ed.), *Regulation of Cytokine Gene Expression in Immunity and Diseases*, Advances in Experimental Medicine and Biology 941,
DOI 10.1007/978-94-024-0921-5_6

Keywords IL-12 • Antigen-presenting cells • Macrophage • Dendritic cell • Interferon-γ • Transcription

6.1 IL-12 Composition, Producer Cells, Receptors, and Target Cells

Interleukin-12 (IL-12) was discovered originally as a natural killer cell stimulatory factor (NKSF) [51] and cytotoxic lymphocyte maturation factor (CLMF) [92], respectively, by two different groups. IL-12, composed of p35 and p40 chains, is the first member of a family of heterodimeric cytokines identified [51], which also includes IL-23 consisted of p19 and p40 [72], IL-27 consisted of p28 and EBI3 [73], and IL-35 composed of p35 and EBI3 [15]. The main physiological producers of IL-12 are phagocytes (monocytes/macrophages and neutrophils) and dendritic cells (DCs) in response to pathogens (bacteria, fungi, intracellular parasites and viruses) through Toll-like receptors (TLRs) [97].

Other "non-typical" cell types than professional APCs have been reported to produce low levels of biologically active IL-12 such as TLR1/2 ligand-stimulated mouse liver endothelial cells [57]. It was shown that stimulation of primary human donor islets and murine beta cell lines with a proinflammatory cytokine cocktail (TNF-α, IL-1β, and IFN-γ) induced islet dysfunction and potently increased the expression and production of IL-12 and IL-12 receptor. Further, IL-12 induced upregulation of IFN-γ and impaired glucose-stimulated insulin secretion [94]. These data identify beta cells as a novel and local source of IL-12 and suggest a direct role of IL-12 in mediating beta cell pathology.

IL-12 is composed of p35 and p40 chains, and it principally activates NK cells and induces the differentiation of naïve $CD4^+$ T lymphocytes to become IFN-γ-producing Th1 effectors in cell-mediated immune responses to intracellular pathogens[2]. IL-12 signals through the IL-12 receptor (IL-12R) comprised of the IL-12Rβ1 and IL-12Rβ2 subunits that are expressed on T cells, NK cells, and DCs [21, 34]. IL-12 stimulates nonreceptor JAK2 and TYK2 activities, leading to phosphorylation of signal transduction and transcription factors (STATs), in particular, STAT4 homodimers [97, 106]. It is widely believed that the majority of IL-12's immunological activities are mediated through IFN-γ produced by activated NK and Th1 cells that have been exposed to APC-derived IL-12. However, significant levels of IFN-γ-independent activities of IL-12 have been demonstrated in many infectious disease and cancer models [2, 37, 78, 84, 95, 105, 112].

Production of IL-12 in vitro results in the secretion of a 5- to 500-fold excess of p40 relative to IL-12 [19, 76]. There is evidence that the p40 subunit plays a critical role in enhancing the stability, intracellular trafficking, and export of the p35 subunit. This posttranslational regulation mediated by the p40 subunit is conserved in mammals [44]. In addition to form IL-12 and IL-23, the p40 monomer has been recently suggested to function as an adaptor that is able to generate multiple de novo composites in combination with other locally available polypeptide partners after secretion, although the function of these novel forms of p40-containing heterodimers remains obscure [1].

6.2 IL-12's Immunological Activities

The physiologically important target cells of IL-12 are hematopoietic progenitors for which IL-12 induces increased proliferation and colony formation in synergy with other colony-stimulating factors; NK cells, NKT cells, and T cells, for which IL-12 induces proliferation, enhancement of cytotoxicity and of the expression of cytotoxic mediators, and the production of cytokines, particularly IFN-γ, as well as favoring differentiation to cells that produce type-1 cytokines (Th1, T_C1, and NK1 cells); and B cells, for which IL-12, directly or through the effects of type-1 cytokines such as IFN- γ, enhances the activation and production of Th1-associated classes of immunoglobulin (e.g., IgG2a in the mouse).

The importance of IL-12 in host defense is underscored by observations that IL-12-deficient mice are susceptible to infection with regular doses of intracellular pathogens, including *Leishmania major* [62], *Leishmania donovani* [81], *Toxoplasma gondii* [24], *Trypanosoma cruzi* [67], *Cryptococcus neoformans* [47], *mycobacteria* [16, 102], or *Listeria monocytogenes* [12], displaying impaired effector functions of NK, $CD4^+$, and $CD8^+$ T cells. In humans, recessive mutations in the IFN-γ receptor ligand-binding chain, IFN-γ receptor signaling chain, IL-12p40 subunit, and IL-12 receptor β1 chain genes have been identified in a number of patients with disseminated mycobacterial infection [4]. Although genetically distinct, these conditions are immunologically related and highlight the essential role of IFN-γ-mediated immunity in the control of mycobacteria. More frequently in human populations, however, is the opposite problem: having overly active IL-12-producing and/or signaling activities, which contributes to the occurrence and/or pathogenesis of inflammatory autoimmune disorders such as psoriasis, arthritis, Crohn's disease, multiple sclerosis, graft vs host disease, atopic dermatitis, hidradenitis suppurativa, primary biliary cirrhosis, sarcoidosis, systemic lupus erythematosus, etc. [96]. Thus, understanding how IL-12 production is regulated in various physiological and pathological conditions is vitally important for human health.

IL-12's immunological activities are complicated by the existence of IL-12p40 homodimer, IL-12p80, which acts as an IL-12 antagonist by binding to the IL-12R but which does not mediate a biological response [27, 54]. Secretion of IL-12 is associated with excess production of IL-12p80 [19]. For example, in contrast to the dogma about the restrictive nature of IL-12-producing cell types, meaningful amounts of IL-12p40 monomer and IL-12p80 have been observed in human breast cancer cells [40], which could potentially thwart the IL-12-induced antitumor responses in vivo. Approximately 20–40 % of the p40 in the serum of normal and endotoxin-treated mice is in the form of IL-12p80 [41]. In IL-12-dependent shock models, exogenous IL-12p80 inhibits IL-12-induced cell-mediated immune response and protects mice from sepsis-associated death [63]. However, IL-12p80 has also been reported to stimulate, rather than inhibit, the differentiation of $CD8^+$ Tc1 (type I cytotoxic T) cells in vitro contrary to its suppressive activity on Th1 function [74]. The divergent functions of the various forms of p40 highlight our lack of a complete understanding of its true range of biological activities.

Furthermore, several studies have found evidence of immunological activities of the p40 monomer [45]. A mathematical model for IL-12 bioactivity was created by incorporating the production of IL-12p70, IL-12p40 (monomer), and IL-12$(p40)^2$ (homodimer) by mature human DCs and the interaction of these species with the IL-12 receptor. Using this model, Klinke simulated the effects of IFN-γ, IL-4, and PGE2 concentrations on the bioactivity of IL-12. The simulations suggest that the concentration of IL-12p70 alone is not indicative of IL-12 bioactivity; rather, the bioactivity of IL-12 produced by mature DCs depends on IL-12p70, IL-12p40, and IL-12(p40)2 production and their competitive interaction with the IL-12 receptor [50]. Thus the ratio of IL-12p40 to IL-12(p40)2 is an important yet underestimated determinant of IL-12 bioactivity.

6.3 IL-12-Encoding Genes

The two IL-12-encoding genes *Il12a* (for p35) and *Il12b* (for p40) are located on separate chromosomes (3p12–q13.2 and 5q31–33, respectively, in humans; chromosomes 6 and 11, respectively, in the mouse) [70, 83, 85], and they do not share sequence homology. The *Il12a* cDNA encodes a 209-amino-acid polypeptide corresponding to a mature protein of 27.5 kDa. It contains seven cysteine residues and three potential N-glycosylation sites. The *Il12b* cDNA sequence encodes a 328-amino-acid polypeptide with a 22-amino-acid signal peptide generating a mature protein of 34.7 kDa. Heterodimers of p40 and p35 are formed via disulfide bonds and secreted, usually upon stimulation of producer cells.

Both *Il12a* and *Il12b* genes must be expressed coordinately in the same cells to produce biologically activate IL-12 [108]. Paradoxically, the mRNA of *Il12a* is widely expressed in many cell types, albeit at low levels in some cells, most of which do not even produce IL-12. The *Il12b* mRNA is restricted to cells that can produce biologically active heterodimer [19]. Synthesis of the p35 chain was proposed to be a rate-limiting step for IL-12 production for its low abundance transcripts in cells under steady-state conditions [88]. The regulation of *Il12a* and *Il12b* genes is primarily exerted at the transcriptional level, although post-transcriptional mechanisms play minor roles.

6.4 Transcriptional Regulation of *Il12b*

Unlike for IL-12p35, over the past two decades, a large number of molecular studies have explored numerous mechanistic aspects that control the production of IL-12p40 encoded by *Il12b*. The *Il12b* promoter has been extensively studied, and numerous transcriptional factors have been identified as direct regulators for its transcription. Murphy et al. first demonstrated binding of a proximal promoter sequence of the murine *Il12b* gene to NF-κB (p50/p65 and p50/c-Rel) complexes in a functional manner in macrophages activated by several pathogens. Further, IFN-γ treatment of cells enhanced this binding interaction, providing a mechanism for IFN-γ augmentation of IL-12 production by macrophages [68]. The activating role of NF-κB was also confirmed with the human *IL12B* promoter with an additional finding for a molecular role for Ets2 [33].

When murine macrophages are stimulated with LPS, nucleosome 1 is selectively remodeled so that the transcription factor CCAAT enhancer-binding protein β (C/EBPβ)/LAP could gain access to this region [75]. However, remodeling of nucleosome 1 alone is not sufficient for *Il12b* transcription and more factors are required for its induced expression. These factors include NF-κB [68, 80], PU.1 [8], interferon regulatory factor 1 (IRF-1) [61], nuclear factor in activated T cells (NFAT) [111], and IFN consensus sequence binding protein (ICSBP), also called IRF-8 [103] in human and/or murine macrophages. Activation protein 1 (AP-1) has been reported to be an activator of *Il12b* transcription in LPS-stimulated macrophages [110]. However, this finding is in contradiction with an earlier study showing that LPS-stimulated macrophages from mice deficient in c-fos, a component of AP-1 with c-Jun, *Il12a/b* mRNA expression, and IL-12 production were greatly enhanced, suggesting that c-fos is a negative regulator of IL-12 expression [79]. This notion is supported by the observation that overexpression of c-fos in macrophages completed blocked *Il12b* transcription [65].

An important pathway in IL-12 induction is the requirement for "priming" of LPS-activated macrophages and DCs by IFN-γ for the expression of maximal amounts of *Il12a* and *Il12b* mRNAs and for IL-12 production [38, 56, 59]. The IFN-γ priming is a positive feedback mechanism for more robust IL-12 production in certain immune responses as the priming agent IFN-γ is principally derived from NK cells and activated Th1 lymphocytes, cells that are initially activated by APC-derived IL-12 upon pathogen infection. The priming effects by and large involve IRFs and STATs induced by IFN-γ. The role of interferons in the regulation of IL-12 production through activation of IRFs has also been demonstrated in myeloid DCs infected with *Leishmania* species. Addition of a neutralizing type I IFN decoy receptor blocked the expression of IRF-7 and IL-12p40 during infection with *L. major* but not *L. donovani*, indicating the parasite-induced expression of IL-12p40 is dependent upon the type I IFN signaling pathway [25].

Not all IRFs are activators of IL-12 production. IRF-4 plays pivotal roles in the differentiation and function of T and B lymphocytes. Mice with IRF4-deficient DCs showed reduced parasite burden upon *Leishmania major* infection, and their CD4$^+$

T cells produced higher levels of IFN-γ in response to *L. major* Ag. In the draining lymph nodes, the proportion of IFN-γ-producing cells was increased, suggesting a Th1 bias in the immune response. Moreover, the numbers of migrating Langerhans cells and other migratory DCs in the draining lymph nodes produced higher levels of IL-12 in IRF4-deficient DCs [3]. These results suggest that IRF-4 expression in DCs inhibits their ability to produce IL-12, thereby regulating CD4⁺ T cell responses against local infection with *L. major*.

The host immune system responds to CpG motifs in bacterial DNA by rapidly producing proinflammatory cytokines. The response to CpG DNA via TLR9 resembles, in many ways, the response to bacterial LPS via TLR4. While both agents can induce a powerful inflammatory response, CpG DNA promotes Th1 and suppresses Th2 immunity. Interestingly, it was observed that CpG DNA induced IL-12p40 secretion by macrophages from mice with either disrupted STAT1 or IRF-8 genes, unlike the LPS-induced response. Thus, CpG DNA-induced IL-12p40 production is mediated by one or more signaling elements distinct from those induced by either LPS or IFN-γ [10].

While TLR4 and TLR9 are primarily involved in pathogen-induced IL-12 production, TLR2, in contrast, plays principally a negative role. Gravina et al. showed that when TLR2-deficient mice were infected with the parasite *Trypanosome cruzi*, an increased production of IL-12p40 and IFN-γ and enhanced Th1 response were noted. The differential use of TLR2 and TLR9 by the immune cells during the acute phase of the infection explains why TLR9- but not TLR2-deficient mice are susceptible to *T. cruzi* infection [31].

Figure 6.1 summarizes our current understanding of the transcriptional mechanisms regulating the human IL-12p40 promoter.

6.5 Transcriptional Regulation of *Il12a*

Il12a and *Il12b* gene expression can be coordinately as well as differentially regulated. Grumont et al. first demonstrated that in contrast to macrophages which require c-Rel for *Il12b* but not *Il12a* transcription, in CD8⁺ DCs, the induced expression of *Il12a* rather than *Il12b* by inactivated *Staphylococcus aureus* (SAC), bacterial DNA, or LPS is c-Rel dependent and regulated directly by c-Rel complexes binding to the *Il12a* promoter [35].

Subsequently, Liu et al. provided evidence that IFN regulatory factor (IRF)-1-deficient macrophages have a selective impairment in mRNA synthesis of *Il12a* but not *Il12b* and a strong deficiency in the production of IL-12 p70 but not p40 [56]. IRF-1 physically interacts with an inverted IRF element within the *Il12a* promoter upon IFN-γ stimulation. Moreover, IRF-1-mediated transcriptional activation of the *Il12a* promoter requires the cooperation of two adjacent Sp1 elements. Thus, IRF-1 acts as a critical component of IFN-γ signaling in the selective activation of *Il12a* transcription in synergy with LPS-mediated events. The study was borne out by Nigishi et al., showing that MyD88-associated IRF-1 migrates into the nucleus

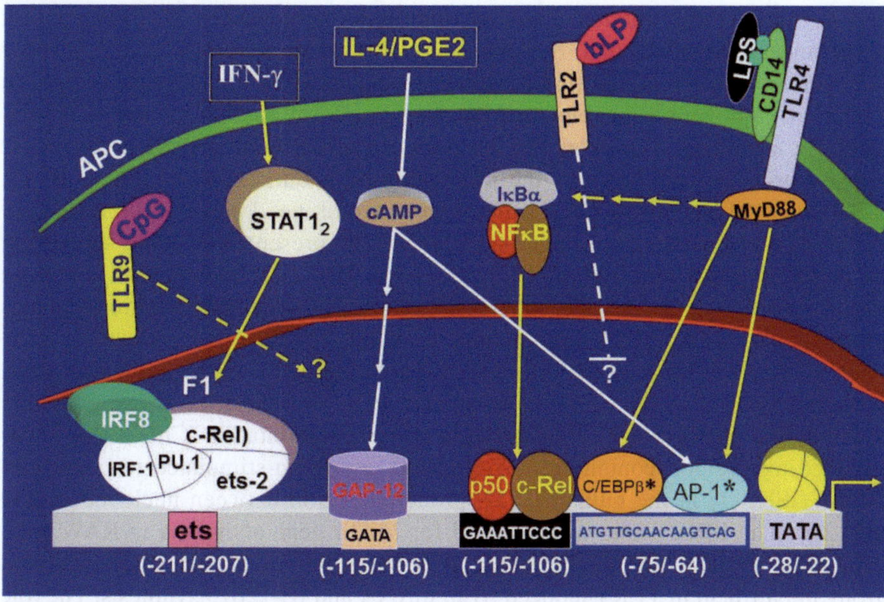

Fig. 6.1 Transcriptional regulation of IL-12p40 (IL12B) in antigen-presenting cells (*APCs*). The data draws primarily from macrophage studies. In DCs, c-Rel is not required for IL12B transcription. *Broken arrows* indicate unconfirmed aspects. *Continuous short arrows* denote multiple steps involved that are not specified in details. *Dashed lines* indicate undetermined signaling pathway. *bLP* bacterial lipoprotein. *IRF-1* interferon regulatory factor 1. The promoter coordinates are with respect to the transcription start site, which is designated +1. GAP-12 is a putative transcriptional repressor of unidentified nature that is induced by IL-4 or prostaglandin E2 (PGE2) treatment of human monocytes. * denotes controversial transcriptional factors that are defined as repressors by mouse knockout studies, but as activators in some in vitro studies. *Arrows* denote activating signals while horizontal lines indicate inhibitory signals

more efficiently than non-MyD88-associated IRF-1. The critical role of MyD88-dependent "IRF-1 licensing" is underscored by the observation that the induction of a specific gene subset downstream of the TLR-MyD88 pathway, such as IFN-β, inducible NO synthase, and IL-12p35, is impaired in IRF1-deficient cells [69]. These studies provide mechanistic insights into the enhancement of the TLR-dependent *Il12a* induction program by IFN-γ. In fact, the phenomenon of IFN-γ-augmented IL-12 production was first observed in 1996 [59]. It was thought to be a mechanism for a positive feedback that amplifies the production of IL-12 for robust immune responses against intracellular pathogens. The observation was extended more broadly to a greater number of genes in human monocytes and macrophages exposed to IFN-γ priming followed by LPS activation through TLR4 [9]. The study found that IFN-γ enhanced mRNA and surface or intracellular expression of TLR4, the accessory component MD-2 and the adapter protein MyD88, respectively. IFN-γ counteracted the LPS-induced downregulation of TLR4. IFN-γ-primed monocytes showed increased responsiveness to LPS with respect to phosphorylation of the interleukin-1 receptor-associated kinase (IRAK), NF-kB DNA-binding activity,

TNF-α, and IL-12 production. The study highlights the enhanced TLR4 expression as the underlying effect of IFN-γ priming effect on mononuclear phagocytes for pathogen recognition and killing as well as its synergism with LPS in macrophage activation. Pathogens can explore the IRF/IL-12 pathway for their own invasive and surviving advantage. For example, the presence of helminth infection coincident with malaria profoundly alters the production of malaria-specific IFN-γ, IL-12, CXCL9, CXCL10, and CXCL11, cytokines/chemokines known to be critical in mediating malaria-specific immunity. Metenou et al. presented evidence supporting the notion that the inhibition of IRF expression (IRF-1, 2, 3 particularly) seen in filarial infection underlies the suppression of malaria-specific cytokines/chemokines that play a crucial role in immunity to malaria [64].

To address the issue of the differential regulation of IL-12p35 and p40 mRNA through TLR3 or TLR4 vs TLR2 in human DCs, Goriely et al. investigated the potential role of IRF-3 [30]. By chromatin immunoprecipitation (ChIP), IRF-3 was recruited to a region containing a site named interferon-stimulated response element-1 (ISRE-1) in the human *IL12A* promoter region between nucleotides -251 and -240. The ISRE-1 site was required for *IL12A* gene activation in RAW264.7 cells stimulated by LPS or Poly I:C. DCs from IRF-3-deficient mice exhibited strong impairment in TLR4-induced *Il12a* mRNA and IL-12p70 synthesis. The role of IRF-3 in the differential regulation of TLR-induced *Il12a* and *Il12b* transcription was corroborated in a study using a specific inhibitor of conventional protein kinase C α (PKCα) (Gö6976) targeting IRF-3 activation and IFN-β synthesis [46].

A well-illustrated example of the differential transcriptional level regulation of *Il12a* and *Il12b* genes was shown in macrophages derived from C/EBPβ-deficient mice. In contrast to the enhanced induction of *Il12b* mRNA, C/EBPβ-deficient primary macrophages derived from both the bone marrow and the peritoneal cavity displayed totally defective expression of *Il12a* mRNA. This may explain the decreased production of bioactive IL-12 and the impaired Th1 responses of C/EBPβ-deficient mice to *Candida albicans* infection, a pathogen that requires Th1-mediated control [29]. However, the enhanced p40 production in C/EBPβ-deficient macrophages is in contradiction to an earlier in vitro molecular study which described C/EBPβ as a direct transcription activator for *Il12b* [75].

Alternative transcription initiation may also play a role in the regulation of *Il12a* expression and IL-12 production. Transcription of the gene for murine *Il12a* initiates within the first exon, an alternate first exon (exon 1a), or second exon. Vaidyanathan et al. showed that LPS and IFN-γ/CD40 ligation increased the amount of total *Il12a* mRNA in splenic adherent cells (primarily macrophages) to a similar extent. Nevertheless, the exon 1 transcript was a smaller fraction of total *Il12a* mRNA in IFN-γ/CD40-stimulated cells than in unstimulated or LPS-stimulated cells. Despite comparable levels of total p35 mRNA, LPS-induced *Il12a* exon 1 transcripts led to significantly more bioactive IL-12 than IFN-γ/CD40-induced exon 1a/exon 2 transcripts. The data suggest that LPS-inducible *Il12a* synthesis from exon 1 transcripts leads to greater amount of bioactive IL-12 than IFN-γ/CD40-induced p35 expression from alternate exon 1a/exon 2 transcripts [99]. In addition to this alternative transcriptional regulatory mechanism, there is evidence that murine p35 expression may be regulated at the level of translation in the B cell lymphoma line A20 and in bone marrow-derived DCs through the use of different transcription start sites with or without the translation initiation signal ATG [6].

As the saying goes: "What goes up must come down." The existence of various mechanisms that control IL-12 production ensures appropriate immune responses against pathogens and limits "collateral damage."

Fatty acids and their metabolites can modulate various functions of DCs including their differentiation and cytokine production. Epidermal-type fatty acid-binding protein (E-FABP) is exclusively expressed in splenic DCs among FABP family. To investigate the mechanisms underlying E-FABP's cellular functions, Kitanaka et al. examined the phenotype of E-FABP-null mutant mice. Although the mutant mice showed no apparent abnormalities in the population density and subset distribution of DCs as well as the microscopic morphology in the spleen, DCs isolated from E-FABP-null spleen showed enhanced production of IL-12, compared with wild-type controls, in response to appropriate stimuli. Mutant DCs displayed excessive expression level of *Il12a* mRNA after LPS stimulation, with no apparent change in *Il12b* mRNA level. Phosphorylated forms of p38 mitogen-activated protein kinase (MAPK) and IkB-α, molecules critical for IL-12 production, were detected at higher levels in LPS-stimulated E-FABP-null mutant DCs [49]. Thus, E-FABP may be a novel negative regulator of IL-12 production in DCs, via its involvement in the p38 MAPK-mediated transcription of *Il12a* in a differential manner. Further elucidation of the molecular mechanism underlying the negative regulation of IL-12 transcription by E-FABP may lead to a better understanding of the anti-inflammatory actions of polyunsaturated long chain fatty acids (PUFAs) to which E-FABP bind as ligands [60].

Bordetella pertussis, the causative agent of whooping cough, possesses an array of virulence factors, including adenylate cyclase toxin (ACT), important in the establishment of infection. Spensieri et al. assessed the impact of cyclic AMP (cAMP) intoxication due to the action of ACT on DC-induced immune response, by infecting monocyte-derived DC (MDDC) with a strain of *B. pertussis* (WT18323) or its ACT-deficient mutant. Infection with the mutant strain induced the expression of the IRF-1 and IRF-8 and of IFN-β involved in IL-12p35 regulation, and the expression of these genes was inhibited by the cAMP analogous D-butyril-cAMP (D-cAMP) in WT18323-infected DCs. The enhanced expression of IL-12 induced by the ACT mutant strain triggered a more pronounced Th1 polarization compared to WT18323 [90]. The study suggests that ACT-dependent cAMP induction selectively targets the *Il12a* expression pathway, thus representing a potential mechanism for *B. pertussis* to escape the host immune response mediated by IL-12.

The Th2-promoting transcription factor c-Maf is pivotal for tissue-specific IL-4 expression [42]. Homma et al. identified a novel NF-κB element within the proximal *Il12a* promoter and show that c-Maf inhibits *Il12a* transcription by antagonizing the effects of NF-κB, especially c-Rel, on *Il12a* activation. It does so not by directly interacting with the target DNA but by interfering with the nuclear localization of c-Rel [43]. This study extends our understanding of the molecular basis of the homeostatic regulation of IL-12 production by c-Maf, which plays a dual role both in the function of APCs and in T helper cell differentiation.

The rationale for the differential regulation of IL-12p35 and p40 expression is thought to be due to the requirement for varying levels of "optimal production" of their composite cytokines IL-12, IL-23, and IL-35 in different types of immune responses to exogenous and endogenous ligands.

In addition to forming the IL-12 heterodimer with the p40 chain, *Il12a* gene encoding the p35 chain has been found to have at least one splice variant derived from an insert corresponding to an intron in the *Il12a* gene that would result in a truncated form of p35 if translated [86]. The truncated form of p35 constructed into a fusion protein with a p40-encoding cDNA. The truncated IL-12 fusion protein inhibited the action of the full-length IL-12p40/IL-12p35 fusion in a T cell proliferation assay and also blocked the induction of IFN-γ. These findings raise the possibility that alternative splicing may provide an additional regulatory mechanism for IL-12's biological activities. In reality, this type of "aberrant" forms of IL-12 is yet to be identified in primary cells or animals.

6.6 Signal Transduction in the Regulation of IL-12 Production

Following TLR ligation on APCs, intracellular signaling activities are initiated involving cascades of protein kinases and/or phosphatases, which ultimately impact specific transcriptional regulators for the induction or suppression of IL-12 gene expression. Several studies have demonstrated that activation of p38 MAPK is required for IL-12 regulation. Mice deficient in MKK3, an upstream kinase of p38, were defective in the production of IL-12 by APCs [58]. LPS-stimulated macrophages isolated from MKK3-deficient mice produced normal levels of IL-6, IL-1α, TNF-α, and IL-1β but sharply reduced levels of *Il12a* and *Il12b* mRNAs and IL-12 protein (in both macrophages and DCs). It was further shown that MKK3 activated the *Il12b* promoter primarily through p38a. Nonetheless, the *Il12a/b*-activating role of p38 MAPK was countered by a later study showing that compared to WT macrophages, LPS-induced expression of IL-12p40 was lower in primary mouse peritoneal and bone marrow-derived macrophages from mice deficient in mitogen-activated protein kinase phosphatase-1 (MKP-1), a nuclear tyrosine/threonine phosphatase that inhibits p38 MAPK activity [53]. The apparent contradiction could probably be explained in part by the lack of target exclusivity of MKP-1 or p38 MAPK.

In contrast to p38 MAPK, extracellular signal-regulated kinase (ERK) signaling mediates negative feedback regulation of p40 (but not p35) production. Such ERK activation is downstream of calcium influx and targets LPS-induced *Il12b* transcription by suppressing the synthesis of IRF-1. On the other hand, negative regulation of the p35 subunit of IL-12 occurs via a calcium-dependent, but ERK-independent, mechanism, which was thought to involve NF-κB signaling [28].

Helminth infection leads to the local proliferation and accumulation of macrophages in tissues. Hadidi et al. identified a critical role for SH2-containing inositol 5′-phosphatase 1 (Ship1, Inpp5d) in the negative regulation of p40 production by macrophages during infection with the intestinal helminth parasite *Trichuris muris* [36]. Mice with myeloid cell-specific deletion of Ship1 (Ship1(ΔLysM) mice) developed a non-protective Th1 cell response and failed to expel parasites. Ship1-deficient macrophages produced increased levels of p40 in vitro and in vivo and antibody blockade of p40 renders Ship1(ΔLysM) mice resistant to *Trichuris* infection.

Emerging evidence has demonstrated that mammalian target of rapamycin (mTOR) is an important regulator of immunity by modulating the differentiation, activation, and function of lymphocytes and APCs [77]. In exploring the long-held "puzzle" of low levels of IL-12 induced through TLR4 signaling in macrophages and DCs, which implied the existence of stringent regulatory mechanisms, He et al. identified the critical regulatory roles of three protein kinases, mTOR, phosphoinositide-3 kinase (PI3K), and ERK, in TLR-induced Th1 responses by reciprocally controlling IL-12 and IL-10 production in innate immune cells of murine origin [39]. Moreover, it was further revealed that mTOR was a key molecule that mediated the kinase-modulated IL-12 and IL-10 expression in TLR4 signaling by regulating c-fos expression and NF-κB binding to the promoters of IL-12 and IL-10 in a differential manner [39], confirming the role of c-fos as a negative regulator of IL-12 expression reported in two previous studies [65, 79]. These findings were also corroborated by a similar study in human DCs with a further delineation of the opposing activities of the two components of the mTOR complex, mTORC1 and mTORC2, in this signaling pathway [107]. Thus, by controlling the balance between IL-12 and IL-10, mTOR can specifically regulate the TLR-induced T cell response in vivo. Indeed, blockade of mTOR by rapamycin efficiently boosted TLR-induced antigen-specific T and B cell responses to hepatitis B virus (HBV) and hepatitis C virus (HCV) vaccines [39]. This study links a ubiquitously present and fundamentally important pathway of cellular survival, proliferation, and function with the production of a highly restricted specialist molecule in the immune system. Yet, it is presently not clear if the induction of IL-10 via mTOR signaling is directly responsible for the inhibition of IL-12 production.

Sphingosine-1-phosphate (S1P) modulates multiple cellular functions such as lymphocyte trafficking and signaling as well as keratinocyte proliferation relevant to psoriasis. Schaper et al. showed that S1P decreased the production of the proinflammatory cytokines IL-12 and IL-23 in LPS-stimulated DCs via the common subunit p40 as well as in the cross talk with activated keratinocytes through S1P receptor 1 primarily in a phosphoinositide 3-kinase (PI3K)-dependent manner [82]. Notably, the PI3K pathway also suppresses TLR-induced IL-12 production as an autoinhibitory regulator in DCs in a context-dependent manner [66]. The role of PI3K in the regulation of IL-12 production seems controversial as a previous study found that PI3K p110β positively controls LPS-induced IL-12 production through the JNK1-dependent pathway in human macrophages and DCs [98].

6.7 Regulation of IL-12 Production by Immunosuppressive Cytokines

The potency of IL-12 in host defense makes it a target for stringent regulation. Indeed, the temporal, spatial, and quantitative expression of IL-12 during an immune response in a microenvironment contributes critically to the determination of the type, extent, and ultimate resolution of the reaction. Breaching of the delicate control and balance frequently leads to immunologic disorders and pathogenesis. Some of the most important and well-studied negative regulators of TLR-induced IL-12 production are IL-10 [18], TGF-β [23], and prostaglandin E2 (PGE2) [100].

IL-10 suppression of both *IL12a* and *Il12b* genes is primarily seen at the transcriptional level and that the induction of the two genes has different requirements for de novo protein synthesis [5]. How IL-10 suppresses *Il12a* transcription is presently unknown. IL-10 targets an enhancer 10 kb upstream of the *Il12b* transcriptional start site that is bound by nuclear factor, interleukin 3-regulated (NFIL3), a B-ZIP transcription factor. Myeloid cells lacking NFIL3 produce excessive IL-12p40 and increased IL-12p70 [87]. Thus, the STAT3-dependent expression of NFIL3 is a key component of a negative feedback pathway in myeloid cells that suppresses proinflammatory responses. Kobayashi et al. observed that acetylated histone H4 transiently associated with the *Il12b* promoter in WT bone marrow-derived macrophages (BMDMs), whereas association of these factors was prolonged in IL-10-deficient BMDMs. Experiments using histone deacetylase (HDAC) inhibitors and HDAC3 short hairpin RNA indicate that HDAC3 is involved in histone deacetylation of the *Il12b* promoter by IL-10 [52]. These results suggest that histone deacetylation on the *Il12b* promoter by HDAC3 mediates the homeostatic effect of IL-10 in macrophages. More detailed work needs to be performed to understand the important homeostatic regulation of IL-12 production by IL-10. In this context, the IL-4-inducing transcription factor c-Maf is an interesting molecule, which can directly and conversely regulate IL-12 and IL-10 gene expression in activated macrophages [13]. Certain viruses such as hepatitis viruses B and C appear to tap into the IL-10 signaling pathway to inhibit IL-12 production and cell-mediated control of viral infections. This was shown by Liu et al. in human monocytes treated with non-γ interferons, IFN-α, β, and IL-29, a member of the type III IFN family. Increased sensitivity of monocytes to IL-10 signaling and, as a result, decreased TLR-induced IL-12 production by IFN-pretreated cells were observed. These findings may explain partially why only about 30–50 % of chronic hepatitis C virus (HCV) and hepatitis B virus (HBV) patients respond to IFN-based therapy and suggest new ways to improve the therapy for patients with viral hepatitis [55].

Ogawa et al. characterized the negative regulation of the IL-12p40 expression by TGF-β in macrophages. NF-κB p65 and IRF-1-induced murine *Il12b* promoter activity was suppressed by the expression of a constitutively active TGF-β type I receptor in the presence of Smad3 and Smad4. Further evidence indicated that both the NF-κB and IRF-1 sites are required for the repression of promoter activity of *Il12b* by TGF-β in an indirect manner [71]. In addition, Nodal, a member of the TGF-β superfamily and an embryonic morphogen that is upregulated in different types of tumors, increases the tumorigenesis by inducing angiogenesis and promoting metastasis. Treatment with recombinant Nodal (rNodal) decreased the expression of IL-12 in murine macrophages. However, the mechanism is not known. Furthermore, rNodal promoted macrophage polarization to an alternatively activated macrophage-like/tumor-associated macrophage (TAM) phenotype and modulated its function [104]. It is noted that TGF-β's effect on IL-12 production in macrophages seems to depend on how the cells are activated. Tada et al. observed that TGF-β inhibited IL-12p40 production by LPS-stimulated macrophages, whereas it enhanced p40 production by anti-CD40/IFN-γ-stimulated cells [93].

To explore the role of human immunodeficiency virus (HIV)-induced PGE2 in IL-12 synthesis, it was observed that LPS-induced IL-12 synthesis in human whole blood cultures was almost completed inhibited by PGE2, whereas IL-6 production was only partially inhibited. In contrast, the production of IL-10 was approximately twofold enhanced under these conditions. The effects of PGE2 were due to its cAMP-inducing capacity, since they could be mimicked by other cAMP inducers. The inhibitory effect of PGE2 on IL-12 production was independent of IL-10 since neutralizing anti-IL-10 antibodies were unable to reverse this inhibition. These results suggest that the capacity of an antigen to induce PGE2 synthesis may play a crucial role in the development of either a Th1 or Th2 response [100]. HIV has been postulated to induce Th2-dominant responses that favor their persistence [11]. Mitsuhashi et al. further demonstrated that murine mammary tumor-derived PGE2 regulated IL-12p40 expression with the functional cooperation of AP-1, which strongly suppressed *Il12b* transcription [65]. Correa et al. showed that anandamide regulated IL-12p40 expression by acting on the repressor site GA-12 [GATA sequence in *Il12b* promoter first described by Becker et al. [8]]. Prostamide E2 (prostaglandin E2 ethanolamide), a product considered to be a putative metabolite of anandamide by COX-2 (cyclooxygenase 2) oxygenation, was also able to inhibit the activity of the *Il12b* promoter by acting at the repressor site, providing potential mechanistic insights into the activities of anandamide in immune-related disorders [17].

It's interesting that IL-23 has been noted to have a direct inhibitory activity on *Il12a* and *Il12b* mRNA expression in murine bone marrow-derived DCs, consistent with increased IL-12 production in DCs and Th1-mediated inflammatory colitis observed in mice deficient in IL-23p19 [7]. This is the first example that a member of the IL-12 family has a direct inhibitory role in the production of another member within the family, suggesting that IL-12 and IL-23 may have evolved a "competitive" relationship.

6.8 Regulation of IL-12 Production by Apoptotic Cells

Apoptosis and the rapid clearance of apoptotic cells (ACs) by professional or non-professional phagocytes are normal and coordinated processes that ensure controlled cell growth with a nonpathological outcome. Defects in clearance of ACs by macrophages have serious consequences often resulting in autoimmune disorders. Voll et al. first reported that the presence of ACs during monocyte activation increased their secretion of the anti-inflammatory and immunoregulatory cytokine IL-10 and decreased secretion of the proinflammatory cytokines TNF-α, IL-1, and IL-12. This may inhibit inflammation and contribute to impaired cell-mediated immunity in conditions associated with increased apoptosis, such as viral infections, pregnancy, cancer, and exposure to radiation [101]. Kim et al. showed that cell-cell contact with ACs is sufficient to induce profound inhibition of IL-12 production by activated macrophages. Phosphatidylserine-coated liposomes could mimic the inhibitory effect. The inhibition does not involve autocrine or paracrine actions of IL-10 and TGF-β and has no impact on the state of NF-κB activation [48]. The inhibitory effect was much more pronounced on *Il12a* than *Il12b* gene expression. The investigators further identified a novel nuclear protein called GC-binding protein (GC-BP) found in macrophages that engulf ACs via phagocytosis. GC-BP is activated via tyrosine phosphorylation induced by interactions between the phagocyte and the AC expressing externalized phosphatidylserine. GC-BP has a direct and selective inhibitory activity on the transcription of the *Il12a* gene (but not *Il12b*) and IL-12 production via binding to a sequence just downstream of the transcription initiation site [48]. In contrast, the converse observation was made of the induction of IL-10 production during phagocytosis of ACs via CD36 in a p38 MAPK-dependent mechanism involving the transcription factor pre-B cell leukemia transcription factor-1b (Pbx-1) [14]. The inhibition of IL-12 during clearance of ACs was also observed in human monocyte-derived DCs by blocking surface CD47, a thrombospondin receptor and a "Don't eat me" signal [89], via either intact or F(ab′)2 of CD47 mAb or 4N1K, a peptide derived from the CD47-binding site of thrombospondin. However, unlike in the AC-phagocytosis system, IL-10 production was also inhibited by CD47 mAb in such an experimental setting [20]. It is postulated that following exposure to microorganisms, CD47 ligation may limit the intensity and duration of the inflammatory response by preventing inflammatory cytokine production by DCs and favoring their maintenance in an immature state.

In contrast to conventional DCs, in "tolerogenic" myeloid DCs (TolDCs) generated in vitro from monocytes by GM-CSF/IL-4 and dexamethasone [26, 91], AC uptake resulted in diminished production of IL-12p40 and IL-12p70 and increased expression of IL-12p35 and EBI3, the constituents of IL-35 [22]. The differential regulation of the IL-12 family by TolDCs may represent an additional mechanism in determining the immune response to dying cells by producing the T-suppressing cytokine IL-35.

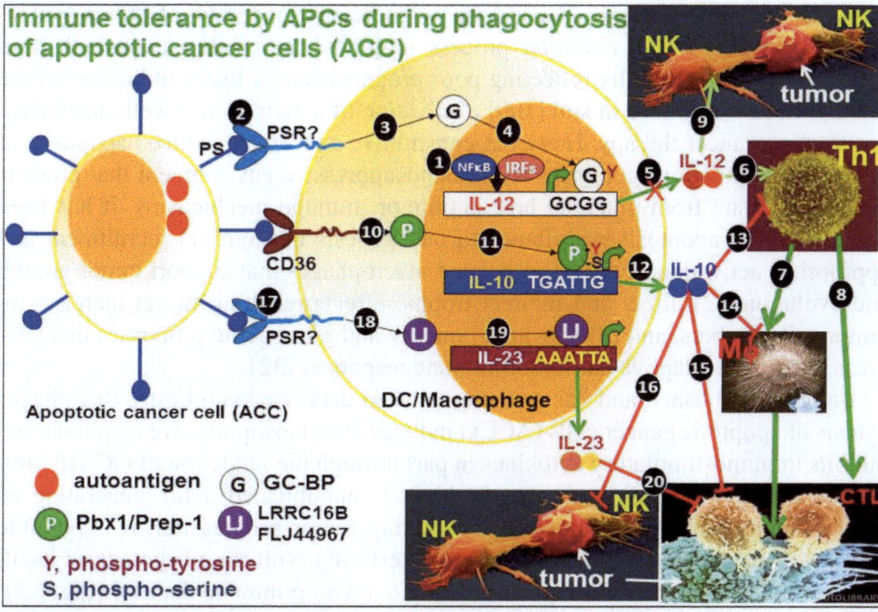

Fig. 6.2 Model of apoptotic cancer cell-induced inhibition of IL-12 production, induction of IL-10 and IL-23 synthesis, and inhibition of cell-mediated immunity against cancer. IL-12 gene transcription is stimulated in professional antigen-presenting cells (dendritic cells and macrophages) by innate immune cues, such as TLR-mediated signaling, and by adaptive immune signals such as CD40L through activation of NF-kB and interferon regulatory factors (IRFs) 1 and 8 that induce IL-12p35 gene transcription (*#1*). Apoptotic cancer cells (ACCs), principally via the "eat me" signal phosphatidylserine (PS) on their surface interacting with certain phagocytic receptors (PSR) on DCs (*#2*), induce GC-BP tyrosine phosphorylation (*#3*). Phosphorylated GC-BP translocates to the nucleus (*#4*) and blocks IL-12 production by binding to the proximal p35 promoter region at the "apoptotic cell response element" (ACRE) (*#5*). The lack of IL-12 results in the block of Th1 differentiation and activation from naïve T (Th0) cells (*#6*), which limits cell-mediated immune responses against malignant tumors via two major cell types: macrophages (*#7*) and CTL (*#8*). Lack of IL-12 also directly dampens NK activation (*#9*). This process also induces the production of IL-10, a highly immunosuppressive cytokine, via the activation of homeoproteins Pbx-1 and Prep-1. The induction of IL-10 by ACCs involves the scavenger receptor CD36 on DCs (*#10*). The signaling triggered by the ACC-DC interaction induces serine and tyrosine phosphorylation of Pbx-1 and Prep-1, respectively (*#11*). The activated Pbx1/Prep1 complex binds to the IL-10 promoter at the "TGATTG" motif (*#12*), driving its transcription. IL-10 produced in this context can inhibit all four effector cell types critical for antitumor immunity: Th1, macrophages, CTL, and NK (*#13–16*). The ACC-phagocyte interaction can induce IL-23 through a yet to be defined phagocytic receptor (*#17*), which activates LRRC16B/FLJ44967 (LJ) (*#18*). These two transcription factors bind to the proximal IL-23p40 promoter region at the "AAATTA" motif (*#19*) driving its transcription. IL-23 can inhibit CTL infiltration into tumors and upregulates IL-17 and IL-22 and matrix metalloproteinase MMP9 and increases angiogenesis (*#20*). Thus, blocking these transcriptional regulators during ACC-DC interaction may benefit antitumor immunity without interfering with the process of clearance of ACCs by phagocytes

Cell loss in malignant disease is a very significant component of tumor dynamics, and apoptosis is a common process in high-grade malignancy, with high apoptosis indices generally reflecting poor prognosis and a likely indication of the rapidity of AC clearance in situ [109]. Activation of apoptosis is a well-established approach to cancer therapy. However, constitutive or therapy-induced apoptosis of tumor cell populations generates an immunosuppressive environment that protects malignant tissue from potential host antitumor immune mechanisms. It has been postulated that apoptosis contributes to oncogenesis through (1) recruitment and appropriate activation of tumor-associated macrophages that support tumor growth and evolution, (2) direct and indirect trophic effects resulting in net increases in tumor cell numbers, and (3) anti-inflammatory and tolerogenic properties that suppress innate and adaptive antitumor immune responses [32].

Based on published and our own unpublished data, we hypothesize that phagocytosis of apoptotic cancer cells (ACCs) induces immunosuppressive cytokines and inhibits immunostimulatory cytokines in part through the induction of GC-BP [48], Pbx-1/Prep-1 [14], LRRC16B, and FLJ44967 (unpublished data), generating an immunosuppressive microenvironment favoring tumor progression. It is possible that targeting these molecules, instead of interfering with the phagocytosis itself, may have antitumor benefits without causing overt autoimmune damage (Fig. 6.2).

Understanding the fundamentally important mechanisms involved in the generation of highly immunosuppressive cytokine microenvironment during clearance of ACCs by phagocytes carries the potential of uncovering novel inner workings that underlie some of the most profound networks of nature and evolution. It could move the field forward conceptually and inspire the development of innovative strategies to overcome immune tolerance to self-tumor antigens in cancer therapeutic modality and vaccination.

References

1. Abdi K, Singh NJ, Spooner E, Kessler BM, Radaev S, Lantz L, Xiao TS, Matzinger P, Sun PD, Ploegh HL. Free IL-12p40 monomer is a polyfunctional adaptor for generating novel IL-12-like heterodimers extracellularly. J Immunol. 2014;192:6028–36.
2. Afanasyeva M, Wang Y, Kaya Z, Stafford EA, Dohmen KM, Sadighi Akha AA, Rose NR. Interleukin-12 receptor/STAT4 signaling is required for the development of autoimmune myocarditis in mice by an interferon-gamma-independent pathway. Circulation. 2001;104:3145–51.
3. Akbari M, Honma K, Kimura D, Miyakoda M, Kimura K, Matsuyama T, Yui K. IRF4 in dendritic cells inhibits IL-12 production and controls Th1 immune responses against Leishmania major. J Immunol. 2014;192:2271–9.
4. Altare F, Jouanguy E, Lamhamedi S, Doffinger R, Fischer A, Casanova JL. Mendelian susceptibility to mycobacterial infection in man. Curr Opin Immunol. 1998;10:413–7.
5. Aste-Amezaga M, Ma X, Sartori A, Trinchieri G. Molecular mechanisms of the induction of IL-12 and its inhibition by IL-10. J Immunol. 1998;160:5936–44.
6. Babik JM, Adams E, Tone Y, Fairchild PJ, Tone M, Waldmann H. Expression of murine IL-12 is regulated by translational control of the p35 subunit. J Immunol. 1999;162:4069–78.

7. Becker C, Dornhoff H, Neufert C, Fantini MC, Wirtz S, Huebner S, Nikolaev A, Lehr HA, Murphy AJ, Valenzuela DM, et al. Cutting edge: IL-23 cross-regulates IL-12 production in T cell-dependent experimental colitis. J Immunol. 2006;177:2760–4.
8. Becker C, Wirtz S, Ma X, Blessing M, Galle PR, Neurath MF. Regulation of IL-12 p40 promoter activity in primary human monocytes: roles of NF-kappaB, CCAAT/enhancer-binding protein beta, and PU.1 and identification of a novel repressor element (GA-12) that responds to IL-4 and prostaglandin E(2). J Immunol. 2001;167:2608–18.
9. Bosisio D, Polentarutti N, Sironi M, Bernasconi S, Miyake K, Webb GR, Martin MU, Mantovani A, Muzio M. Stimulation of toll-like receptor 4 expression in human mononuclear phagocytes by interferon-gamma: a molecular basis for priming and synergism with bacterial lipopolysaccharide. Blood. 2002;99:3427–31.
10. Bradford M, Schroeder A, Morse 3rd HC, Vogel SN, Cowdery JS. CpG DNA induced IL-12 p40 gene activation is independent of STAT1 activation or production of interferon consensus sequence binding protein. J Biomed Sci. 2002;9:688–96.
11. Bretscher PA. On the mechanism determining the TH1/TH2 phenotype of an immune response, and its pertinence to strategies for the prevention, and treatment, of certain infectious diseases. Scand J Immunol. 2014;79:361–76.
12. Brombacher F, Dorfmuller A, Magram J, Dai WJ, Kohler G, Wunderlin A, Palmer-Lehmann K, Gately MK, Alber G. IL-12 is dispensable for innate and adaptive immunity against low doses of Listeria monocytogenes. Int Immunol. 1999;11:325–32.
13. Cao S, Liu J, Chesi M, Bergsagel PL, Ho IC, Donnelly RP, Ma X. Differential regulation of IL-12 and IL-10 gene expression in macrophages by the basic leucine zipper transcription factor c-Maf fibrosarcoma. J Immunol. 2002;169:5715–25.
14. Chung EY, Liu J, Homma Y, Zhang Y, Brendolan A, Saggese M, Han J, Silverstein R, Selleri L, Ma X. Interleukin-10 expression in macrophages during phagocytosis of apoptotic cells is mediated by homeodomain proteins Pbx1 and Prep-1. Immunity. 2007;27:952–64.
15. Collison LW, Workman CJ, Kuo TT, Boyd K, Wang Y, Vignali KM, Cross R, Sehy D, Blumberg RS, Vignali DA. The inhibitory cytokine IL-35 contributes to regulatory T-cell function. Nature. 2007;450:566–9.
16. Cooper AM, Magram J, Ferrante J, Orme IM. Interleukin 12 (IL-12) is crucial to the development of protective immunity in mice intravenously infected with mycobacterium tuberculosis. J Exp Med. 1997;186:39–45.
17. Correa F, Docagne F, Clemente D, Mestre L, Becker C, Guaza C. Anandamide inhibits IL-12p40 production by acting on the promoter repressor element GA-12: possible involvement of the COX-2 metabolite prostamide E(2). Biochem J. 2008;409:761–70.
18. D'Andrea A, Aste-Amezaga M, Valiante NM, Ma X, Kubin M, Trinchieri G. Interleukin 10 (IL-10) inhibits human lymphocyte interferon gamma-production by suppressing natural killer cell stimulatory factor/IL-12 synthesis in accessory cells. J Exp Med. 1993;178:1041–8.
19. D'Andrea A, Rengaraju M, Valiante NM, Chehimi J, Kubin M, Aste M, Chan SH, Kobayashi M, Young D, Nickbarg E, et al. Production of natural killer cell stimulatory factor (interleukin 12) by peripheral blood mononuclear cells. J Exp Med. 1992;176:1387–98.
20. Demeure CE, Tanaka H, Mateo V, Rubio M, Delespesse G, Sarfati M. CD47 engagement inhibits cytokine production and maturation of human dendritic cells. J Immunol. 2000;164:2193–9.
21. Desai BB, Quinn PM, Wolitzky AG, Mongini PK, Chizzonite R, Gately MK. IL-12 receptor. II. Distribution and regulation of receptor expression. J Immunol (Baltimore, Md: 1950). 1992;148:3125–32.
22. Dixon KO, O'Flynn J, van der Kooij SW, van Kooten C. Phagocytosis of apoptotic or necrotic cells differentially regulates the transcriptional expression of IL-12 family members in dendritic cells. J Leukoc Biol. 2014;96:313–24.
23. Du C, Sriram S. Mechanism of inhibition of LPS-induced IL-12p40 production by IL-10 and TGF-beta in ANA-1 cells. J Leukoc Biol. 1998;64:92–7.

24. Ely KH, Kasper LH, Khan IA. Augmentation of the CD8+ T cell response by IFN-gamma in IL-12-deficient mice during Toxoplasma gondii infection. J Immunol. 1999;162:5449–54.
25. Favila MA, Geraci NS, Zeng E, Harker B, Condon D, Cotton RN, Jayakumar A, Tripathi V, McDowell MA. Human dendritic cells exhibit a pronounced type I IFN signature following Leishmania major infection that is required for IL-12 induction. J Immunol. 2014;192:5863–72.
26. Finkelman FD, Lees A, Birnbaum R, Gause WC, Morris SC. Dendritic cells can present antigen in vivo in a tolerogenic or immunogenic fashion. J Immunol. 1996;157:1406–14.
27. Gillessen S, Carvajal D, Ling P, Podlaski FJ, Stremlo DL, Familletti PC, Gubler U, Presky DH, Stern AS, Gately MK. Mouse interleukin-12 (IL-12) p40 homodimer: a potent IL-12 antagonist. Eur J Immunol. 1995;25:200–6.
28. Goodridge HS, Harnett W, Liew FY, Harnett MM. Differential regulation of interleukin-12 p40 and p35 induction via Erk mitogen-activated protein kinase-dependent and -independent mechanisms and the implications for bioactive IL-12 and IL-23 responses. Immunology. 2003;109:415–25.
29. Gorgoni B, Maritano D, Marthyn P, Righi M, Poli V. C/EBP beta gene inactivation causes both impaired and enhanced gene expression and inverse regulation of IL-12 p40 and p35 mRNAs in macrophages. J Immunol. 2002;168:4055–62.
30. Goriely S, Molle C, Nguyen M, Albarani V, Haddou NO, Lin R, De Wit D, Flamand V, Willems F, Goldman M. Interferon regulatory factor 3 is involved in Toll-like receptor 4 (TLR4)- and TLR3-induced IL-12p35 gene activation. Blood. 2006;107:1078–84.
31. Gravina HD, Antonelli L, Gazzinelli RT, Ropert C. Differential use of TLR2 and TLR9 in the regulation of immune responses during the infection with Trypanosoma cruzi. PLoS One. 2013;8, e63100.
32. Gregory CD, Pound JD. Cell death in the neighbourhood: direct microenvironmental effects of apoptosis in normal and neoplastic tissues. J Pathol. 2011;223:177–94.
33. Gri G, Savio D, Trinchieri G, Ma X. Synergistic regulation of the human interleukin-12 p40 promoter by NFkappaB and Ets transcription factors in Epstein-Barr virus-transformed B cells and macrophages. J Biol Chem. 1998;273:6431–8.
34. Grohmann U, Belladonna ML, Bianchi R, Orabona C, Ayroldi E, Fioretti MC, Puccetti P. IL-12 acts directly on DC to promote nuclear localization of NF-kappaB and primes DC for IL-12 production. Immunity. 1998;9:315–23.
35. Grumont R, Hochrein H, O'Keeffe M, Gugasyan R, White C, Caminschi I, Cook W, Gerondakis S. c-Rel regulates interleukin 12 p70 expression in CD8(+) dendritic cells by specifically inducing p35 gene transcription. J Exp Med. 2001;194:1021–32.
36. Hadidi S, Antignano F, Hughes MR, Wang SK, Snyder K, Sammis GM, Kerr WG, McNagny KM, Zaph C. Myeloid cell-specific expression of Ship1 regulates IL-12 production and immunity to helminth infection. Mucosal Immunol. 2012;5:535–43.
37. Hafner M, Falk W, Echtenacher B, Mannel DN. Interleukin-12 activates NK cells for IFN-gamma-dependent and NKT cells for IFN-gamma-independent antimetastatic activity. Eur Cytokine Netw. 1999;10:541–8.
38. Hayes MP, Wang J, Norcross MA. Regulation of interleukin-12 expression in human monocytes: selective priming by interferon-gamma of lipopolysaccharide-inducible p35 and p40 genes. Blood. 1995;86:646–50.
39. He L, Zang A, Du M, Ma D, Yuan C, Zhou C, Mu J, Shi H, Li D, Huang X, et al. mTOR regulates TLR-induced c-fos and Th1 responses to HBV and HCV vaccines. Virol Sin. 2015;30:174–89.
40. Heckel MC, Wolfson A, Slachta CA, Schwarting R, Salgame P, Katsetos CD, Platsoucas CD. Human breast tumor cells express IL-10 and IL-12p40 transcripts and proteins, but do not produce IL-12p70. Cell Immunol. 2011;266:143–53.
41. Heinzel FP, Hujer AM, Ahmed FN, Rerko RM. In vivo production and function of IL-12 p40 homodimers. J Immunol. 1997;158:4381–8.
42. Ho IC, Hodge MR, Rooney JW, Glimcher LH. The proto-oncogene c-maf is responsible for tissue-specific expression of interleukin-4. Cell. 1996;85:973–83.

43. Homma Y, Cao S, Shi X, Ma X. The Th2 transcription factor c-Maf inhibits IL-12p35 gene expression in activated macrophages by targeting NF-kappaB nuclear translocation. J Interferon Cytokine Res. 2007;27:799–808.
44. Jalah R, Rosati M, Ganneru B, Pilkington GR, Valentin A, Kulkarni V, Bergamaschi C, Chowdhury B, Zhang GM, Beach RK, et al. The p40 subunit of interleukin (IL)-12 promotes stabilization and export of the p35 subunit: implications for improved IL-12 cytokine production. J Biol Chem. 2013;288:6763–76.
45. Jana M, Dasgupta S, Saha RN, Liu X, Pahan K. Induction of tumor necrosis factor-alpha (TNF-alpha) by interleukin-12 p40 monomer and homodimer in microglia and macrophages. J Neurochem. 2003;86:519–28.
46. Johnson J, Molle C, Aksoy E, Goldman M, Goriely S, Willems F. A conventional protein kinase C inhibitor targeting IRF-3-dependent genes differentially regulates IL-12 family members. Mol Immunol. 2011;48:1484–93.
47. Kawakami K, Koguchi Y, Qureshi MH, Miyazato A, Yara S, Kinjo Y, Iwakura Y, Takeda K, Akira S, Kurimoto M, Saito A. IL-18 contributes to host resistance against infection with Cryptococcus neoformans in mice with defective IL-12 synthesis through induction of IFN-gamma production by NK cells. J Immunol. 2000;165:941–7.
48. Kim S, Elkon KB, Ma X. Transcriptional suppression of interleukin-12 gene expression following phagocytosis of apoptotic cells. Immunity. 2004;21:643–53.
49. Kitanaka N, Owada Y, Okuyama R, Sakagami H, Nourani MR, Aiba S, Furukawa H, Watanabe M, Ono M, Ohteki T, Kondo H. Epidermal-type fatty acid binding protein as a negative regulator of IL-12 production in dendritic cells. Biochem Biophys Res Commun. 2006;345:459–66.
50. Klinke 2nd DJ. The ratio of P40 monomer to dimer is an important determinant of IL-12 bioactivity. J Theor Biol. 2006;240:323–35.
51. Kobayashi M, Fitz L, Ryan M, Hewick RM, Clark SC, Chan S, Loudon R, Sherman F, Perussia B, Trinchieri G. Identification and purification of natural killer cell stimulatory factor (NKSF), a cytokine with multiple biologic effects on human lymphocytes. J Exp Med. 1989;170:827–45.
52. Kobayashi T, Matsuoka K, Sheikh SZ, Russo SM, Mishima Y, Collins C, deZoeten EF, Karp CL, Ting JP, Sartor RB, Plevy SE. IL-10 regulates Il12b expression via histone deacetylation: implications for intestinal macrophage homeostasis. J Immunol. 2012;189:1792–9.
53. Korhonen R, Huotari N, Hommo T, Leppanen T, Moilanen E. The expression of interleukin-12 is increased by MAP kinase phosphatase-1 through a mechanism related to interferon regulatory factor 1. Mol Immunol. 2012;51:219–26.
54. Ling P, Gately MK, Gubler U, Stern AS, Lin P, Hollfelder K, Su C, Pan YC, Hakimi J. Human IL-12 p40 homodimer binds to the IL-12 receptor but does not mediate biologic activity. J Immunol. 1995;154:116–27.
55. Liu BS, Janssen HL, Boonstra A. Type I and III interferons enhance IL-10R expression on human monocytes and macrophages, resulting in IL-10-mediated suppression of TLR-induced IL-12. Eur J Immunol. 2012;42:2431–40.
56. Liu J, Cao S, Herman LM, Ma X. Differential regulation of interleukin (IL)-12 p35 and p40 gene expression and interferon (IFN)-gamma-primed IL-12 production by IFN regulatory factor 1. J Exp Med. 2003;198:1265–76.
57. Liu J, Jiang M, Ma Z, Dietze KK, Zelinskyy G, Yang D, Dittmer U, Schlaak JF, Roggendorf M, Lu M. TLR1/2 ligand-stimulated mouse liver endothelial cells secrete IL-12 and trigger CD8+ T cell immunity in vitro. J Immunol. 2013;191:6178–90.
58. Lu HT, Yang DD, Wysk M, Gatti E, Mellman I, Davis RJ, Flavell RA. Defective IL-12 production in mitogen-activated protein (MAP) kinase kinase 3 (Mkk3)-deficient mice. EMBO J. 1999;18:1845–57.
59. Ma X, Chow JM, Gri G, Carra G, Gerosa F, Wolf SF, Dzialo R, Trinchieri G. The interleukin 12 p40 gene promoter is primed by interferon gamma in monocytic cells. J Exp Med. 1996;183:147–57.

60. Makowski L, Hotamisligil GS. Fatty acid binding proteins – the evolutionary crossroads of inflammatory and metabolic responses. J Nutr. 2004;134:2464S–8.
61. Maruyama S, Sumita K, Shen H, Kanoh M, Xu X, Sato M, Matsumoto M, Shinomiya H, Asano Y. Identification of IFN regulatory factor-1 binding site in IL-12 p40 gene promoter. J Immunol. 2003;170:997–1001.
62. Mattner F, Magram J, Ferrante J, Launois P, Di Padova K, Behin R, Gately MK, Louis JA, Alber G. Genetically resistant mice lacking interleukin-12 are susceptible to infection with Leishmania major and mount a polarized Th2 cell response. Eur J Immunol. 1996;26:1553–9.
63. Mattner F, Ozmen L, Podlaski FJ, Wilkinson VL, Presky DH, Gately MK, Alber G. Treatment with homodimeric interleukin-12 (IL-12) p40 protects mice from IL-12-dependent shock but not from tumor necrosis factor alpha-dependent shock. Infect Immun. 1997;65:4734–7.
64. Metenou S, Kovacs M, Dembele B, Coulibaly YI, Klion AD, Nutman TB. Interferon regulatory factor modulation underlies the bystander suppression of malaria antigen-driven IL-12 and IFN-gamma in filaria-malaria co-infection. Eur J Immunol. 2012;42:641–50.
65. Mitsuhashi M, Liu J, Cao S, Shi X, Ma X. Regulation of interleukin-12 gene expression and its anti-tumor activities by prostaglandin E2 derived from mammary carcinomas. J Leukoc Biol. 2004;76:322–32.
66. Miyazawa K. A negative regulator or just an unconcerned passerby: phosphoinositide 3-kinase signalling in IL-12 production. J Biochem. 2012;152:497–9.
67. Muller U, Kohler G, Mossmann H, Schaub GA, Alber G, Di Santo JP, Brombacher F, Holscher C. IL-12-independent IFN-gamma production by T cells in experimental Chagas' disease is mediated by IL-18. J Immunol. 2001;167:3346–53.
68. Murphy TL, Cleveland MG, Kulesza P, Magram J, Murphy KM. Regulation of interleukin 12 p40 expression through an NF-kappa B half-site. Mol Cell Biol. 1995;15:5258–67.
69. Negishi H, Fujita Y, Yanai H, Sakaguchi S, Ouyang X, Shinohara M, Takayanagi H, Ohba Y, Taniguchi T, Honda K. Evidence for licensing of IFN-gamma-induced IFN regulatory factor 1 transcription factor by MyD88 in Toll-like receptor-dependent gene induction program. Proc Natl Acad Sci U S A. 2006;103:15136–41.
70. Noben-Trauth N, Schweitzer PA, Johnson KR, Wolf SF, Knowles BB, Shultz LD. The interleukin-12 beta subunit (p40) maps to mouse chromosome 11. Mamm Genome. 1996;7:392.
71. Ogawa K, Funaba M, Tsujimoto M. Suppression of NF-kappaB and IRF-1-induced transcription of the murine IL-12 p40 by transforming growth factor-beta Smad pathway in macrophages. Mol Cell Biochem. 2008;308:9–15.
72. Oppmann B, Lesley R, Blom B, Timans JC, Xu Y, Hunte B, Vega F, Yu N, Wang J, Singh K, et al. Novel p19 protein engages IL-12p40 to form a cytokine, IL-23, with biological activities similar as well as distinct from IL-12. Immunity. 2000;13:715–25.
73. Pflanz S, Timans JC, Cheung J, Rosales R, Kanzler H, Gilbert J, Hibbert L, Churakova T, Travis M, Vaisberg E, et al. IL-27, a heterodimeric cytokine composed of EBI3 and p28 protein, induces proliferation of naive CD4+ T cells. Immunity. 2002;16:779–90.
74. Piccotti JR, Chan SY, Li K, Eichwald EJ, Bishop DK. Differential effects of IL-12 receptor blockade with IL-12 p40 homodimer on the induction of CD4+ and CD8+ IFN-gamma-producing cells. J Immunol. 1997;158:643–8.
75. Plevy SE, Gemberling JH, Hsu S, Dorner AJ, Smale ST. Multiple control elements mediate activation of the murine and human interleukin 12 p40 promoters: evidence of functional synergy between C/EBP and Rel proteins. Mol Cell Biol. 1997;17:4572–88.
76. Podlaski FJ, Nanduri VB, Hulmes JD, Pan YC, Levin W, Danho W, Chizzonite R, Gately MK, Stern AS. Molecular characterization of interleukin 12. Arch Biochem Biophys. 1992;294:230–7.
77. Powell JD, Pollizzi KN, Heikamp EB, Horton MR. Regulation of immune responses by mTOR. Annu Rev Immunol. 2012;30:39–68.
78. Roilides E, Tsaparidou S, Kadiltsoglou I, Sein T, Walsh TJ. Interleukin-12 enhances antifungal activity of human mononuclear phagocytes against Aspergillus fumigatus: implications for a gamma interferon-independent pathway. Infect Immun. 1999;67:3047–50.

79. Roy S, Charboneau R, Melnyk D, Barke RA. Interleukin-4 regulates macrophage interleukin-12 protein synthesis through a c-fos mediated mechanism. Surgery. 2000;128:219–24.
80. Sanjabi S, Hoffmann A, Liou HC, Baltimore D, Smale ST. Selective requirement for c-Rel during IL-12 P40 gene induction in macrophages. Proc Natl Acad Sci U S A. 2000;97:12705–10.
81. Satoskar AR, Rodig S, Telford 3rd SR, Satoskar AA, Ghosh SK, von Lichtenberg F, David JR. IL-12 gene-deficient C57BL/6 mice are susceptible to Leishmania donovani but have diminished hepatic immunopathology. Eur J Immunol. 2000;30:834–9.
82. Schaper K, Kietzmann M, Baumer W. Sphingosine-1-phosphate differently regulates the cytokine production of IL-12, IL-23 and IL-27 in activated murine bone marrow derived dendritic cells. Mol Immunol. 2014;59:10–8.
83. Schweitzer PA, Noben-Trauth N, Pelsue SC, Johnson KR, Wolf SF, Shultz LD. Genetic mapping of the IL-12 alpha chain gene (Il12a) on mouse chromosome 3. Mamm Genome. 1996;7:394–5.
84. Shi X, Liu J, Xiang Z, Mitsuhashi M, Wu RS, Ma X. Gene expression analysis in Interleukin-12-induced suppression of mouse mammary carcinoma. Int J Cancer. 2004;110:570–8.
85. Sieburth D, Jabs EW, Warrington JA, Li X, Lasota J, LaForgia S, Kelleher K, Huebner K, Wasmuth JJ, Wolf SF. Assignment of genes encoding a unique cytokine (IL12) composed of two unrelated subunits to chromosomes 3 and 5. Genomics. 1992;14:59–62.
86. Skrombolas D, Wylie I, Maharaj S, Frelinger JG. Characterization of an IL-12 p40/p35 truncated fusion protein that can inhibit the action of IL-12. J Interferon Cytokine Res. 2015;35:690–7.
87. Smith AM, Qualls JE, O'Brien K, Balouzian L, Johnson PF, Schultz-Cherry S, Smale ST, Murray PJ. A distal enhancer in Il12b is the target of transcriptional repression by the STAT3 pathway and requires the basic leucine zipper (B-ZIP) protein NFIL3. J Biol Chem. 2011;286:23582–90.
88. Snijders A, Hilkens CM, van der Pouw Kraan TC, Engel M, Aarden LA, Kapsenberg ML. Regulation of bioactive IL-12 production in lipopolysaccharide-stimulated human monocytes is determined by the expression of the p35 subunit. J Immunol. 1996;156:1207–12.
89. Sosale NG, Spinler KR, Alvey C, Discher DE. Macrophage engulfment of a cell or nanoparticle is regulated by unavoidable opsonization, a species-specific 'Marker of Self' CD47, and target physical properties. Curr Opin Immunol. 2015;35:107–12.
90. Spensieri F, Fedele G, Fazio C, Nasso M, Stefanelli P, Mastrantonio P, Ausiello CM. Bordetella pertussis inhibition of interleukin-12 (IL-12) p70 in human monocyte-derived dendritic cells blocks IL-12 p35 through adenylate cyclase toxin-dependent cyclic AMP induction. Infect Immun. 2006;74:2831–8.
91. Steptoe RJ, Thomson AW. Dendritic cells and tolerance induction. Clin Exp Immunol. 1996;105:397–402.
92. Stern AS, Podlaski FJ, Hulmes JD, Pan YC, Quinn PM, Wolitzky AG, Familletti PC, Stremlo DL, Truitt T, Chizzonite R, et al. Purification to homogeneity and partial characterization of cytotoxic lymphocyte maturation factor from human B-lymphoblastoid cells. Proc Natl Acad Sci U S A. 1990;87:6808–12.
93. Tada Y, Asahina A, Fujita H, Mitsui H, Torii H, Watanabe T, Tamaki K. Differential effects of LPS and TGF-beta on the production of IL-6 and IL-12 by Langerhans cells, splenic dendritic cells, and macrophages. Cytokine. 2004;25:155–61.
94. Taylor-Fishwick DA, Weaver JR, Grzesik W, Chakrabarti S, Green-Mitchell S, Imai Y, Kuhn N, Nadler JL. Production and function of IL-12 in islets and beta cells. Diabetologia. 2013;56:126–35.
95. Taylor AP, Murray HW. Intracellular antimicrobial activity in the absence of interferon-gamma: effect of interleukin-12 in experimental visceral leishmaniasis in interferon-gamma gene-disrupted mice. J Exp Med. 1997;185:1231–9.
96. Teng MW, Bowman EP, McElwee JJ, Smyth MJ, Casanova JL, Cooper AM, Cua DJ. IL-12 and IL-23 cytokines: from discovery to targeted therapies for immune-mediated inflammatory diseases. Nat Med. 2015;21:719–29.

97. Trinchieri G. Interleukin-12 and the regulation of innate resistance and adaptive immunity. Nat Rev Immunol. 2003;3:133–46.
98. Utsugi M, Dobashi K, Ono A, Ishizuka T, Matsuzaki S, Hisada T, Shimizu Y, Kawata T, Aoki H, Kamide Y, Mori M. PI3K p110beta positively regulates lipopolysaccharide-induced IL-12 production in human macrophages and dendritic cells and JNK1 plays a novel role. J Immunol. 2009;182:5225–31.
99. Vaidyanathan H, Gentry JD, Weatherman A, Schwartzbach SD, Petro TM. Differential response of the murine IL-12 p35 gene to lipopolysaccharide compared with interferon-gamma and CD40 ligation. Cytokine. 2001;16:1–9.
100. van der Pouw Kraan TC, Boeije LC, Smeenk RJ, Wijdenes J, Aarden LA. Prostaglandin-E2 is a potent inhibitor of human interleukin 12 production. J Exp Med. 1995;181:775–9.
101. Voll RE, Herrmann M, Roth EA, Stach C, Kalden JR, Girkontaite I. Immunosuppressive effects of apoptotic cells. Nature. 1997;390:350–1.
102. Wakeham J, Wang J, Magram J, Croitoru K, Harkness R, Dunn P, Zganiacz A, Xing Z. Lack of both types 1 and 2 cytokines, tissue inflammatory responses, and immune protection during pulmonary infection by Mycobacterium bovis bacille Calmette-Guerin in IL-12-deficient mice. J Immunol. 1998;160:6101–11.
103. Wang IM, Contursi C, Masumi A, Ma X, Trinchieri G, Ozato K. An IFN-gamma-inducible transcription factor, IFN consensus sequence binding protein (ICSBP), stimulates IL-12 p40 expression in macrophages. J Immunol. 2000;165:271–9.
104. Wang XF, Wang HS, Zhang F, Guo Q, Wang H, Wang KF, Zhang G, Bu XZ, Cai SH, Du J. Nodal promotes the generation of M2-like macrophages and downregulates the expression of IL-12. Eur J Immunol. 2014;44:173–83.
105. Wang ZE, Zheng S, Corry DB, Dalton DK, Seder RA, Reiner SL, Locksley RM. Interferon gamma-independent effects of interleukin 12 administered during acute or established infection due to Leishmania major. Proc Natl Acad Sci U S A. 1994;91:12932–6.
106. Watford WT, Hissong BD, Bream JH, Kanno Y, Muul L, O'Shea JJ. Signaling by IL-12 and IL-23 and the immunoregulatory roles of STAT4. Immunol Rev. 2004;202:139–56.
107. Wei WC, Liu CP, Yang WC, Shyur LF, Sheu JH, Chen SS, Yang NS. Mammalian target of rapamycin complex 2 (mTORC2) regulates LPS-induced expression of IL-12 and IL-23 in human dendritic cells. J Leukoc Biol. 2015;97:1071–80.
108. Wolf SF, Temple PA, Kobayashi M, Young D, Dicig M, Lowe L, Dzialo R, Fitz L, Ferenz C, Hewick RM, et al. Cloning of cDNA for natural killer cell stimulatory factor, a heterodimeric cytokine with multiple biologic effects on T and natural killer cells. J Immunol. 1991;146:3074–81.
109. Wyllie AH. Apoptosis and the regulation of cell numbers in normal and neoplastic tissues: an overview. Cancer Metastasis Rev. 1992;11:95–103.
110. Zhu C, Gagnidze K, Gemberling JH, Plevy SE. Characterization of an activation protein-1-binding site in the murine interleukin-12 p40 promoter. Demonstration of novel functional elements by a reductionist approach. J Biol Chem. 2001;276:18519–28.
111. Zhu C, Rao K, Xiong H, Gagnidze K, Li F, Horvath C, Plevy S. Activation of the murine interleukin-12 p40 promoter by functional interactions between NFAT and ICSBP. J Biol Chem. 2003;278:39372–82.
112. Zilocchi C, Stoppacciaro A, Chiodoni C, Parenza M, Terrazzini N, Colombo MP. Interferon gamma-independent rejection of interleukin 12-transduced carcinoma cells requires CD4+ T cells and granulocyte/macrophage colony-stimulating factor. J Exp Med. 1998;188:133–43.

Chapter 7
Regulation of Interleukin-17 Production

Wenjuan Dong and Xiaojing Ma

Abstract Innate immunity represents the first line of host defense against extracellular infections. Timely initiation of inflammatory responses to invading pathogens confers host the ability to prevent, restrict, and eliminate microorganisms and further trigger adaptive immunity to provide extended eradication of the pathogens as well as long-term protection of the host from recurrent infections. During this process, various types of cytokines are required for efficient immune responses, and interleukin 17 (IL-17) is one of them. The IL-17 family of cytokines bridges innate and adaptive immune responses and play an essential role in defining the Th17 cell lineage differentiation and function. Th17 cells have pivotal roles in protecting the host against extracellular pathogens as well as in promoting inflammatory pathology in autoimmune diseases. Strict regulation of the IL-17 family of cytokines at different levels ensures appropriate controls of their abilities to shape immune responses. This chapter provides the recent progress in the field of IL-17 studies for the understanding of the regulatory mechanisms in the expression and signaling of the IL-17 family cytokines. IL-17A (IL-17) is the most well-studied member in this family, while information on the other members is relatively sketchy. Thus, we mainly focus on the study of IL-17A.

Keywords IL-17 • Transcription regulation • IL-17 receptor signaling • Th17 cell • Infection • Autoimmune/inflammatory diseases

---

The authors declare no conflict of interest.

W. Dong
Department of Microbiology and Immunology, Weill Cornell Medical College,
1300 York Avenue, New York, NY 10065, USA

X. Ma (✉)

State Key Laboratory of Microbial Metabolism, Sheng Yushou Center of Cell Biology and Immunology, School of Life Science and Biotechnology, Shanghai Jiaotong University, 800 Dongchuan Road, Shanghai 200240, China

Department of Microbiology and Immunology, Weill Cornell Medical College,
1300 York Avenue, New York, NY 10065, USA
e-mail: xim2002@sjtu.edu.cn; xim2002@med.cornell.edu

X. Ma (ed.), *Regulation of Cytokine Gene Expression in Immunity and Diseases*,
Advances in Experimental Medicine and Biology 941,
DOI 10.1007/978-94-024-0921-5_7

7.1 The Source of IL-17

It has been reported that many types of innate and adaptive immune cells produce IL-17, among which, T helper 17 (Th17) cells, a subset of helper CD 4^+ T cells that is distinct from Th1 and Th2 lineages, are defined as the major cellular source of IL-17.

Earlier studies showed that naïve CD 4^+ T cells could differentiate into the Th1 or Th2 lineage upon stimulation, with different cytokine production profiles. The development of the two lineages requires distinct set of cytokines, respectively [122]. However, a number of other observations could not be explained based on the Th1 and Th2 subtypes, indicating the existence of a separate helper cell subset [29]. For example, IL-23p19-deficient mice are more susceptible to certain autoimmune diseases but not IL-12p35-deficient mice. Furthermore, IL-17-producing $CD4^+$ T cells are generated independently of Th1 and Th2 cell development, and the differentiation of naïve $CD4^+$ T cells to this subset is triggered and tightly regulated by a cytokine network which is different from the ones both for Th1 and Th2 lineages [132]. These studies eventually led to the discovery of the Th17 subset as a new branch of the Th cell family tree, which plays both protective and pathogenic roles in clearance of invading pathogens and organ-specific autoimmune and chronic inflammatory diseases [57, 92, 194].

Although IL-17 is characterized as a signature cytokine of the Th17 lineage, it can also be derived from other innate and adaptive immune cell populations [26]. γδ T cells express a T cell receptor consisting of a γ-chain and a δ-chain. Similar to $CD4^+$ T cells, γδ T cells can also be subdivided into different populations based on their cytokine production. The subset of γδ T cells with Scart-2 and CCR6 expression is further defined as the functional phenotype of IL-17-producing γδ T cells [48, 143].

These IL-17-producing γδ T cells are mainly localized to mucosal tissues including the skin, lung, and intestine. γδ T cells express pattern recognition receptors like Toll-like receptor 2 (TLR2) and C-type lectin domain family 7 member A (CLEC7A), which allow for rapid production of IL-17 in response to invading microbes [114, 142]. The respond speed is crucial in the combat with infections especially in the mucosal tissues that constantly interact with outside environment. The rapid reaction of γδ T cells to antigen provides immediate activated immune response to restrict and eventually eradicate the pathogens by producing pro-inflammatory cytokines and recruitment of immune cells to the site of infection. Early reports have provided strong evidence on the important role of IL-17-producing γδ T cells in response to bacteria and fungi encounter [49, 58, 107, 116, 164, 178].

In *Toxoplasma gondii*-infected mice, it has been demonstrated that IL-17 can be produced by natural killer cells in an IL-23- and IL-6-induced manner, whereas natural killer T cells which lack the NK1.1 marker secrete IL-17 following the stimulation of anti-CD3 and IL-23, independently of IL-6 [118, 133, 137]. iNKT cells that form murine spleen and human peripheral blood which express CD161 are also able to produce IL-17 in the presence of IL-1β, TGF-β, and IL-23, but not IL-6 [121, 143].

Lymphoid tissue inducer (LTi) cells are key components of the lymphoid structures. LTi and LTi-like cells have been defined as cells express that lymphotoxin-α (LT-α) and lymphotoxin-β (LT-β) and the chemokine receptors CCR7 and CXCR5 (CXC receptor 5). Those cells inhibit fungal and bacterial infection by releasing IL-17, and IFNγ in intestinal mucosa provides the protection of intestinal homeostasis [42, 154].

A subset of CD8+ T cells, also known as Tc17 cells, participates in host defense against foreign microorganisms also by releasing IL-17 [125]. B cells were found to become a major source of IL-17 during *Trypanosoma cruzi* infection and enhance the eradication of this parasite [9]. Neutrophils, the early immune activity responding cells and hallmark effector cells for acute inflammation, are also found to contribute to adaptive immunity by production of IL-17, which induces other inflammatory mediators, resulting to further activation and recruitment of immune cells [98]. Mast cells are tissues resident which, upon activation, secrete a broad array of biologically active products including IL-17 and play a protective role in host defense against pathogens, wound healing, and angiogenesis. Macrophage is another important source of IL-17. Lacking of IL-17 signaling affects monocyte recruitment and adherence and homeostasis and prevents macrophage undergoing apoptosis [33, 152, 205].

In addition to immune cells, Paneth cells which are specialized epithelial cells located in small intestine secrete cytokines including IL-17 into the lumen of the intestinal gland, thereby contributing to the maintenance of gastrointestinal barrier [26].

7.2 IL-17 Receptor and Responds Cells

The IL-17 receptor family is composed of five receptor subunits including IL-17RA, IL-17RB, IL-17RC, IL-17RD, and IL-17RE [1, 39]. They are all containing single transmembrane domain, ranging in size from 499 to 866 amino acids [39]. IL-17Rs showed little homology to other cytokine receptors and thus have been classified as a new cytokine receptor family. Despite the divergent sequence between IL-17R family members, their encoding genes show a cluster pattern on chromosomes with exception of IL-17RA that is located on human chromosome 22; human IL-17RB, IL-17RC, IL-17RD, and IL-17RE are clusters on chromosome 3. IL-17RA was first identified in 1995 as the receptor for IL-17A and was later proved to be the most commonly used signaling subunit among the IL-17 receptor family [192]. IL-17RA binds to IL-17A and also IL-17B, IL-17C, IL-17D, and IL-17E with weaker affinities. The binding between IL-17A and IL-17F is the weakest, which is approximately 100–1000 times weaker than IL-17A binding with other IL-17Rs. Even the binding of IL-17RA with IL-17A is the highest among other IL-17 family members; the binding affinity of IL-17RA for IL-17A was still too low ($4 \times 10^7\ M^{-1}$) to initiate immediate respond, indicating the cooperation of an additional subunit. Further studies toward IL-17 receptor structure and function showed that IL-17RA pairs with IL-17RC to efficiently induce responses to IL-17A and IL-17F at a very low concentration. The underlying mechanism is the binding of ligand to the first IL-17 receptor subunit that changes the affinity and specificity of the second binding event, thus promoting the formation of a heterodimeric receptor complex [176]. In addition, IL-17RA can also form a complex with IL-17B and IL-17E, suggesting that IL-RA is the most commonly used subunit in the signaling transduction of IL-17 by participating in heterodimeric complexes [39]. Furthermore, IL-17RA can also be shared as a receptor by other cytokine family molecules such as IL-6 family, suggesting its significant roles in signal transduction [131].

IL-17RC is known to form a heterodimer with IL-17RA for IL-17A- and IL-17F-mediated signaling, but not without the presence of IL-17RA. IL-17RC is highly spliced at sites in the extracellular domain with more than 90 spliced isoforms that were identified in human. Human17RC shows similar binding affinity for both IL-17A and IL-17F, while mouse IL-17RC has higher binding affinity to IL-17F than IL-17A [77, 90]. There are also other isoforms that showed a different preference in the binding of IL-17A and IL-17F, while some isoforms do not bind either of them, suggesting that a regulatory mechanism of IL-17RC induced a different response to IL-17 family cytokines stimuli [55]. Similarly to IL-17RC, IL-17RB is also highly spliced in extracellular domain, which has the ability to bind with both IL-17B and IL-17E, but not IL-17A or IL-17C. Compared to the rest of the members of IL-17 receptor family, the structure and function of IL-17RD and IL-17RE are the least well understood. IL-17C is speculated to be the ligand of IL-17RE. IL-17RD homologues have been found in sea lamprey, *C. elegans*, and frogs; however, its binding ligand has not been identified yet.

All cell types and tissues which express IL-17 receptor family molecules could become the targeting cells of IL-17 cytokine family membranes, thereby initiating downstream signaling transduction. However, the different expression level and distinct tissue distribution of IL-17 receptor subunits and their capacity for various splice variations confer the responder cell with different signal strength, respond degree, and width [54, 90]. Among IL-17Rs, IL-17RA is the most ubiquitously expressed member; it has been reported that IL-17RA is expressed on various tissues and cell types with particularly high levels in hematopoietic tissues [82, 129]. However, further studies showed that the main responsive cells to IL-17A are epithelial cells, endothelial cells, fibroblasts, and innate immune cells such as macrophages and DCs; although IL-17-inducible genes have been detected in lymphocytes, these genes are distinct from those expressed in other cell types induced by IL-17 [60, 71, 158]. IL-17RB is detected in a variety of cell types, and the highest expression level is detected in the kidney, liver, and brain [96]. The expression profile showed that IL-17RD was expressed on many different tissues including the GI tract, breast, lung, and kidney. In contrast to IL-17RD, IL-17RE showed a limited expression profile; the tissue which has the highest expression of IL-17RE is the kidney. IL-17RC has a lower expression in hemopoietic tissues and a higher expression in the liver, kidney, and thyroid compared to IL-17RA, which hint the tissue specificity of IL-17 receptor family in IL-17 signaling transduction [54, 90]. Due to the distinct expression profiles of IL-17 receptor family and different binding affinities to their ligands, the

IL-17/IL-R axis achieved accurately modulation of the complex network by selectively regulating the signal respond strength, location, and extensity.

7.3 IL-17 Signaling Pathways

The IL-17/IL-17R axis is a delicate signaling pathway, regulated by various kinases and transcriptional factors. IL-17A and IL-17F can form both homodimer and heterodimer with each other to bind IL-17RA and IL-17RC complex, thereby activating downstream cascade including NF-kB (nuclear factor kappa-light-chain-enhancer of activated B cells), MAP kinases (mitogen-activated protein kinases), and C/EBPs (CCAAT/enhancer-binding proteins) which lead to the upregulation of pro-inflammatory genes as shown in Fig. 7.1. NF-kB, a hallmark transcription factor associated with inflammation induction, is activated by IL-17A through distinct proximal signaling pathways [11]. There is a conserved region at the C-terminus of the IL-17 receptors known as the SEFIR (similar expression of fibroblast growth factor genes and IL-17Rs) domain that is also present in Act1 which has been known as NF-kB activator. Act1 was recruited to IL-17 receptors through the interaction of SEFIR domain upon IL-17 stimulation. After Act1 binding to the receptor complex, it further recruited TRAF6 (TNF receptor-associated factor 6) through the binding motifs of Act1. Act1 also serves as an E3 ubiquitin ligase; it mediates Lys^{63}-linked ubiquitination of TARF6. TARF6 is a critical adaptor in TLR (toll-like receptor) and also exerts the function of E3 ubiquitin ligase. Polyubiquitinated TRAF6 further activates downstream TAK1 (TRAF6-dependent TGF-β-activated kinase 1) which is also by ubiquitination of TAK1. Polyubiquitinated TAK1 complex activates IKKs (IkB kinase) for NF-kB activation [15, 126, 136]. These results are supported by recent reports, where TARF6-deficient mice showed significantly reduced production of pro-inflammatory cytokines upon IL-17 stimulation [196]. However, in *ACT1*$^{-/-}$ cells, ERK1 (extracellular signal-regulated kinase 1) and ERK2 can be activated in response to IL-17A, indicating that IL-17RA-mediated activation of NF-kB could be different from those of TLRs and IL-1Rs due to the unique structure of IL-17RA [136].

IL-17A can also augment NF-kB activation by the enhancement of mRNA stability. IL-17A induces weak activation of NF-kB, while TNFα can strongly activate NF-kB. However, the pro-inflammatory mRNAs induced by TNFα are highly unstable. IL-17A exerts the ability to promote pro-inflammatory response by synergizing with TNFα and stabilizing pro-inflammatory genes' mRNA such as Cxcl1 (CXC chemokine ligand 1), Cxcl2 (CXC chemokine ligand 2), and MIP2 (macrophage inflammatory protein 2) [13, 53, 170]. IKKi, a kinase involved in type I IFN production for antivirus, is recruited to the IL-17/IL-17R/Act-1 complex to specifically phosphorylate Act1 at Ser^{311}, which allow the complex to further recruit TRAF2 and TRAF5, but not TRAF6, the newly formed complex dissociate ASF (alternative splicing factor) from mRNA [13, 170]. The synergistic effect of IL-17

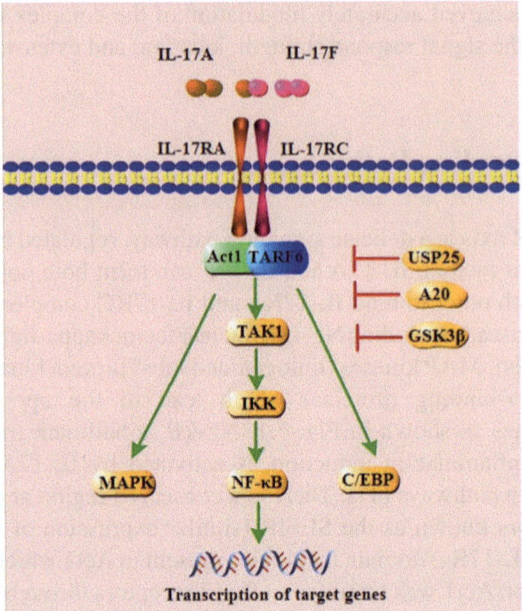

Fig. 7.1 IL-17A- and IL-17F-induced signaling pathways. IL-17RA forms heterodimeric complexes with IL-17RC, which can bind with IL-17A and IL-17F homodimers or IL-17A/IL-17F heterodimers. After the binding, IL-17 receptors recruit Act1 to the binding site as an adaptor molecule. IL-17A and IL-17F signal through the IL-17RA/IL-17RC complex and trigger TRAF6 expression to form Act1/TRAF6 complex, which further activates downstream signal molecules. IL-17 signaling can be suppressed by deubiquitinating enzymes like USP25 and A20 as well as phosphorylase GSK3β, which regulate the ubiquitination status of TRAFs and phosphorylated status of C/EBP-β to blockade the signaling cascade

and TNFα which regulates pro-inflammatory molecules production not only exists in posttranscriptional level but also in transcription level. It has been shown that they increase IL-6 production through cooperative induction of C/EBPs at the promoter level. In addition to TNFα, IL-17 has been reported that it can have synergistic effects with many other cytokines, such as IL-1β, IL-22, and IFNγ [129].

The MAPK signaling pathway plays a critical role in induction of pro-inflammatory mediators' activation. All MAPK members can be activated by IL-17 including ERK, JNK, and p38. Among which, ERK is generally the most rapidly activated kinase. IL-17 efficiently induces responding gene expression by controlling the stability of mRNA transcripts by activation of MAPK family. MAPKs keep mRNA stabilization by inhibiting the activity of destabilizing protein like tristetraprolin. Tristetraprolin delivers mRNA into exosome complex to promote their degradation by binding to their AU-rich elements in the 3′-untranslated regions (UTRs) of the mRNAs [5]. It has been reported that IL-17 can activate JAK/STAT pathway for induction of host defense genes [62]. In this cascade, the involvement of STAT3 is critical for IL-17-induced chemokine expression [153].

Besides the role of activated Src/PI3K, MAPKs, STAT3, and PKC by IL-17 in pro-inflammatory gene expression, it can also result in carcinogenesis and change of the microenvironment of tumors and promote autoimmune disease [12, 17, 43, 47, 89, 91]. C/EBP which is known to be a critical regulator of IL-6 transcription was found to be upregulated by IL-17 [148]. The promoters of genes stimulated by IL-17A are enriched for C/EBP-binding elements, and for some genes, it needs both C/EBP and NF-kB to activate transcription [159, 189]. There are six membranes in C/EBP family, but only C/EBP-δ and C/EBP-β are reported to respond to IL-17 stimulation [15, 161]. Despite both C/EBP-δ and C/EBP-β which are regulated by IL-17 in the transcriptional/posttranscriptional level, the underlying molecular mechanisms between them are different. The induction of C/EBP-δ requires SEFIR, ACT1, and TILL domains of IL-17RA; in addition, *C/EBP-δ* gene is capable of auto-enhanced, as it contains a functional C/EBP-binding element. In contrast, the regulation of C/EBP-β expression is more complex. C/EBP-β has three isoforms, LAP*(liver-enriched activator protein), the largest isoform of all; LAP, the shortened form; and LIP (liver-enriched inhibitory protein) which serves as a dominant-negative inhibitor of transcription. IL-17 preferentially induces LAP*. Induction of LAP* requires the SEFIR, TILL, and CBA (C/EBP-β-activation) domains of IL-17RA. C/EBP-β and the reconstitution experiment showed that C/EBP-δ shares some overlapping functions. In the cells deficient in both C/EBP-β and C/EBP-δ, reconstituting with either transcription factor can restore IL-17A-dependent induction of IL-6 and lipocalin-2 expressions [148]. C/EBP is a relatively late transcriptional factor activated by IL-17A in the comparison of NF-kB. In terms of IL-6 activation induced by IL-17A, C/EBP-β and C/EBP-δ seem functionally redundant, but it is not a universal event in all IL-17A-induced genes. The recent research showed that mutant IL-RA can activate C/EBP-δ but not C/EBP-β, resulting in impaired induction of some but not all IL-17 target genes, indicating that some target gene transcription may have a preference for C/EBP-β.

AP-1 (activator protein-1) is a sequence-specific transcriptional activator which is a heterodimeric protein composed of proteins belonging to the c-Fos, c-Jun, and ATF families. Numerous literatures have delineated the critical role of AP-1 in response to various stimuli, especially in the transduction of MAPK signal. Recently, studies showed that AP-1 binding site is also enriched in IL-17 target gene promoters; however, in the case of IL-6 activation, AP-1 site in the promoter is dispensable for IL-17-mediated activity [148, 159].

Since the IL-17/IL-17R signaling pathway has strong effect on induction of many pro-inflammatory cytokines, its activity needs to be strictly controlled to prevent inflammatory disorders. The search for negative regulators in IL-17/IL-17R cascade revealed the mechanisms of controlling its signal transduction. Activation of ERK upon IL-17 stimulation phosphorylates C/EBP-β at Thr^{188} which generated a docking site for GSK3β (glycogen synthase kinase 3-β). Following the second phosphorylation of C/EBP-β at Thr^{179} by GSK3β, the dual phosphorylated C/EBP-β activates itself to suppress IL-17-induced downstream gene activation [160]. TRAF3 downregulates IL-17-mediated gene expression by interfering with the formation of IL-17R/Act1-TRAF6 complex; the function of TRAF3 which serves as a negative regulator of IL-17 signaling pathway has also been illustrated in vivo by an EAE model [203].

PI3K is well known as an upstream kinase in PI3K/AKT/mTOR pathway; studies also show its ability on inhibiting IL-17 signaling transduction. Blockade of PI3K pathway could lead to elevated expression of IL-17RA, which additionally strengthened IL-17 signaling pathway [105]. USP25 and A20 can both restrict the ubiquitination of TRAF6, thereby natively regulating IL-17 signaling pathway. USP25 is a deubiquitinating enzyme; deficiency in USP25 resulted markedly upregulation of IL-17-inducible genes in vitro and pulmonary inflammation in vivo [201]. A20 is encoded by *TNFAIP* (TNFα-induced protein 3) and can directly interact with IL-17RA and restrict IL-17-dependent activation of NF-kB and MAPKs [41].

In addition to proteins, microRNA also participates in the controlling of IL-17 signaling pathway; miR-13b interacts with upstream activators, such as IKKα, Tab2, and Tab3, to impair the activity of NF-kB [61, 204]. When under long-period presence of IL-17, it led to the induction of TrCP (β-transducin repeat-containing protein), which will subsequently mediate the ubiquitination of Act1 for its degradation [162]. Interestingly, IL-17RD is also involved in the regulation of IL-17 signaling. IL-17RD disrupts the binding between Act1 and TRAF6, thus suppressing NF-kB activation. In addition to IL-17RD, IL-17 itself also exerts the modulation activity to control its signaling pathway by triggering the ubiquitylation and degradation of IL-17RA [145].

7.4 IL-17 Structure and Function

The IL-17 family consists of a subset of cytokines that participate in both host defense and autoimmune diseases. IL-17A was discovered by a subtractive hybridization screen in a rodent T cell library, based on amino acid sequence homology; genomic sequencing analysis identified five other members of this family, IL-17B, IL-17C, IL-17D, IL-17E, and IL-17F [66, 96, 100, 146, 167]. It also shows that an open reading frame of T cell-tropic gammaherpesvirus, in herpesvirus saimiri, shared 58 % homology with IL-17A, was then named vIL-17.

The IL-17 family members have unique amino acid sequence that shares no similarity with other known cytokines; among the members, IL-17F has the highest homology to IL-17A (60 %), and IL-17E has the least (17 %), suggesting the functional difference of these members [7]. Despite the sequence difference, they are all well conserved in mammalians; the homologues in mouse show 62–88 % similarity compared to those in human.

The gene-coded IL-17A is located at chromosome 6, except IL-17F gene, other member's gene is located at different chromosomes, and gene-coded IL-17B,IL-17C, IL-17D, and IL-17E are located at chromosomes 5, 16, 13, and 14, respectively, suggesting the different activation and regulation mechanisms in IL-17 family cytokines.

IL-17A contains 155 amino acids; they form homodimer which is linked by disulfide bond. Each subunit of the homodimer is approximately 15–20 kDa, with a molecular mass of 35 kDa in total. The structure of IL-17 consists of an N-linked glycosylation site, a signal peptide of 23 amino acids and 123 amino acid chain region characteristic of the IL-17 family. Subsequent study compared the different members of the IL-17 family and revealed that there are four conserved cysteines that form two disulfide bonds [88, 193]. The crystal structure of IL-17A and IL-17F showed that they both present cysteine-knot fold architecture [66]. In IL-17 family members, the cysteine-knot fold is characterized by two sets of paired β-strands stabilized by two disulfide interactions. IL-17A and IL-17F not only form homodimer, but they also can form heterodimer with each other; in some cell types like human peripheral blood mononuclear cells, IL-17A/IL-17F heterodimer is produced at a higher level compared to IL-17A homodimer [183]. Either IL-17A and IL-17F homodimer or IL-17A/IL-17F heterodimer, they all activate downstream signaling pathway through the same receptor subunit composed of IL-17RA and IL-17RC [90, 182].

Owing to the different similarities among the IL-17 family, they exert overlapped yet distinct functions. The IL-17 family of cytokines play essential roles in host defense against microbial pathogens by rapidly induction of pro-inflammatory cascade of chemokines and cytokines that further activate and recruit neutrophils and monocytes required for timely control of the pathogen by the immune system [27]. IL-17RA knockout mice showed high mortality in intrapulmonary *Klebsiella pneumoniae* infections and demonstrate that the IL-17 family is a major contributor in eradication of invading pathogens [194]. Subsequent characterization showed that the effect of IL-17 on *K. pneumoniae* is through expansion and maintenance of Th17 population by the induction of IL-23 [50, 51]. The efficient response to IL-17 builds up the antibacterial barriers at epithelial and mucosal, for example, IL-17 efficiently restricted *S. aureus* and *C. rodentium* which infect the skin and colon as well as other bacteria including *Salmonella enterica* [138], *Streptococcus pneumoniae* [108], *Listeria monocytogenes* [184], *Staphylococcus aureus* [20], *Helicobacter pylori* [3], and *Citrobacter rodentium* [102]. In the infection of *S. aureus* model, IL-17A and IL-17F showed redundant function; animals become more susceptible to the bacterium only when both cytokines are defective [71]. However, in the case of *C. rodentium* infection, the two cytokines are both required to trigger effective protection effects. Deficiency in IL-17A or IL-17RA led to increased bacterial dissemination, correlating with decreased inflammatory gene expression and neutrophil recruitment.

IL-17 not only exerts antibacterial activity, but it is also involved in antiviral signaling through activation of Act1 [150]. Utilizing gene modulating and neutralizing antibody technologies, it showed that IL-17A is required in controlling fungal infection. Under the infection of *Pneumocystis carinii*, blockade of IL-17A resulted in increased pathogen burden, exacerbated the disease, and lowered the survival rate [149, 151]. Genetic mutation in *STAT3* gene develops hyper-IgE syndrome in human due to impaired IL-17A and IL-17F production. Those patients suffer from high susceptibility to extrocellular pathogen infection such as S. *aureus*, *C. albicans, and Streptococcus pneumoniae*. These observations provide more evidence on the important role of IL-17 in the immune response against foreign pathogens [119]. Different from the main source of IL-17 come from innate immune cells when encounter to bacteria and viruses, B cells becoming a major source in the combat of parasites. In *Trypanosoma cruzi* parasite-infected mice, principal IL-17-producing cells found in spleen are B cells, which mediate pathogen clearance [34]. The protective role of IL-17 also extends against the infection of other parasites such as *Leishmania infantum* and *Nippostrongylus brasiliensis* [124, 200].

The protective role of IL-17 is mainly through induction of pro-inflammatory cytokines and chemokines on the infection site. The induction of IL-6 facilitates Th17 differentiation and suggested a positive feedback circuit induced by IL-17 [128]. Furthermore, induction of other cytokines, such as TNFα and IL-1β, helps to amplify the signal and induce more inflammatory factors to the infection site. NO and PGE_2 are two mediators of inflammation, which could also be increased by IL-17 via its induction of NOS (NO synthase) and COX (cyclooxygenase) [97, 177].

Upon stimulation of IL-17, a set of chemokines can be induced including CXCL1 [GRO (growth-related oncogene)-α], CXCL5, CXCL8 (IL-8), and CCL20 (MIP3a) [56, 81, 94, 157]. CXCL1, CXCL5, and CXCL8 have the ability to attract neutrophils, while CCL20 recruits more IL-17-producing cells. In addition to induce the production of cytokines and chemokines against pathogens, IL-17 can also induce the expression of many antimicrobial peptides, such as β-defensins and S100 proteins to further recruit DCs and T cells [40, 83, 102, 188].

Although crucial in host defense against various invading pathogens, dysregulated IL-17 production can lead to acute chronic inflammation, which further resulted in tissue damage and autoimmunity. The IL-17 family of cytokines have been link to many autoimmune diseases, such as MS (multiple sclerosis), RA (rheumatoid arthritis), systemic lupus erythematosus (SLE), IBD (inflammatory bowel disease), and psoriasis.

MS is characterized by autoimmune inflammation in the CNS. IL-17A/IL-17F disrupts the blood-brain barrier tight junction, which results in the local migration of CD4+ T cells [93]. IL-17A and IL-17F enhance the production of IL-23 and other cytokines that promote Th17 cell development and accumulation, causing severe pathogenesis [172]. Th17 cell as major cause of central nervous system inflammation and lesion formation has also been confirmed in an EAE model [92, 132]. Furthermore, IL-17-deficient mice show resistance to the induction of EAE [190].

The pathogenesis of RA is characterized by infiltration of CD4+ T cells and autoantibody-producing plasma cells. IL-17A can be detected in the synovial fluids and synovium of RA patients [111]. In a collagen-induced arthritis model, the critical role of IL-17A in the progression of disease has been demonstrated. γδ T cells and CD4+ T cell secrete IL-17 which affect innate immune cells and Th17 cells. Instead of Th17 cells, mast cells are the main source of IL-17 in humans synovium affected by rheumatoid arthritis [63]. Neutralizing IL-17 in CIA mouse models decreases joint inflammation, cartilage destruction, and bone erosion [110], whereas overexpression of IL-17 promotes collagen arthritis and accelerates joint destruction [109]. In addition, IL-17-deficient mice display resistance to CIA, confirming the essential role of IL-17 in the generation and development of RA [123].

SLE is characterized by autoantibody production. SLE patients develop immune responses against self-antigens which cause local inflammation and tissue damage [23, 25]. In SLE patients, the IL-17 level in serum is elevated; the pathological role of IL-17 has also been confirmed in a spontaneously lupus-like disease development MRL/lpr mice mode [30, 181, 199]. IL-17 synergizes with IFNγ and IL-1β causing inflammation; it can also synergize with BAFF to prolong B cell survival, thus increasing the number of autoantibody-producing cells, causing severe SLE pathogenesis [24, 46].

Psoriasis is a chronic skin disease, characterized by dermal hyperplasia. IL-17 cooperative with IFNγ, TNFα, IL-22, and IL-23 increases the production of inflammatory cytokines, chemokines, and antimicrobial peptides, leading to severe infiltration of immune cells.

IBD is a chronic relapsing inflammatory disorder of the gastrointestinal tract, developed with chronic inflammation. In TNBS-induced colitis, using gene deletion and overexpression methods, IL-17 shows a correlation with colitis pathogenesis, indicating the crucial role of IL-17 in TNBS-induced colitis pathogenesis. IL-17 pathogenic role has also been confirmed in dextran sulfate sodium (DSS)-induced colitis [72, 127]. In humans, the expression of IL-17 was elevated in patients with either CD or UD compared to healthy subjects or to patients with infectious or ischemic colitis, suggesting the essential role of IL-17 in IBD [37, 169, 185]. Studies showed that Th17 and IL-17 are involved in autoimmune diabetes in animal models, while IL-17 neutralization experiment further confirmed its key role in the development of autoimmune diabetes [32, 75]. IL-17 also promotes the pathology in viral-induced inflammation. HSV-1 infection of the cornea can lead to SK (stromal keratitis), a blinding immune-inflammatory lesion of the eye. In IL-17R knockout mice or IL-17 neutralization in wild-type mice, HSV infection showed diminished SK severity, indicating the responsive role of IL-17/IL-17R in the pathogenesis of SK.

Besides the activity to induce the cytokines and chemokines, IL-17 has also the ability to induce the production of other genes to fulfill other IL-17-mediated physiological functions. For example, IL-17 not only protects mucosal barrier during pathogen infections but also homeostasis by enhancing its integrity. IL-17 treatment increases the synthesis of tight junction proteins like claudin-1 and claudin-2 to forming tight structures, thereby connecting epithelial cells to form a network [59]. IL-17 also plays a pathological role in tissue damage, bone destruction, and tumorigenesis through induction of tissue-remodeling factors MMPs (matrix metalloproteinases) including MMP1, MMP3, MMP9, and MMP13 [36, 45, 120].

IL-17 not only plays critical roles in host defense but also in the pathogenesis of other inflammation-related diseases such as allergy, transplantation, obesity, and malignancy [52, 84, 156, 166, 187].

The IL-17 family of cytokines exert critical functions associated with many responses and diseases related to whole organism. In addition, the members in this family could have both redundant and distinct roles. For example, IL-17A and IL-17F showed the overlapping function in reducing certain cytokines; however, they play a distinct role in causing autoimmune disease. In the EAE model, only IL-17A knockout mice display significant reduction in disease score, but not IL-17F knockout mice. In addition, IL-17A promotes inflammation in an asthma model, while IL-17F showed suppressive function [71, 190]. The regulation of such a delicate signal transduction network requires strict control and precise modulation.

7.5 Transcription Regulation of IL-17

Due to the fundamental role of IL-17 in immune cell development and its association with immune diseases, increasing studies aim to illustrate the regulation mechanisms of IL-17 transcription. RORγt (retinoic acid-related orphan nuclear receptor) which is encoded by the *RORc* gene is a unique lineage-specific transcription factor [73, 191]. RORγt is highly expressed on CD4+CD8+ thymocytes and is required for the development of Th17 cells, LTi, and LTi-like cells [31, 171]. ROR response elements (ROREs) contain the consensus core motif AGGTCA preceded by a 5 bp A-/T-rich sequence [76]; the search for potential ROREs within the IL-17 locus revealed a conserved RORγt binding site within the 200 bp core promoter region, suggesting RORγt could be a direct activator to IL-17 [73]. Recent research has been shown that RORγt is a strong activator for both IL-17A and IL-17F transcription. Mice with RORγt-deficient T cells showed attenuated symptoms of autoimmune disease and lack of infiltrating Th17 cells in the tissue, confirming the essential role of RORγt in IL-17 production and Th17 cell development. Another member of ROR family, RORα, also participates in the activation of IL-17 transcription; however, subsequent study shows that it seems and plays redundant and synergistic roles with RORγt. Different Th cells show a signature expression of their definite regulation factor. For example, Th1 cells are defined by the expression of T-bet (Tbx21), Th2 cells by GATA3. In the case of Th17 cells, RORγt is its master regulator. RORγt plays a determining role in IL-17 transcription that is also due to lots of other regulators which indirectly regulate IL-17 transcription by modulating RORγt transcription as shown in Fig. 7.2.

1.1-kb promoter fragment containing the RORγt binding site did not efficiently initiate IL-17 transcription compared to a 2-kb fragment; the analysis toward this discrepancy discovered runt-related transcription factor 1 (Runx1). Runx1 regulates IL-17 transcription by direct binding to and acting synergically with RORγt to promote IL-17 transcription [106, 198]. Subsequently study showed that Runx2 can also promote IL-17 transcription by binding to the promoter and enhance regions. The optimal transcription of gene encoding IL-17 not only requires the 2-kb promoter but also CNS5. Both *cis*-regulatory elements contained regions have the ability to bind the transcription factors RORγt and Runx1, further confirmed the central role of this two transcriptional factors in the activation of IL-17 transcription [101].

SIRT1 is protein deacetylase which serves as an epigenetic regulator that has effects on the modulation of several immune function-associated transcription factors. In the case of IL-17 transcription, SIRT1 promotes the transcription by enhancing RORγt transcriptional activity. SIRT1-deficient mice or mice treated with SIRT1 inhibitor displayed a less severe disease in MS mice model and suppressed Th17 differentiation [104].

BATF is a member of the AP-1 family of transcription factors, has been shown is a positive regulator of IL-17 transcription. BATF promotes IL-17 production and controls Th17 cell differentiation by upregulating RORγt. Mice with BATF deficiency fail

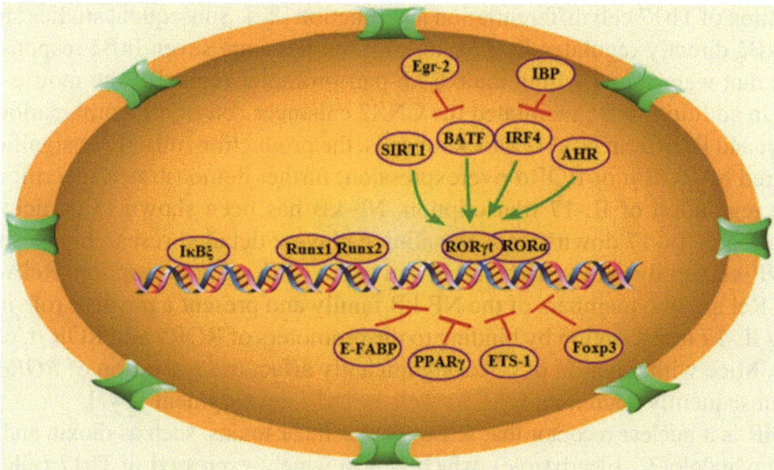

Fig. 7.2 Transcriptional and regulatory factors involved in IL-17 transcription. In *IL17* gene enhancer and promoter regions, it contains transcriptional factors binding site. RORγt, Runx1, and IkBξ directly regulate IL-17 transcription by binding at the enhancer and promoter regions. SIRT1, BATF, IRF4, and AHR enhance IL-17 transcription by increasing RORγt expression and activity, whereas Egr-2, IBP, E-FABP, PPARγ, ETS-1, and Foxp3 suppress IL-17 transcription by decreasing RORγt expression and blocking the binding of it to *IL17* gene enhancer and promoter

to induce RORγt and IL-17 expression. Reconstitutive experiment further showed that reconstitution of RORγt in BATF-deficient mice can only partially restore IL-17 expression, suggesting that optimized transcription of IL-17 requires the participation of BATF [70]. In addition, IRF4 can cooperatively bind with BATF; the complex interacts with STAT3 and modulates the transcription of IL-17 [21]. The role of IRF4 has also been demonstrated in vivo; in IRF4-deficient mice, RORα and RORγt are largely decreased, indicating impaired transcription of IL-17. The activation of IRF4 is through phosphorylation by ROCK; activated IRF4 translocates into nuclei and binds to IL-17 promoter [10].

STAT3 can be activated by cytokines including IL-23 and IL-6; activated STAT3 directly binds to IL-17 promoter. The binding site of STAT3 is present in both promoter regions of both IL-17A and IL-17F. The activation of STAT3 is negatively regulated by Socs3; in the absence of Socs3, the phosphorylation of STAT3 is markedly enhanced. STAT3 can also be activated by IL-6, to form IL-6/IL-6R/STAT3 complex regulating IL-17 transcription, partially through RORγt. In the collagen-induced arthritis mice model, blockade of JAK-STAT3 pathway shows decreased Th17 differentiation. KLF4 (Kruppel-like factor 4) showed a RORγt-independent regulation role in IL-17 transcription by direct binding with IL-17 promoter [4, 95].

IkBξ is a nuclear protein encoded by *Nfkbiz* gene, expressed in many immune cells such as, macrophages, DCs, and Th17 cells. Among these cells, Th17 cells showed the highest expression of *Nfkbiz* gene, indicating an important role in the regulation of Th17 cell differentiation and function [87]. Subsequent studies showed that IkBξ directly regulates IL-17 transcription; there are seven IkBξ response elements that were found within the 6.6-kb-pair promoter region of the mouse *IL17a* gene. In addition, IkBξ facilitated the CNS2 enhancer activity in combination with RORγt and RORα. In Nfkbiz knockout cells, the production of IL-17 is significantly impaired by RORγt or RORα overexpression, further demonstrated its critical role in the regulation of IL-17 transcription. NF-kB has been shown to participate in regulation of IL-17 downstream signaling pathway; detailed research also showed that NF-kB family is also involved in activation of IL-17 transcription. RelA (p65) and c-Rel are two members of the NF-kB family and present a positive role in activating IL-17 transcription by binding to the promoters of RORγ and RORγt, respectively. Mice with deletion of c-Rel significantly affect the activation of *RORc* gene and subsequently decreased EAE as well as Th17 development [147].

AHR is a nuclear receptor that senses extracellular toxins, such as dioxin and FICZ (6-formylindolo[3,2-b]carbazole), which is also widely expressed in Th17 cells, γδ T cells, LTi cells, and NK cells, suggesting a potential role in regulation of pro-inflammatory gene transcription [140, 155, 195]. Subsequent studies showed that AHR has cooperation activity with RORγt in the induction of maximal amount of IL-17 [202]. The function of AHR is further illustrated by its activity to suppress STAT1- and STAT5-mediated pathway, thereby strengthening IL-17/IL-23 and RORγt-dependent signaling.

Foxp3 is a major negative regulator of IL-17 transcription which downregulates IL-17 transcription by directly interacting with RORγt and suppressing RORγt-mediated IL-17 promoter activation. The exon 2 region and FKH (forkhead) domain were required for inhibiting the activity of RORγt activation of IL-17 transcription. Overexpressed foxp3 induces significant reduction of IL-17 expression, further confirmed the fundamental role of foxp3 which negatively regulates IL-17 transcription [19, 67]. In addition, foxp3 can also downregulate IL-17 transcription by interacting with Runx1 [198]. Another negative regulator T-bet which is known as a Th1-lineage-specific transcriptional factor can also suppress IL-17 production by binding to Runx1. The complex blocks the transactivation of *RORc* gene, thus impairing IL-17 secretion. It showed a dual regulation effect of Runx1; as described earlier, Runx1 has the ability to upregulate IL-17 when associated with RORγt, whereas downregulate IL-17 when complexed with T-bet/Foxp3 [44].

ETS-1 (v-ets erythroblastosis virus E26 oncogene homologue 1) exerts negative regulation of IL-17 by its ability to suppress Th17 differentiation through interacting with T-bet. Detailed study showed that ETS-1 knockout mice have increased IL-17 expression without affecting RORγt, suggesting its suppressive role in IL-17 expression. IRF8 interferes with IL-17 production by interacting with RORγt and also decreased Th17-associated gene expression, causing a suppressed efficiency in amplifying IL-17 production signal [130, 135].

PPARγ (peroxisome proliferator-activated receptor γ) also acts as an intrinsic suppressor of IL-17 transcription by affecting RORγt expression. It downregulates RORγt by preventing the removal of repressor complexes from RORγt promoter, resulting in decreased IL-17 production and Th17 differentiation [86]. Egr-2 (early growth response gene-2) is also involved in the regulation of IL-17 expression by interacting with BATF that suppresses its binding with IL-17 promoter [117]. IBP (IRF-4-binding protein) decreases IL-17 transcription by resistance of released IRF4, an efficient IL-17 expression activator. The role of IBP has also been illustrated in its deficiency in mice; the deficiency leads to a rapid development of rheumatoid arthritis-like joint disease [18]. It has been illustrated that IL-6 can induce IL-21 production, which together stimulates the expression of RORγt and RORα, to activate IL-17 transcription. E-FABP blocks the IL-21 expression upon IL-6 stimulation and further reduced expression of RORγt and RORα in CD4+ T cells, causing reduced production of IL-17 [65, 99].

The regulation of differentiation of THi cells is modulated by epigenetic modifications on *IL-17* locus. During IL-17 transcription, chromatin undergoes extensive changes, and remodeled chromatin spreads through the entire locus, thereby providing accessibility of IL-17 gene promoter and enhancer regions. Under THi-polarizing conditions, histone acetylation initiation is enhanced by TGF-β and IL-6, followed by histone H3 at the IL-17 promoter regions which becomes hyperacetylated and trimethylated at Lys^4 [2]. Etv5, an Ets-family transcription factor, plays a positive regulation of IL-17 transcription. Etv5 upregulates IL-17 transcription by recruiting histone-modifying enzymes to the IL-17 locus, creating an increased active histone and decreased repressive histone environment. Etv5-deficient mice showed reduced IL-17 production, suggesting that it is required by efficient IL-17 secretion [134]. In addition, overexpression of S1P1 (sphingosine-1-phosphate) showed increased IL-17 production as well as Th17 differentiation and showed its positive regulation role in IL-17 production [103]. There are also several conserved noncoding sequences within the IL-17 locus associated with hyperacetylated histone 3 and thus may act as potential regulatory regions. The transcription of IL-17 can also be regulated by microRNAs, CNSs, and cytokines [113, 115, 179].

A set of cytokines such as IL-6, IL-1β, IL-21, IL-23, and TGF-β can activate RORγt and lead to IL-17 transcription. The regulation of IL-17 expression also exerts on posttranscriptional level. It has been shown that HuR decreases IL-17 steady-state mRNA and protein levels by directly binding to the IL-17 mRNA 39 untranslated regions, causing reduced frequency of IL-17-positive cells in Th17 cells which are from HuR knockout mice.

7.6 Other Members of the IL-17 Family

IL-17E, also known as IL-25 is the most distinct cytokine in the IL-17 family and exerts unique functions in regulating immune responses [96]. IL-17E is widely expressed in many immune and nonimmune cells such as macrophage, dendritic cells, and mast cells and eosinophil, epithelial, and endothelial cells [8, 28, 68, 85, 180]. IL-17E is also involved in host defense, however, different from the function of IL-17A and IL-17F playing a protective role in eliminating bacterial and fungal infections; IL-17E is mostly involved in Th2-mediated host defense against helminth and parasitic infections by inducing cytokines including IL-4,IL-5, and IL-13 for the recruitment of eosinophil [35, 64].

Similar to IL-17 A and IL-17F, IL-17E can also potentiate allergic inflammation and involved in the pathogenesis of multiple autoimmune diseases [85]. The innate immune responder cells of IL-17E include monocytes and NKT cells; however, the mostly targeted cells of IL-17E are hematopoietic cells. These cells produce an amount of Th2 cytokines upon stimulation [14, 168, 175]. IL-17E initiates downstream signaling pathway through binding to IL-17RA/IL-17RB heterodimeric receptor complex [6, 144]. Similarly, Act1 is participated in the signal transduction by homotypic interactions of the SEFIR domain [22, 173, 174]. The activation of MAPK family pathways is in TRAF6-dependent and TRAF6-independent manners [112, 197]. It has been shown that IL-17E induces caspase-mediated apoptosis through the death domain (DD) within IL-RB. In breast cancer cells, IL-17E treatment induces FAS-associated protein with death domain (FADD) and TNF-R1-associated death domain protein (TRADD) to interact with IL-17RB, confirming IL-17E-induced apoptosis function [38].

IL-17C was found mainly secreted by epithelial cells, which play a protective role in the gastrointestinal tract by inducing cytokines and antimicrobial peptide production [139, 165]. Mice lacking IL-17C showed impaired anti-pathogen activity due to reduction of cytokine, chemokine, and antimicrobial peptide secretion, suggesting the importance of IL-17C in maintaining intestinal barrier and homeostasis [141].

Similar to IL-17A, in addition of host defense functions, IL-17C also causes autoimmune disease such as psoriasis and MS [16, 78–80]. Mice lacking IL-17C are partially resistant to EAE and attenuated imiquimod-induced skin inflammation [16]. IL-17C binds to the IL-17RA/IL-17RE receptor complex to trigger the downstream signal transduction. The cascades involving IL-17C signaling are similar to those of IL-17A such as the activation of Act1, NF-kB, and MAPK [16, 78].

IL-17B and IL-17D are less studied as IL-17A. IL-17B is highly expressed in chondrocytes and neurons and is also detected in immune cell line THP-1 [74, 100, 163]. IL-17B manipulates immune response and accelerates autoimmune disease progress by inducing the production of TNFα, IL-1β, and other pro-inflammatory cytokine production through binding to IL-17RB [96, 186]. IL-17D was in endothelial cells, and resting CD4+T cell and CD19+B cells, the major function of IL-17D, are the induction of pro-inflammatory genes.

The fast-growing number of literature characterized the cellular and molecular mechanisms toward IL-17 family cytokine production, regulation, and signaling transduction, which provide an insight for the development of therapeutics and drugs for IL-17-mediated diseases. It has been shown that many molecules targeted IL-17-cascade such as IL-17 antagonists and monoclonal antibodies are currently under evaluation for the effects on IL-17-associated disease [69]. Further dissection of the functional roles in mediating immune response and the regulation of IL-17 family of cytokines' expression and signaling mechanisms may provide us detailed information for understanding of cross talk of innate and adaptive immune response and more specific targets and candidates to fulfill the future drug pipeline in infectious and inflammatory diseases.

References

1. Aggarwal S, Gurney AL. IL-17: prototype member of an emerging cytokine family. J Leukoc Biol. 2002;71:1–8.
2. Akimzhanov AM, Yang XO, Dong C. Chromatin remodeling of interleukin-17 (IL-17)-IL-17F cytokine gene locus during inflammatory helper T cell differentiation. J Biol Chem. 2007;282:5969–72.
3. Algood HM, Allen SS, Washington MK, Peek Jr RM, Miller GG, Cover TL. Regulation of gastric B cell recruitment is dependent on IL-17 receptor A signaling in a model of chronic bacterial infection. J Immunol. 2009;183:5837–46.
4. An J, Golech S, Klaewsongkram J, Zhang Y, Subedi K, Huston GE, Wood 3rd WH, Wersto RP, Becker KG, Swain SL, et al. Kruppel-like factor 4 (KLF4) directly regulates proliferation in thymocyte development and IL-17 expression during Th17 differentiation. FASEB J. 2011;25:3634–45.
5. Anderson P. Post-transcriptional control of cytokine production. Nat Immunol. 2008;9:353–9.
6. Angkasekwinai P, Chang SH, Thapa M, Watarai H, Dong C. Regulation of IL-9 expression by IL-25 signaling. Nat Immunol. 2010;11:250–6.
7. Auger KR, Serunian LA, Soltoff SP, Libby P, Cantley LC. PDGF-dependent tyrosine phosphorylation stimulates production of novel polyphosphoinositides in intact cells. Cell. 1989;57:167–75.
8. Becker S, Mundandhara S, Devlin RB, Madden M. Regulation of cytokine production in human alveolar macrophages and airway epithelial cells in response to ambient air pollution particles: further mechanistic studies. Toxicol Appl Pharmacol. 2005;207:269–75.
9. Bermejo DA, Jackson SW, Gorosito-Serran M, Acosta-Rodriguez EV, Amezcua-Vesely MC, Sather BD, Singh AK, Khim S, Mucci J, Liggitt D, et al. Trypanosoma cruzi trans-sialidase initiates a program independent of the transcription factors RORgammat and Ahr that leads to IL-17 production by activated B cells. Nat Immunol. 2013;14:514–22.
10. Biswas PS, Gupta S, Chang E, Song L, Stirzaker RA, Liao JK, Bhagat G, Pernis AB. Phosphorylation of IRF4 by ROCK2 regulates IL-17 and IL-21 production and the development of autoimmunity in mice. J Clin Invest. 2010;120:3280–95.
11. Blonska M, Lin X. CARMA1-mediated NF-kappaB and JNK activation in lymphocytes. Immunol Rev. 2009;228:199–211.
12. Brereton CF, Sutton CE, Lalor SJ, Lavelle EC, Mills KH. Inhibition of ERK MAPK suppresses IL-23- and IL-1-driven IL-17 production and attenuates autoimmune disease. J Immunol. 2009;183:1715–23.
13. Bulek K, Liu C, Swaidani S, Wang L, Page RC, Gulen MF, Herjan T, Abbadi A, Qian W, Sun D, et al. The inducible kinase IKKi is required for IL-17-dependent signaling associated with neutrophilia and pulmonary inflammation. Nat Immunol. 2011;12:844–52.
14. Caruso R, Stolfi C, Sarra M, Rizzo A, Fantini MC, Pallone F, MacDonald TT, Monteleone G. Inhibition of monocyte-derived inflammatory cytokines by IL-25 occurs via p38 Map kinase-dependent induction of Socs-3. Blood. 2009;113:3512–9.
15. Chang SH, Park H, Dong C. Act1 adaptor protein is an immediate and essential signaling component of interleukin-17 receptor. J Biol Chem. 2006;281:35603–7.
16. Chang SH, Reynolds JM, Pappu BP, Chen G, Martinez GJ, Dong C. Interleukin-17C promotes Th17 cell responses and autoimmune disease via interleukin-17 receptor E. Immunity. 2011;35:611–21.
17. Chen J. The Src/PI3K/Akt pathway may play a key role in the production of IL-17 in obesity. J Leukoc Biol. 2010;87:355. author reply 357.
18. Chen Q, Yang W, Gupta S, Biswas P, Smith P, Bhagat G, Pernis AB. IRF-4-binding protein inhibits interleukin-17 and interleukin-21 production by controlling the activity of IRF-4 transcription factor. Immunity. 2008;29:899–911.

19. Chen Z, Lin F, Gao Y, Li Z, Zhang J, Xing Y, Deng Z, Yao Z, Tsun A, Li B. FOXP3 and RORgammat: transcriptional regulation of Treg and Th17. Int Immunopharmacol. 2011;11:536–42.
20. Cho JS, Pietras EM, Garcia NC, Ramos RI, Farzam DM, Monroe HR, Magorien JE, Blauvelt A, Kolls JK, Cheung AL, et al. IL-17 is essential for host defense against cutaneous Staphylococcus aureus infection in mice. J Clin Invest. 2010;120:1762–73.
21. Ciofani M, Madar A, Galan C, Sellars M, Mace K, Pauli F, Agarwal A, Huang W, Parkurst CN, Muratet M, et al. A validated regulatory network for Th17 cell specification. Cell. 2012;151:289–303.
22. Claudio E, Sonder SU, Saret S, Carvalho G, Ramalingam TR, Wynn TA, Chariot A, Garcia-Perganeda A, Leonardi A, Paun A, et al. The adaptor protein CIKS/Act1 is essential for IL-25-mediated allergic airway inflammation. J Immunol. 2009;182:1617–30.
23. Crispin JC, Liossis SN, Kis-Toth K, Lieberman LA, Kyttaris VC, Juang YT, Tsokos GC. Pathogenesis of human systemic lupus erythematosus: recent advances. Trends Mol Med. 2010;16:47–57.
24. Crispin JC, Tsokos GC. Human TCR-alpha beta+ CD4–CD8- T cells can derive from CD8+ T cells and display an inflammatory effector phenotype. J Immunol. 2009;183:4675–81.
25. Crispin JC, Tsokos GC. IL-17 in systemic lupus erythematosus. J Biomed Biotechnol. 2010;2010:943254.
26. Cua DJ, Tato CM. Innate IL-17-producing cells: the sentinels of the immune system. Nat Rev Immunol. 2010;10:479–89.
27. Curtis MM, Way SS. Interleukin-17 in host defence against bacterial, mycobacterial and fungal pathogens. Immunology. 2009;126:177–85.
28. Dolgachev V, Petersen BC, Budelsky AL, Berlin AA, Lukacs NW. Pulmonary IL-17E (IL-25) production and IL-17RB+ myeloid cell-derived Th2 cytokine production are dependent upon stem cell factor-induced responses during chronic allergic pulmonary disease. J Immunol. 2009;183:5705–15.
29. Dong C, Nurieva RI. Regulation of immune and autoimmune responses by ICOS. J Autoimmun. 2003;21:255–60.
30. Doreau A, Belot A, Bastid J, Riche B, Trescol-Biemont MC, Ranchin B, Fabien N, Cochat P, Pouteil-Noble C, Trolliet P, et al. Interleukin 17 acts in synergy with B cell-activating factor to influence B cell biology and the pathophysiology of systemic lupus erythematosus. Nat Immunol. 2009;10:778–85.
31. Eberl G, Littman DR. Thymic origin of intestinal alphabeta T cells revealed by fate mapping of RORgammat+ cells. Science. 2004;305:248–51.
32. Emamaullee JA, Davis J, Merani S, Toso C, Elliott JF, Thiesen A, Shapiro AM. Inhibition of Th17 cells regulates autoimmune diabetes in NOD mice. Diabetes. 2009;58:1302–11.
33. Erbel C, Akhavanpoor M, Okuyucu D, Wangler S, Dietz A, Zhao L, Stellos K, Little KM, Lasitschka F, Doesch A, et al. IL-17A influences essential functions of the monocyte/macrophage lineage and is involved in advanced murine and human atherosclerosis. J Immunol. 2014;193:4344–55.
34. Erdmann H, Rossnagel C, Bohme J, Iwakura Y, Jacobs T, Schaible UE, Holscher C. IL-17A promotes macrophage effector mechanisms against Trypanosoma cruzi by trapping parasites in the endolysosomal compartment. Immunobiology. 2013;218:910–23.
35. Fallon PG, Ballantyne SJ, Mangan NE, Barlow JL, Dasvarma A, Hewett DR, McIlgorm A, Jolin HE, McKenzie AN. Identification of an interleukin (IL)-25-dependent cell population that provides IL-4, IL-5, and IL-13 at the onset of helminth expulsion. J Exp Med. 2006;203:1105–16.
36. Feng M, Wang Y, Chen K, Bian Z, Jinfang W, Gao Q. IL-17A promotes the migration and invasiveness of cervical cancer cells by coordinately activating MMPs expression via the p38/NF-kappaB signal pathway. PLoS One. 2014;9:e108502.
37. Fujino S, Andoh A, Bamba S, Ogawa A, Hata K, Araki Y, Bamba T, Fujiyama Y. Increased expression of interleukin 17 in inflammatory bowel disease. Gut. 2003;52:65–70.

38. Furuta S, Jeng YM, Zhou L, Huang L, Kuhn I, Bissell MJ, Lee WH. IL-25 causes apoptosis of IL-25R-expressing breast cancer cells without toxicity to nonmalignant cells. Sci Transl Med. 2011;3:78ra31.
39. Gaffen SL. Structure and signalling in the IL-17 receptor family. Nat Rev Immunol. 2009;9:556–67.
40. Ganz T. Defensins and host defense. Science. 1999;286:420–1.
41. Garg AV, Ahmed M, Vallejo AN, Ma A, Gaffen SL. The deubiquitinase A20 mediates feedback inhibition of interleukin-17 receptor signaling. Sci Signal. 2013;6:ra44.
42. Geremia A, Arancibia-Carcamo CV, Fleming MP, Rust N, Singh B, Mortensen NJ, Travis SP, Powrie F. IL-23-responsive innate lymphoid cells are increased in inflammatory bowel disease. J Exp Med. 2011;208:1127–33.
43. Gislette T, Chen J. The possible role of IL-17 in obesity-associated cancer. Sci World J. 2010;10:2265–71.
44. Gocke AR, Cravens PD, Ben LH, Hussain RZ, Northrop SC, Racke MK, Lovett-Racke AE. T-bet regulates the fate of Th1 and Th17 lymphocytes in autoimmunity. J Immunol. 2007;178:1341–8.
45. Greenlee KJ, Werb Z, Kheradmand F. Matrix metalloproteinases in lung: multiple, multifarious, and multifaceted. Physiol Rev. 2007;87:69–98.
46. Groom J, Mackay F. B cells flying solo. Immunol Cell Biol. 2008;86:40–6.
47. Gu FM, Li QL, Gao Q, Jiang JH, Zhu K, Huang XY, Pan JF, Yan J, Hu JH, Wang Z, et al. IL-17 induces AKT-dependent IL-6/JAK2/STAT3 activation and tumor progression in hepatocellular carcinoma. Mol Cancer. 2011;10:150.
48. Haas JD, Gonzalez FH, Schmitz S, Chennupati V, Fohse L, Kremmer E, Forster R, Prinz I. CCR6 and NK1.1 distinguish between IL-17A and IFN-gamma-producing gammadelta effector T cells. Eur J Immunol. 2009;39:3488–97.
49. Hamada S, Umemura M, Shiono T, Tanaka K, Yahagi A, Begum MD, Oshiro K, Okamoto Y, Watanabe H, Kawakami K, et al. IL-17A produced by gammadelta T cells plays a critical role in innate immunity against listeria monocytogenes infection in the liver. J Immunol. 2008;181:3456–63.
50. Happel KI, Dubin PJ, Zheng M, Ghilardi N, Lockhart C, Quinton LJ, Odden AR, Shellito JE, Bagby GJ, Nelson S, et al. Divergent roles of IL-23 and IL-12 in host defense against Klebsiella pneumoniae. J Exp Med. 2005;202:761–9.
51. Happel KI, Zheng M, Young E, Quinton LJ, Lockhart E, Ramsay AJ, Shellito JE, Schurr JR, Bagby GJ, Nelson S, et al. Cutting edge: roles of Toll-like receptor 4 and IL-23 in IL-17 expression in response to Klebsiella pneumoniae infection. J Immunol. 2003;170:4432–6.
52. Harley IT, Stankiewicz TE, Giles DA, Softic S, Flick LM, Cappelletti M, Sheridan R, Xanthakos SA, Steinbrecher KA, Sartor RB, et al. IL-17 signaling accelerates the progression of nonalcoholic fatty liver disease in mice. Hepatology. 2014;59:1830–9.
53. Hartupee J, Liu C, Novotny M, Li X, Hamilton T. IL-17 enhances chemokine gene expression through mRNA stabilization. J Immunol. 2007;179:4135–41.
54. Haudenschild D, Moseley T, Rose L, Reddi AH. Soluble and transmembrane isoforms of novel interleukin-17 receptor-like protein by RNA splicing and expression in prostate cancer. J Biol Chem. 2002;277:4309–16.
55. Haudenschild DR, Curtiss SB, Moseley TA, Reddi AH. Generation of interleukin-17 receptor-like protein (IL-17RL) in prostate by alternative splicing of RNA. Prostate. 2006;66:1268–74.
56. Herjan T, Yao P, Qian W, Li X, Liu C, Bulek K, Sun D, Yang WP, Zhu J, He A, et al. HuR is required for IL-17-induced Act1-mediated CXCL1 and CXCL5 mRNA stabilization. J Immunol. 2013;191:640–9.
57. Higgins SC, Jarnicki AG, Lavelle EC, Mills KH. TLR4 mediates vaccine-induced protective cellular immunity to Bordetella pertussis: role of IL-17-producing T cells. J Immunol. 2006;177:7980–9.

58. Hirota K, Duarte JH, Veldhoen M, Hornsby E, Li Y, Cua DJ, Ahlfors H, Wilhelm C, Tolaini M, Menzel U, et al. Fate mapping of IL-17-producing T cells in inflammatory responses. Nat Immunol. 2011;12:255–63.
59. Howe KL, Reardon C, Wang A, Nazli A, McKay DM. Transforming growth factor-beta regulation of epithelial tight junction proteins enhances barrier function and blocks enterohemorrhagic Escherichia coli O157:H7-induced increased permeability. Am J Pathol. 2005;167:1587–97.
60. Hsu HC, Yang P, Wang J, Wu Q, Myers R, Chen J, Yi J, Guentert T, Tousson A, Stanus AL, et al. Interleukin 17-producing T helper cells and interleukin 17 orchestrate autoreactive germinal center development in autoimmune BXD2 mice. Nat Immunol. 2008;9:166–75.
61. Hu R, O'Connell RM. MiR-23b is a safeguard against autoimmunity. Nat Med. 2012;18:1009–10.
62. Huang F, Kao CY, Wachi S, Thai P, Ryu J, Wu R. Requirement for both JAK-mediated PI3K signaling and ACT1/TRAF6/TAK1-dependent NF-kappaB activation by IL-17A in enhancing cytokine expression in human airway epithelial cells. J Immunol. 2007;179:6504–13.
63. Hueber AJ, Asquith DL, Miller AM, Reilly J, Kerr S, Leipe J, Melendez AJ, McInnes IB. Mast cells express IL-17A in rheumatoid arthritis synovium. J Immunol. 2010;184:3336–40.
64. Hurst SD, Muchamuel T, Gorman DM, Gilbert JM, Clifford T, Kwan S, Menon S, Seymour B, Jackson C, Kung TT, et al. New IL-17 family members promote Th1 or Th2 responses in the lung: in vivo function of the novel cytokine IL-25. J Immunol. 2002;169:443–53.
65. Hwang ES. Transcriptional regulation of T helper 17 cell differentiation. Yonsei Med J. 2010;51:484–91.
66. Hymowitz SG, Filvaroff EH, Yin JP, Lee J, Cai L, Risser P, Maruoka M, Mao W, Foster J, Kelley RF, et al. IL-17s adopt a cystine knot fold: structure and activity of a novel cytokine, IL-17F, and implications for receptor binding. EMBO J. 2001;20:5332–41.
67. Ichiyama K, Yoshida H, Wakabayashi Y, Chinen T, Saeki K, Nakaya M, Takaesu G, Hori S, Yoshimura A, Kobayashi T. Foxp3 inhibits RORgammat-mediated IL-17A mRNA transcription through direct interaction with RORgammat. J Biol Chem. 2008;283:17003–8.
68. Ikeda K, Nakajima H, Suzuki K, Kagami S, Hirose K, Suto A, Saito Y, Iwamoto I. Mast cells produce interleukin-25 upon Fc epsilon RI-mediated activation. Blood. 2003;101:3594–6.
69. Isailovic N, Daigo K, Mantovani A, Selmi C. Interleukin-17 and innate immunity in infections and chronic inflammation. J Autoimmun. 2015;60:1–11.
70. Ise W, Kohyama M, Schraml BU, Zhang T, Schwer B, Basu U, Alt FW, Tang J, Oltz EM, Murphy TL, et al. The transcription factor BATF controls the global regulators of class-switch recombination in both B cells and T cells. Nat Immunol. 2011;12:536–43.
71. Ishigame H, Kakuta S, Nagai T, Kadoki M, Nambu A, Komiyama Y, Fujikado N, Tanahashi Y, Akitsu A, Kotaki H, et al. Differential roles of interleukin-17A and -17F in host defense against mucoepithelial bacterial infection and allergic responses. Immunity. 2009;30:108–19.
72. Ito R, Kita M, Shin-Ya M, Kishida T, Urano A, Takada R, Sakagami J, Imanishi J, Iwakura Y, Okanoue T, et al. Involvement of IL-17A in the pathogenesis of DSS-induced colitis in mice. Biochem Biophys Res Commun. 2008;377:12–6.
73. Ivanov II, McKenzie BS, Zhou L, Tadokoro CE, Lepelley A, Lafaille JJ, Cua DJ, Littman DR. The orphan nuclear receptor RORgammat directs the differentiation program of proinflammatory IL-17+ T helper cells. Cell. 2006;126:1121–33.
74. Iwakura Y, Ishigame H, Saijo S, Nakae S. Functional specialization of interleukin-17 family members. Immunity. 2011;34:149–62.
75. Jain R, Tartar DM, Gregg RK, Divekar RD, Bell JJ, Lee HH, Yu P, Ellis JS, Hoeman CM, Franklin CL, et al. Innocuous IFNgamma induced by adjuvant-free antigen restores normoglycemia in NOD mice through inhibition of IL-17 production. J Exp Med. 2008;205:207–18.

76. Jetten AM, Kurebayashi S, Ueda E. The ROR nuclear orphan receptor subfamily: critical regulators of multiple biological processes. Prog Nucleic Acid Res Mol Biol. 2001;69:205–47.
77. Jin W, Dong C. IL-17 cytokines in immunity and inflammation. Emerg Microbes Infect. 2013;2:e60.
78. Johansen C, Riis JL, Gedebjerg A, Kragballe K, Iversen L. Tumor necrosis factor alpha-mediated induction of interleukin 17C in human keratinocytes is controlled by nuclear factor kappaB. J Biol Chem. 2011;286:25487–94.
79. Johansen C, Usher PA, Kjellerup RB, Lundsgaard D, Iversen L, Kragballe K. Characterization of the interleukin-17 isoforms and receptors in lesional psoriatic skin. Br J Dermatol. 2009;160:319–24.
80. Johnston A, Fritz Y, Dawes SM, Diaconu D, Al-Attar PM, Guzman AM, Chen CS, Fu W, Gudjonsson JE, McCormick TS, et al. Keratinocyte overexpression of IL-17C promotes psoriasiform skin inflammation. J Immunol. 2013;190:2252–62.
81. Jones CE, Chan K. Interleukin-17 stimulates the expression of interleukin-8, growth-related oncogene-alpha, and granulocyte-colony-stimulating factor by human airway epithelial cells. Am J Respir Cell Mol Biol. 2002;26:748–53.
82. Jovcic G, Bugarski D, Krstic A, Vlaski M, Petakov M, Mojsilovic S, Stojanovic N, Milenkovic P. The effect of interleukin-17 on hematopoietic cells and cytokine release in mouse spleen. Physiol Res/Acad Sci Bohemoslovaca. 2007;56:331–9.
83. Kao CY, Chen Y, Thai P, Wachi S, Huang F, Kim C, Harper RW, Wu R. IL-17 markedly up-regulates beta-defensin-2 expression in human airway epithelium via JAK and NF-kappaB signaling pathways. J Immunol. 2004;173:3482–91.
84. Kappel LW, Goldberg GL, King CG, Suh DY, Smith OM, Ligh C, Holland AM, Grubin J, Mark NM, Liu C, et al. IL-17 contributes to CD4-mediated graft-versus-host disease. Blood. 2009;113:945–52.
85. Kleinschek MA, Owyang AM, Joyce-Shaikh B, Langrish CL, Chen Y, Gorman DM, Blumenschein WM, McClanahan T, Brombacher F, Hurst SD, et al. IL-25 regulates Th17 function in autoimmune inflammation. J Exp Med. 2007;204:161–70.
86. Klotz L, Burgdorf S, Dani I, Saijo K, Flossdorf J, Hucke S, Alferink J, Nowak N, Beyer M, Mayer G, et al. The nuclear receptor PPAR gamma selectively inhibits Th17 differentiation in a T cell-intrinsic fashion and suppresses CNS autoimmunity. J Exp Med. 2009;206:2079–89.
87. Kobayashi S, Hara A, Isagawa T, Manabe I, Takeda K, MaruYama T. The nuclear IkappaB family protein IkappaBNS influences the susceptibility to experimental autoimmune encephalomyelitis in a murine model. PLoS Onc. 2014;9:e110838.
88. Kolls JK, Linden A. Interleukin-17 family members and inflammation. Immunity. 2004;21:467–76.
89. Krstic A, Ilic V, Mojsilovic S, Jovcic G, Milenkovic P, Bugarski D. p38 MAPK signaling mediates IL-17-induced nitric oxide synthase expression in bone marrow cells. Growth Factors. 2009;27:79–90.
90. Kuestner RE, Taft DW, Haran A, Brandt CS, Brender T, Lum K, Harder B, Okada S, Ostrander CD, Kreindler JL, et al. Identification of the IL-17 receptor related molecule IL-17RC as the receptor for IL-17F. J Immunol. 2007;179:5462–73.
91. Laan M, Lotvall J, Chung KF, Linden A. IL-17-induced cytokine release in human bronchial epithelial cells in vitro: role of mitogen-activated protein (MAP) kinases. Br J Pharmacol. 2001;133:200–6.
92. Langrish CL, Chen Y, Blumenschein WM, Mattson J, Basham B, Sedgwick JD, McClanahan T, Kastelein RA, Cua DJ. IL-23 drives a pathogenic T cell population that induces autoimmune inflammation. J Exp Med. 2005;201:233–40.
93. Larochelle C, Alvarez JI, Prat A. How do immune cells overcome the blood-brain barrier in multiple sclerosis? FEBS Lett. 2011;585:3770–80.

94. le Rolle AF, Chiu TK, Fara M, Shia J, Zeng Z, Weiser MR, Paty PB, Chiu VK. The prognostic significance of CXCL1 hypersecretion by human colorectal cancer epithelia and myofibroblasts. J Transl Med. 2015;13:199.
95. Lebson L, Gocke A, Rosenzweig J, Alder J, Civin C, Calabresi PA, Whartenby KA. Cutting edge: the transcription factor Kruppel-like factor 4 regulates the differentiation of Th17 cells independently of RORgammat. J Immunol. 2010;185:7161–4.
96. Lee J, Ho WH, Maruoka M, Corpuz RT, Baldwin DT, Foster JS, Goddard AD, Yansura DG, Vandlen RL, Wood WI, et al. IL-17E, a novel proinflammatory ligand for the IL-17 receptor homolog IL-17Rh1. J Biol Chem. 2001;276:1660–4.
97. LeGrand A, Fermor B, Fink C, Pisetsky DS, Weinberg JB, Vail TP, Guilak F. Interleukin-1, tumor necrosis factor alpha, and interleukin-17 synergistically up-regulate nitric oxide and prostaglandin E2 production in explants of human osteoarthritic knee menisci. Arthritis Rheum. 2001;44:2078–83.
98. Ley K, Laudanna C, Cybulsky MI, Nourshargh S. Getting to the site of inflammation: the leukocyte adhesion cascade updated. Nat Rev Immunol. 2007;7:678–89.
99. Li B, Reynolds JM, Stout RD, Bernlohr DA, Suttles J. Regulation of Th17 differentiation by epidermal fatty acid-binding protein. J Immunol. 2009;182:7625–33.
100. Li H, Chen J, Huang A, Stinson J, Heldens S, Foster J, Dowd P, Gurney AL, Wood WI. Cloning and characterization of IL-17B and IL-17C, two new members of the IL-17 cytokine family. Proc Natl Acad Sci U S A. 2000;97:773–8.
101. Li L, Patsoukis N, Petkova V, Boussiotis VA. Runx1 and Runx3 are involved in the generation and function of highly suppressive IL-17-producing T regulatory cells. PLoS One. 2012;7:e45115.
102. Liang SC, Tan XY, Luxenberg DP, Karim R, Dunussi-Joannopoulos K, Collins M, Fouser LA. Interleukin (IL)-22 and IL-17 are coexpressed by Th17 cells and cooperatively enhance expression of antimicrobial peptides. J Exp Med. 2006;203:2271–9.
103. Liao JJ, Huang MC, Goetzl EJ. Cutting edge: alternative signaling of Th17 cell development by sphingosine 1-phosphate. J Immunol. 2007;178:5425–8.
104. Lim HW, Kang SG, Ryu JK, Schilling B, Fei M, Lee IS, Kehasse A, Shirakawa K, Yokoyama M, Schnolzer M, et al. SIRT1 deacetylates RORgammat and enhances Th17 cell generation. J Exp Med. 2015;212:973.
105. Lindemann MJ, Hu Z, Benczik M, Liu KD, Gaffen SL. Differential regulation of the IL-17 receptor by gammac cytokines: inhibitory signaling by the phosphatidylinositol 3-kinase pathway. J Biol Chem. 2008;283:14100–8.
106. Liu HP, Cao AT, Feng T, Li Q, Zhang W, Yao S, Dann SM, Elson CO, Cong Y. TGF-beta converts Th1 cells into Th17 cells through stimulation of Runx1 expression. Eur J Immunol. 2015;45:1010–8.
107. Lockhart E, Green AM, Flynn JL. IL-17 production is dominated by gammadelta T cells rather than CD4 T cells during Mycobacterium tuberculosis infection. J Immunol. 2006;177:4662–9.
108. Lu YJ, Gross J, Bogaert D, Finn A, Bagrade L, Zhang Q, Kolls JK, Srivastava A, Lundgren A, Forte S, et al. Interleukin-17A mediates acquired immunity to pneumococcal colonization. PLoS Pathog. 2008;4:e1000159.
109. Lubberts E, Joosten LA, van de Loo FA, Schwarzenberger P, Kolls J, van den Berg WB. Overexpression of IL-17 in the knee joint of collagen type II immunized mice promotes collagen arthritis and aggravates joint destruction. Inflamm Res. 2002;51:102–4.
110. Lubberts E, Koenders MI, Oppers-Walgreen B, van den Bersselaar L, Coenen-de Roo CJ, Joosten LA, van den Berg WB. Treatment with a neutralizing anti-murine interleukin-17 antibody after the onset of collagen-induced arthritis reduces joint inflammation, cartilage destruction, and bone erosion. Arthritis Rheum. 2004;50:650–9.
111. Lubberts E, van den Berg WB. Cytokines in the pathogenesis of rheumatoid arthritis and collagen-induced arthritis. Adv Exp Med Biol. 2003;520:194–202.

112. Maezawa Y, Nakajima H, Suzuki K, Tamachi T, Ikeda K, Inoue J, Saito Y, Iwamoto I. Involvement of TNF receptor-associated factor 6 in IL-25 receptor signaling. J Immunol. 2006;176:1013–8.
113. Mai J, Virtue A, Maley E, Tran T, Yin Y, Meng S, Pansuria M, Jiang X, Wang H, Yang XF. MicroRNAs and other mechanisms regulate interleukin-17 cytokines and receptors. Front Biosci. 2012;4:1478–95.
114. Martin B, Hirota K, Cua DJ, Stockinger B, Veldhoen M. Interleukin-17-producing gammadelta T cells selectively expand in response to pathogen products and environmental signals. Immunity. 2009;31:321–30.
115. McGeachy MJ, Bak-Jensen KS, Chen Y, Tato CM, Blumenschein W, McClanahan T, Cua DJ. TGF-beta and IL-6 drive the production of IL-17 and IL-10 by T cells and restrain T(H)-17 cell-mediated pathology. Nat Immunol. 2007;8:1390–7.
116. Meeks KD, Sieve AN, Kolls JK, Ghilardi N, Berg RE. IL-23 is required for protection against systemic infection with Listeria monocytogenes. J Immunol. 2009;183:8026–34.
117. Miao T, Raymond M, Bhullar P, Ghaffari E, Symonds AL, Meier UC, Giovannoni G, Li S, Wang P. Early growth response gene-2 controls IL-17 expression and Th17 differentiation by negatively regulating Batf. J Immunol. 2013;190:58–65.
118. Michel ML, Keller AC, Paget C, Fujio M, Trottein F, Savage PB, Wong CH, Schneider E, Dy M, Leite-de-Moraes MC. Identification of an IL-17-producing NK1.1(neg) iNKT cell population involved in airway neutrophilia. J Exp Med. 2007;204:995–1001.
119. Milner JD, Brenchley JM, Laurence A, Freeman AF, Hill BJ, Elias KM, Kanno Y, Spalding C, Elloumi HZ, Paulson ML, et al. Impaired T(H)17 cell differentiation in subjects with autosomal dominant hyper-IgE syndrome. Nature. 2008;452:773–6.
120. Moran EM, Mullan R, McCormick J, Connolly M, Sullivan O, Fitzgerald O, Bresnihan B, Veale DJ, Fearon U. Human rheumatoid arthritis tissue production of IL-17A drives matrix and cartilage degradation: synergy with tumour necrosis factor-alpha, Oncostatin M and response to biologic therapies. Arthritis Res Ther. 2009;11:R113.
121. Moreira-Teixeira L, Resende M, Coffre M, Devergne O, Herbeuval JP, Hermine O, Schneider E, Rogge L, Ruemmele FM, Dy M, et al. Proinflammatory environment dictates the IL-17-producing capacity of human invariant NKT cells. J Immunol. 2011;186:5758–65.
122. Mosmann TR, Cherwinski H, Bond MW, Giedlin MA, Coffman RL. Two types of murine helper T cell clone. I. Definition according to profiles of lymphokine activities and secreted proteins. J Immunol. 1986;136:2348–57.
123. Nakae S, Nambu A, Sudo K, Iwakura Y. Suppression of immune induction of collagen-induced arthritis in IL-17-deficient mice. J Immunol. 2003;171:6173–7.
124. Nascimento MS, Carregaro V, Lima-Junior DS, Costa DL, Ryffel B, Duthie MS, de Jesus A, de Almeida RP, da Silva JS. Interleukin 17A acts synergistically with interferon gamma to promote protection against Leishmania infantum infection. J Infect Dis. 2015;211:1015–26.
125. Nigam P, Kwa S, Velu V, Amara RR. Loss of IL-17-producing CD8 T cells during late chronic stage of pathogenic simian immunodeficiency virus infection. J Immunol. 2011;186:745–53.
126. Novatchkova M, Leibbrandt A, Werzowa J, Neubuser A, Eisenhaber F. The STIR-domain superfamily in signal transduction, development and immunity. Trends Biochem Sci. 2003;28:226–9.
127. Ogawa A, Andoh A, Araki Y, Bamba T, Fujiyama Y. Neutralization of interleukin-17 aggravates dextran sulfate sodium-induced colitis in mice. Clin Immunol. 2004;110:55–62.
128. Ogura H, Murakami M, Okuyama Y, Tsuruoka M, Kitabayashi C, Kanamoto M, Nishihara M, Iwakura Y, Hirano T. Interleukin-17 promotes autoimmunity by triggering a positive-feedback loop via interleukin-6 induction. Immunity. 2008;29:628–36.
129. Onishi RM, Gaffen SL. Interleukin-17 and its target genes: mechanisms of interleukin-17 function in disease. Immunology. 2010;129:311–21.

130. Ouyang X, Zhang R, Yang J, Li Q, Qin L, Zhu C, Liu J, Ning H, Shin MS, Gupta M, et al. Transcription factor IRF8 directs a silencing programme for TH17 cell differentiation. Nat Commun. 2011;2:314.
131. Ozaki K, Leonard WJ. Cytokine and cytokine receptor pleiotropy and redundancy. J Biol Chem. 2002;277:29355–8.
132. Park H, Li Z, Yang XO, Chang SH, Nurieva R, Wang YH, Wang Y, Hood L, Zhu Z, Tian Q, et al. A distinct lineage of CD4 T cells regulates tissue inflammation by producing interleukin 17. Nat Immunol. 2005;6:1133–41.
133. Passos ST, Silver JS, O'Hara AC, Sehy D, Stumhofer JS, Hunter CA. IL-6 promotes NK cell production of IL-17 during toxoplasmosis. J Immunol. 2010;184:1776–83.
134. Pham D, Sehra S, Sun X, Kaplan MH. The transcription factor Etv5 controls TH17 cell development and allergic airway inflammation. J Allergy Clin Immunol. 2014;134:204–14.
135. Qi CF, Li Z, Raffeld M, Wang H, Kovalchuk AL, Morse 3rd HC. Differential expression of IRF8 in subsets of macrophages and dendritic cells and effects of IRF8 deficiency on splenic B cell and macrophage compartments. Immunol Res. 2009;45:62–74.
136. Qian Y, Liu C, Hartupee J, Altuntas CZ, Gulen MF, Jane-Wit D, Xiao J, Lu Y, Giltiay N, Liu J, et al. The adaptor Act1 is required for interleukin 17-dependent signaling associated with autoimmune and inflammatory disease. Nat Immunol. 2007;8:247–56.
137. Rachitskaya AV, Hansen AM, Horai R, Li Z, Villasmil R, Luger D, Nussenblatt RB, Caspi RR. Cutting edge: NKT cells constitutively express IL-23 receptor and RORgammat and rapidly produce IL-17 upon receptor ligation in an IL-6-independent fashion. J Immunol. 2008;180:5167–71.
138. Raffatellu M, Santos RL, Verhoeven DE, George MD, Wilson RP, Winter SE, Godinez I, Sankaran S, Paixao TA, Gordon MA, et al. Simian immunodeficiency virus-induced mucosal interleukin-17 deficiency promotes Salmonella dissemination from the gut. Nat Med. 2008;14:421–8.
139. Ramirez-Carrozzi V, Sambandam A, Luis E, Lin Z, Jeet S, Lesch J, Hackney J, Kim J, Zhou M, Lai J, et al. IL-17C regulates the innate immune function of epithelial cells in an autocrine manner. Nat Immunol. 2011;12:1159–66.
140. Rannug A, Rannug U, Rosenkranz HS, Winqvist L, Westerholm R, Agurell E, Grafstrom AK. Certain photooxidized derivatives of tryptophan bind with very high affinity to the Ah receptor and are likely to be endogenous signal substances. J Biol Chem. 1987;262:15422–7.
141. Reynolds JM, Martinez GJ, Nallaparaju KC, Chang SH, Wang YH, Dong C. Cutting edge: regulation of intestinal inflammation and barrier function by IL-17C. J Immunol. 2012;189:4226–30.
142. Reynolds JM, Pappu BP, Peng J, Martinez GJ, Zhang Y, Chung Y, Ma L, Yang XO, Nurieva RI, Tian Q, et al. Toll-like receptor 2 signaling in CD4(+) T lymphocytes promotes T helper 17 responses and regulates the pathogenesis of autoimmune disease. Immunity. 2010;32:692–702.
143. Ribot JC, deBarros A, Pang DJ, Neves JF, Peperzak V, Roberts SJ, Girardi M, Borst J, Hayday AC, Pennington DJ, et al. CD27 is a thymic determinant of the balance between interferon-gamma- and interleukin 17-producing gammadelta T cell subsets. Nat Immunol. 2009;10:427–36.
144. Rickel EA, Siegel LA, Yoon BR, Rottman JB, Kugler DG, Swart DA, Anders PM, Tocker JE, Comeau MR, Budelsky AL. Identification of functional roles for both IL-17RB and IL-17RA in mediating IL-25-induced activities. J Immunol. 2008;181:4299–310.
145. Rong Z, Cheng L, Ren Y, Li Z, Li Y, Li X, Li H, Fu XY, Chang Z. Interleukin-17F signaling requires ubiquitination of interleukin-17 receptor via TRAF6. Cell Signal. 2007;19:1514–20.
146. Rouvier E, Luciani MF, Mattei MG, Denizot F, Golstein P. CTLA-8, cloned from an activated T cell, bearing AU-rich messenger RNA instability sequences, and homologous to a herpesvirus saimiri gene. J Immunol. 1993;150:5445–56.

147. Ruan Q, Kameswaran V, Zhang Y, Zheng S, Sun J, Wang J, DeVirgiliis J, Liou HC, Beg AA, Chen YH. The Th17 immune response is controlled by the Rel-RORgamma-RORgamma T transcriptional axis. J Exp Med. 2011;208:2321–33.
148. Ruddy MJ, Wong GC, Liu XK, Yamamoto H, Kasayama S, Kirkwood KL, Gaffen SL. Functional cooperation between interleukin-17 and tumor necrosis factor-alpha is mediated by CCAAT/enhancer-binding protein family members. J Biol Chem. 2004;279:2559–67.
149. Rudner XL, Happel KI, Young EA, Shellito JE. Interleukin-23 (IL-23)-IL-17 cytokine axis in murine Pneumocystis carinii infection. Infect Immun. 2007;75:3055–61.
150. Ryzhakov G, Blazek K, Lai CC, Udalova IA. IL-17 receptor adaptor protein Act1/CIKS plays an evolutionarily conserved role in antiviral signaling. J Immunol. 2012;189:4852–8.
151. Saijo S, Ikeda S, Yamabe K, Kakuta S, Ishigame H, Akitsu A, Fujikado N, Kusaka T, Kubo S, Chung SH, et al. Dectin-2 recognition of alpha-mannans and induction of Th17 cell differentiation is essential for host defense against Candida albicans. Immunity. 2010;32:681–91.
152. Sakellariou GT, Anastasilakis AD, Bisbinas I, Oikonomou D, Gerou S, Polyzos SA, Sayegh FE. Circulating periostin levels in patients with AS: association with clinical and radiographic variables, inflammatory markers and molecules involved in bone formation. Rheumatology. 2015;54:908–14.
153. Saleh A, Shan L, Halayko AJ, Kung S, Gounni AS. Critical role for STAT3 in IL-17A-mediated CCL11 expression in human airway smooth muscle cells. J Immunol. 2009;182:3357–65.
154. Sawa S, Lochner M, Satoh-Takayama N, Dulauroy S, Berard M, Kleinschek M, Cua D, Di Santo JP, Eberl G. RORgammat+ innate lymphoid cells regulate intestinal homeostasis by integrating negative signals from the symbiotic microbiota. Nat Immunol. 2011;12:320–6.
155. Schecter A, Birnbaum L, Ryan JJ, Constable JD. Dioxins: an overview. Environ Res. 2006;101:419–28.
156. Serody JS, Hill GR. The IL-17 differentiation pathway and its role in transplant outcome. Biol Blood Marrow Transplant. 2012;18:S56–61.
157. Shahrara S, Pickens SR, Mandelin 2nd AM, Karpus WJ, Huang Q, Kolls JK, Pope RM. IL-17-mediated monocyte migration occurs partially through CC chemokine ligand 2/monocyte chemoattractant protein-1 induction. J Immunol. 2010;184:4479–87.
158. Shen F, Gaffen SL. Structure-function relationships in the IL-17 receptor: implications for signal transduction and therapy. Cytokine. 2008;41:92–104.
159. Shen F, Hu Z, Goswami J, Gaffen SL. Identification of common transcriptional regulatory elements in interleukin-17 target genes. J Biol Chem. 2006;281:24138–48.
160. Shen F, Li N, Gade P, Kalvakolanu DV, Weibley T, Doble B, Woodgett JR, Wood TD, Gaffen SL. IL-17 receptor signaling inhibits C/EBPbeta by sequential phosphorylation of the regulatory 2 domain. Sci Signal. 2009;2:ra8.
161. Shen F, Ruddy MJ, Plamondon P, Gaffen SL. Cytokines link osteoblasts and inflammation: microarray analysis of interleukin-17- and TNF-alpha-induced genes in bone cells. J Leukoc Biol. 2005;77:388–99.
162. Shi P, Zhu S, Lin Y, Liu Y, Liu Y, Chen Z, Shi Y, Qian Y. Persistent stimulation with interleukin-17 desensitizes cells through SCFbeta-TrCP-mediated degradation of Act1. Sci Signal. 2011;4:ra73.
163. Shi Y, Ullrich SJ, Zhang J, Connolly K, Grzegorzewski KJ, Barber MC, Wang W, Wathen K, Hodge V, Fisher CL, et al. A novel cytokine receptor-ligand pair. Identification, molecular characterization, and in vivo immunomodulatory activity. J Biol Chem. 2000;275:19167–76.
164. Shibata K, Yamada H, Hara H, Kishihara K, Yoshikai Y. Resident Vdelta1+ gammadelta T cells control early infiltration of neutrophils after Escherichia coli infection via IL-17 production. J Immunol. 2007;178:4466–72.

165. Song X, Zhu S, Shi P, Liu Y, Shi Y, Levin SD, Qian Y. IL-17RE is the functional receptor for IL-17C and mediates mucosal immunity to infection with intestinal pathogens. Nat Immunol. 2011;12:1151–8.
166. Souwer Y, Szegedi K, Kapsenberg ML, de Jong EC. IL-17 and IL-22 in atopic allergic disease. Curr Opin Immunol. 2010;22:821–6.
167. Starnes T, Broxmeyer HE, Robertson MJ, Hromas R. Cutting edge: IL-17D, a novel member of the IL-17 family, stimulates cytokine production and inhibits hemopoiesis. J Immunol. 2002;169:642–6.
168. Stock P, Lombardi V, Kohlrautz V, Akbari O. Induction of airway hyperreactivity by IL-25 is dependent on a subset of invariant NKT cells expressing IL-17RB. J Immunol. 2009;182:5116–22.
169. Sugihara T, Kobori A, Imaeda H, Tsujikawa T, Amagase K, Takeuchi K, Fujiyama Y, Andoh A. The increased mucosal mRNA expressions of complement C3 and interleukin-17 in inflammatory bowel disease. Clin Exp Immunol. 2010;160:386–93.
170. Sun D, Novotny M, Bulek K, Liu C, Li X, Hamilton T. Treatment with IL-17 prolongs the half-life of chemokine CXCL1 mRNA via the adaptor TRAF5 and the splicing-regulatory factor SF2 (ASF). Nat Immunol. 2011;12:853–60.
171. Sun Z, Unutmaz D, Zou YR, Sunshine MJ, Pierani A, Brenner-Morton S, Mebius RE, Littman DR. Requirement for RORgamma in thymocyte survival and lymphoid organ development. Science. 2000;288:2369–73.
172. Sutton CE, Lalor SJ, Sweeney CM, Brereton CF, Lavelle EC, Mills KH. Interleukin-1 and IL-23 induce innate IL-17 production from gammadelta T cells, amplifying Th17 responses and autoimmunity. Immunity. 2009;31:331–41.
173. Swaidani S, Bulek K, Kang Z, Gulen MF, Liu C, Yin W, Abbadi A, Aronica M, Li X. T cell-derived Act1 is necessary for IL-25-mediated Th2 responses and allergic airway inflammation. J Immunol. 2011;187:3155–64.
174. Swaidani S, Bulek K, Kang Z, Liu C, Lu Y, Yin W, Aronica M, Li X. The critical role of epithelial-derived Act1 in IL-17- and IL-25-mediated pulmonary inflammation. J Immunol. 2009;182:1631–40.
175. Terashima A, Watarai H, Inoue S, Sekine E, Nakagawa R, Hase K, Iwamura C, Nakajima H, Nakayama T, Taniguchi M. A novel subset of mouse NKT cells bearing the IL-17 receptor B responds to IL-25 and contributes to airway hyperreactivity. J Exp Med. 2008;205:2727–33.
176. Toy D, Kugler D, Wolfson M, Vanden Bos T, Gurgel J, Derry J, Tocker J, Peschon J. Cutting edge: interleukin 17 signals through a heteromeric receptor complex. J Immunol. 2006;177:36–9.
177. Trajkovic V, Stosic-Grujicic S, Samardzic T, Markovic M, Miljkovic D, Ramic Z, Mostarica Stojkovic M. Interleukin-17 stimulates inducible nitric oxide synthase activation in rodent astrocytes. J Neuroimmunol. 2001;119:183–91.
178. Umemura M, Yahagi A, Hamada S, Begum MD, Watanabe H, Kawakami K, Suda T, Sudo K, Nakae S, Iwakura Y, et al. IL-17-mediated regulation of innate and acquired immune response against pulmonary Mycobacterium bovis bacille Calmette-Guerin infection. J Immunol. 2007;178:3786–96.
179. Wang X, Zhang Y, Yang XO, Nurieva RI, Chang SH, Ojeda SS, Kang HS, Schluns KS, Gui J, Jetten AM, et al. Transcription of Il17 and Il17f is controlled by conserved noncoding sequence 2. Immunity. 2012;36:23–31.
180. Wang YH, Angkasekwinai P, Lu N, Voo KS, Arima K, Hanabuchi S, Hippe A, Corrigan CJ, Dong C, Homey B, et al. IL-25 augments type 2 immune responses by enhancing the expansion and functions of TSLP-DC-activated Th2 memory cells. J Exp Med. 2007;204:1837–47.
181. Wong CK, Lit LC, Tam LS, Li EK, Wong PT, Lam CW. Hyperproduction of IL-23 and IL-17 in patients with systemic lupus erythematosus: implications for Th17-mediated inflammation in auto-immunity. Clin Immunol. 2008;127:385–93.

182. Wright JF, Bennett F, Li B, Brooks J, Luxenberg DP, Whitters MJ, Tomkinson KN, Fitz LJ, Wolfman NM, Collins M, et al. The human IL-17F/IL-17A heterodimeric cytokine signals through the IL-17RA/IL-17RC receptor complex. J Immunol. 2008;181:2799–805.
183. Wright JF, Guo Y, Quazi A, Luxenberg DP, Bennett F, Ross JF, Qiu Y, Whitters MJ, Tomkinson KN, Dunussi-Joannopoulos K, et al. Identification of an interleukin 17F/17A heterodimer in activated human CD4+ T cells. J Biol Chem. 2007;282:13447–55.
184. Xu S, Han Y, Xu X, Bao Y, Zhang M, Cao X. IL-17A-producing gammadeltaT cells promote CTL responses against Listeria monocytogenes infection by enhancing dendritic cell cross-presentation. J Immunol. 2010;185:5879–87.
185. Yagi Y, Andoh A, Inatomi O, Tsujikawa T, Fujiyama Y. Inflammatory responses induced by interleukin-17 family members in human colonic subepithelial myofibroblasts. J Gastroenterol. 2007;42:746–53.
186. Yamaguchi Y, Fujio K, Shoda H, Okamoto A, Tsuno NH, Takahashi K, Yamamoto K. IL-17B and IL-17C are associated with TNF-alpha production and contribute to the exacerbation of inflammatory arthritis. J Immunol. 2007;179:7128–36.
187. Yang B, Kang H, Fung A, Zhao H, Wang T, Ma D. The role of interleukin 17 in tumour proliferation, angiogenesis, and metastasis. Mediators Inflamm. 2014;2014:623759.
188. Yang D, Chertov O, Bykovskaia SN, Chen Q, Buffo MJ, Shogan J, Anderson M, Schroder JM, Wang JM, Howard OM, et al. Beta-defensins: linking innate and adaptive immunity through dendritic and T cell CCR6. Science. 1999;286:525–8.
189. Yang TT, Chow CW. Transcription cooperation by NFAT.C/EBP composite enhancer complex. J Biol Chem. 2003;278:15874–85.
190. Yang XO, Chang SH, Park H, Nurieva R, Shah B, Acero L, Wang YH, Schluns KS, Broaddus RR, Zhu Z, et al. Regulation of inflammatory responses by IL-17F. J Exp Med. 2008;205:1063–75.
191. Yang XO, Pappu BP, Nurieva R, Akimzhanov A, Kang HS, Chung Y, Ma L, Shah B, Panopoulos AD, Schluns KS, et al. T helper 17 lineage differentiation is programmed by orphan nuclear receptors ROR alpha and ROR gamma. Immunity. 2008;28:29–39.
192. Yao Z, Fanslow WC, Seldin MF, Rousseau AM, Painter SL, Comeau MR, Cohen JI, Spriggs MK. Herpesvirus Saimiri encodes a new cytokine, IL-17, which binds to a novel cytokine receptor. Immunity. 1995;3:811–21.
193. Yao Z, Painter SL, Fanslow WC, Ulrich D, Macduff BM, Spriggs MK, Armitage RJ. Human IL-17: a novel cytokine derived from T cells. J Immunol. 1995;155:5483–6.
194. Ye P, Rodriguez FH, Kanaly S, Stocking KL, Schurr J, Schwarzenberger P, Oliver P, Huang W, Zhang P, Zhang J, et al. Requirement of interleukin 17 receptor signaling for lung CXC chemokine and granulocyte colony-stimulating factor expression, neutrophil recruitment, and host defense. J Exp Med. 2001;194:519–27.
195. Yeste A, Mascanfroni ID, Nadeau M, Burns EJ, Tukpah AM, Santiago A, Wu C, Patel B, Kumar D, Quintana FJ. IL-21 induces IL-22 production in CD4+ T cells. Nat Commun. 2014;5:3753.
196. Zepp JA, Liu C, Qian W, Wu L, Gulen MF, Kang Z, Li X. Cutting edge: TNF receptor-associated factor 4 restricts IL-17-mediated pathology and signaling processes. J Immunol. 2012;189:33–7.
197. Zhang B, Liu C, Qian W, Han Y, Li X, Deng J. Crystal structure of IL-17 receptor B SEFIR domain. J Immunol. 2013;190:2320–6.
198. Zhang F, Meng G, Strober W. Interactions among the transcription factors Runx1, RORgammat and Foxp3 regulate the differentiation of interleukin 17-producing T cells. Nat Immunol. 2008;9:1297–306.
199. Zhang Z, Kyttaris VC, Tsokos GC. The role of IL-23/IL-17 axis in lupus nephritis. J Immunol. 2009;183:3160–9.
200. Zhao A, Urban Jr JF, Sun R, Stiltz J, Morimoto M, Notari L, Madden KB, Yang Z, Grinchuk V, Ramalingam TR, et al. Critical role of IL-25 in nematode infection-induced alterations in intestinal function. J Immunol. 2010;185:6921–9.

201. Zhong B, Liu X, Wang X, Chang SH, Liu X, Wang A, Reynolds JM, Dong C. Negative regulation of IL-17-mediated signaling and inflammation by the ubiquitin-specific protease USP25. Nat Immunol. 2012;13:1110–7.
202. Zhou L, Littman DR. Transcriptional regulatory networks in Th17 cell differentiation. Curr Opin Immunol. 2009;21:146–52.
203. Zhu S, Pan W, Shi P, Gao H, Zhao F, Song X, Liu Y, Zhao L, Li X, Shi Y, et al. Modulation of experimental autoimmune encephalomyelitis through TRAF3-mediated suppression of interleukin 17 receptor signaling. J Exp Med. 2010;207:2647–62.
204. Zhu S, Pan W, Song X, Liu Y, Shao X, Tang Y, Liang D, He D, Wang H, Liu W, et al. The microRNA miR-23b suppresses IL-17-associated autoimmune inflammation by targeting TAB2, TAB3 and IKK-alpha. Nat Med. 2012;18:1077–86.
205. Zizzo G, Cohen PL. IL-17 stimulates differentiation of human anti-inflammatory macrophages and phagocytosis of apoptotic neutrophils in response to IL-10 and glucocorticoids. J Immunol. 2013;190:5237–46.

Chapter 8
Regulation of Interleukin-23 Expression in Health and Disease

Iain Welsby and Stanislas Goriely

Abstract Interleukin (IL)-23 plays a central role in the orchestration of inflammatory responses. Produced by dendritic cells and macrophages, this cytokine promotes the protection of the host against mucosal pathogens through the induction of IL-17 and related cytokines by lymphoid cells. Preclinical disease models and association studies in humans have also clearly demonstrated the implication of IL-23 signalling pathway in inflammatory diseases. Indeed, this cytokine is now considered as a major therapeutic target in immune-based pathologies such as psoriasis, ankylosing spondylitis or Crohn's disease. Furthermore, in the context of inflammation-related cancer, IL-23 is thought to contribute to tumorigenesis and progression to metastatic disease. Herein, we review our current understanding of IL-23 regulation at the transcriptional and post-transcriptional levels. We discuss the relevance of these findings in the context of infection, chronic inflammation and cancer.

Keywords IL-23 • Inflammation • Transcription factors • Dendritic cells • Epigenetic • TLR • mRNA stability • ER stress • Cancer

8.1 Introduction

Since its role in multiple experimental inflammatory models and immune-mediated diseases has been elucidated, there has been tremendous interest in interleukin (IL)-23 as a key therapeutic target. IL-23 is closely related to IL-12 as these heterodimeric cytokines share a common IL-12/IL-23p40 subunit. IL-23-specific subunit (IL-23p19) displays a 4-helix bundle structure that is characteristic of other IL-6- and IL-12-type cytokines and shares about 15 % amino acid sequence identity with

I. Welsby • S. Goriely (✉)
WELBIO and Institute for Medical Immunology (IMI), Université Libre de Bruxelles,
8 rue Adrienne Bolland, B-6041 Charleroi, Belgium
e-mail: stgoriel@ulb.ac.be

X. Ma (ed.), *Regulation of Cytokine Gene Expression in Immunity and Diseases*,
Advances in Experimental Medicine and Biology 941,
DOI 10.1007/978-94-024-0921-5_8

IL-12p35. IL-23 acts through a receptor complex composed of IL-12Rβ1 and IL-23R. The expression of the latter chain is mostly restricted to specific T and innate lymphoid cell subsets [15]. Signalling is dependent on the activation of tyrosine kinase (Tyk)2 and Janus kinase (JAK)2 which predominantly phosphorylates STAT3.

As reviewed extensively [81], polarisation of naïve CD4 T cells by various cytokines including transforming growth factor-β (TGF-β)-1 or (TGF-β)-3, IL-1, IL-21 and IL-6 is dependent on retinoic acid receptor-related orphan receptor-γt (RORγt). This transcription factor promotes expression of IL-23R. Upon IL-23R signalling, STAT3 activation stabilises RORγt expression, establishing a positive feedback loop leading to efficient production of effector cytokines such as IL-17A, IL-17F, IL-21, IL-22 or GM-CSF while suppressing IL-10 expression. IL-23 is therefore essential for the induction of a "pathogenic" Th17 cell gene expression profile [65]. The molecular pathways involved in this process are not fully elucidated, but recent evidences indicate that induction of the salt-sensitive kinase SGK1 downstream of IL-23R is required to promote this pro-inflammatory signature. Importantly, IL-23 also suppresses the development of induced regulatory T cells.

In vivo tracking of IL-23R-expressing cells revealed that a variety of innate and acquired immune cells, other than canonical Th17 cells, are targeted by IL-23 [6]. In particular, RORγt-expressing γδ T cells and type 3 innate lymphoid cells were shown to drive inflammation in psoriasis or colitis models, respectively [26, 89].

Under homeostatic conditions, the IL-23/type 17 axis contributes to the integrity of the mucosal and skin barriers. It is also central for the protective immunity against extracellular bacterial and fungal infections as it is required for optimal induction of chemokines, recruitment of neutrophils and microbicidal activity. Conversely, the importance of IL-23 in promoting autoimmune tissue inflammation has been demonstrated in several experimental models and is further supported by the finding that polymorphisms in genes related to IL-23 and its signalling pathway are associated with several human autoimmune diseases, including Crohn's disease, ankylosing spondylitis, psoriasis and autoimmune thyroiditis. Anti-IL-12/IL-23p40 humanised antibodies (ustekinumab and briakinumab) were evaluated in a number of immune-mediated diseases [121]. They demonstrated substantial benefit for a large fraction of patients with moderate to severe psoriasis. Early data from clinical trials with IL-23-specific antagonists (tildrakizumab and guselkumab) suggest comparable clinical benefit. IL-12/IL-23p40 neutralisation in the context of ankylosing spondylitis is also highly promising. Furthermore, this treatment may lead to clinical improvement in a subgroup of patients with refractory Crohn's disease [121].

In addition to its role in the context of infections and immune-mediated disorders, IL-23-driven inflammation was shown to promote tumorigenesis in several experimental models [62]. Along this line, IL-23p19 mRNA is overexpressed in a majority of human carcinomas [62].

Given the major role of IL-23 in the orchestration of inflammatory responses, it is critical to understand how expression of this cytokine is regulated under homeostatic and pathological conditions, such as infectious, inflammatory or neoplastic disorders.

Production of bioactive IL-23 depends on the availability of both p40 and p19 chains. As IL-12/IL-23p40 can be produced in very large amounts, IL-23p19 is generally considered as the limiting subunit. However, expression of IL-12/IL-23p40 is largely restricted to defined myeloid subpopulations, thereby conferring cell-type specificity for bioactive IL-12 and/or IL-23 production. Hence, while IL-23p19 mRNA can be detected in various tissues and cell types, including epithelial and endothelial cells, IL-23 production is mainly restricted to dendritic cell/macrophages lineages. This is exemplified by the observation that keratinocyte-specific transgenic p40 expression overexpression leads to IL-23-dependent inflammation through dimerisation with endogenous p19 [57]. In contrast, transgenic IL-23p19 expression leads to systemic inflammation only when it occurs in hematopoietic cells and not in other cell types such as hepatocytes, stressing the fact that both p19 and p40 subunits need to be expressed by the same cell [131].

8.2 IL-23 Is Produced by Defined Dendritic Cell Subsets In Vivo

The original article describing IL-23 analysed its mRNA expression in different cell types and found that it was mainly expressed by Th1 cells and activated macrophages. It was not expressed in bone marrow-derived dendritic cells but was also expressed by both mouse and human monocyte-derived DCs [87]. As previously stated, more recent reports have shown that IL-23 is mainly produced by dendritic cells and macrophages, both in humans and in mice. The specific cell type producing IL-23 depends on the model being studied; IL-23 seems to be strongly involved in immunity at epithelial barriers, such as the skin, the gut and the lungs.

Historically, monocytes, macrophages and dendritic cells have been distinguished on the basis of their morphology, phenotype, function or origin. It is now well accepted that many of the proposed markers and functions are shared between these cell types and may vary in different organs or under inflammatory conditions. Recent lineage-tracing studies have revealed that, under steady-state conditions, most macrophages arise from embryonic precursors (except in the gut) and are maintained in the adult independently of the differentiation from blood monocytes [47]. In contrast, classical and plasmacytoid DCs arise from adult haematopoietic stem cell (HSC)-derived common DC precursors (CDPs) that are distinct from classical monocytes. In addition to these steady-state populations, monocytes are recruited to sites of infection or injury and differentiate into cells that have either been termed macrophages, DCs or myeloid-derived suppressor cells. While these monocyte-derived cells are highly heterogeneous, there is a lack of suitable methods to discriminate these different functional states. This diversity is crucial to mount an adequate immune response against a broad range of microbial pathogens while maintaining the integrity of the host [43].

IL-23 has been identified as a key player in the pathogenesis of inflammatory bowel disease, in mice and humans [25, 136], yet the cells that produce IL-23 in the intestinal tract have only recently been characterised. In mice, gut mononuclear phagocytes can be separated into two major populations: classical dendritic cells (cDCs) expressing CD11c and CD103 and monocyte-derived cells expressing CXCR1, CD11b and varying levels of CD11c. cDCs can be further differentiated into $CD11b^+$ and $CD11b^-$ cells [43]. Recent reports have shown that, in mice, production of IL-23 in the gut is clearly context-dependent. In the steady-state, IRF4-dependent $CD103^+CD11b^+$ cDCs express IL-23p19 mRNA and drive homeostatic mucosal Th17 cell differentiation [109]. During attaching and effacing bacterial infections, such as acute *Citrobacter rodentium* infections, $CD103^+CD11b^+$ cDCs are also a critical source of IL-23 and are required for survival after infection [108]. However, others have shown that although IL-23 is important for the control of *C. rodentium* infections, $CD103^+CD11b^+$ DCs are not essential for resistance to either *Citrobacter* or *Salmonella*, but are still required for Th17 cell differentiation, indicating that IL-23 may be produced by other cells [130]. $CD103^+CD11b^+$ DCs can also rapidly produce IL-23 in a TLR5-dependent fashion after systemic flagellin stimulation [55]. Finally, others have shown that during chronic intestinal inflammation induced by *Helicobacter hepaticus* and anti-IL-10R administration, $CD11c^+MHCII^+CX3CR1^+$ monocytes/macrophages are the key mediators of inflammation via production of IL-23, although contribution of $CD103^+CD11b^+$ cDCs could not be formally ruled out [4]. Human $CD1c^+$ DCs, equivalent to murine $CD11b^+$ cDCs, also seem to be an important source of IL-23 in the gut. Indeed, they produce IL-23 following TLR stimulation and control mucosal IL-17 responses [24, 109]. Others have reported that human $CD14^+$ intestinal macrophages from patients with Crohn's disease produce large amounts of IL-23, which surprisingly seems to induce IFN-γ production by lamina propria mononuclear cells rather than IL-17 secretion [53]. Taken together, these observations seem to indicate that $CD103^+CD11b^+$ cells in mice and $CD1c^+$ cells in humans are the main producers of IL-23 in the gut.

IL-23 produced by mononuclear cells also plays a key role in the development of murine and human psoriatic skin lesions. However, the specific subset of cells that secrete IL-23 in the skin is still a matter of debate. A recent report has shown that Langerhans cells (LCs) produce IL-23 in a murine imiquimod (IMQ)-induced model of psoriasis-like disease and are critical for pathogenesis [137]. Depletion of LCs using a Langerin-DTR (diphtheria toxin receptor) system strongly reduced skin inflammation, and chimeras in which IL-23p19 KO mice were reconstituted with WT bone marrow showed that IL-23+ radioresistant (such as LCs) cells were required for pathogenesis. Others have shown that $Langerin^-$ dermal cDCs are required for induction of psoriasis as depletion of Langerhans cells had no effect on the disease [132]. Mature dermal DCs ($CD14^-CD11c^+CD83^+$), monocytes ($CD14^+CD11c^+CD83^-$) and M1 macrophages ($CD14^+CD32^+$) seem to be the main producers of IL-23 in human psoriasis, although it has been reported that keratinocytes also express both the p19 and p40 subunits of IL-23 [64, 92, 135]. A more recent report has shown that human 6-sulfoLacNAc$^+$ DCs (slanDCs), circulating and tissue pro-inflammatory DCs, drive strong Th17 responses in psoriasis through production of IL-23, IL-1β and IL-6 [45].

In the lung, the sources of IL-23 seem also to depend on the stimulus. During LPS-induced acute lung injury, CD11c$^+$ alveolar macrophages produce large quantities of IL-23 that is required for production of IL-17 [9]. In mouse models of invasive pulmonary aspergillosis, fungi-derived β-glucans activate the C-type lectin receptor Dectin-1 on CD103$^+$ lung dendritic cells, which leads to IL-23 production by these cells and a Th17 response [139].

Other cell types have been shown to express *IL23A*. Neutrophils infiltrating colon tissue of paediatric patients with inflammatory bowel disease can express IL-23p19 [61]. Although the authors discuss IL-23, they identified these cells with an antibody recognising only IL-23p19. There is no evidence that these neutrophils express the other subunit of IL-23, p40, to produce the bioactive cytokine. In line with this observation, IL-23p19 is expressed by different cell types, such as keratinocytes and gastric and intestinal epithelial cells [3, 18, 92]. Mouse intestinal epithelial cells can express both IL-23p19 and p40 and secrete IL-23 that promotes wound healing through the induction of IL-22 [74]. On the other hand, human intestinal epithelial cells can express EBI3, IL-12p35 and IL-23p19, but don't express IL-12p40 or IL-27p28 [73]. This would indicate that the different subunits expressed in these cells may possess different functions than when associated to form the canonical dimers. As previously stated, keratinocytes do not express the p40 subunit but do express the p19 subunit [57]. These observations also point to an IL-23-independent function of p19 in keratinocytes. Recent evidence suggests that, in these cells, p19 could interact with EBI3, but whether they are secreted as a dimer remains disputed [100]. Likewise, gastric epithelial cells express IL-23p19 that is regulated by the RUNX3 transcription factor [48]. RUNX3 upregulates p19, but has no effect on other IL-12 family members, and p19 is secreted in a form that is again different from classical heterodimeric IL-23. Whether IL-23p19 associates with other subunits than IL-12/IL-23p40 (such as EBI3) to form a novel cytokine or whether it possesses other independent properties remains unknown, but points out an interesting and potentially IL-23-independent role of the p19 subunit.

In conclusion, IL-23 is produced mainly by antigen-presenting cells that are present or recruited to barrier tissues, such as the gut, skin and lungs, where they recognise pro-inflammatory stimuli. IL-23 produced by these cells then activates IL-17-producing cells (Fig. 8.1).

8.3 Regulation of IL-23 Expression: Role of the Chromatin Environment

The mechanisms that drive this functional specialisation of DCs/macrophages under steady-state conditions and their plasticity in an inflammatory context remain poorly understood. Cell type-specific expression of genes is determined by the activity of distant enhancers during their development. Pioneer transcription factors, such as PU.1 or IRF8, bind to selected genomic regions (enhancers), thereby "marking"

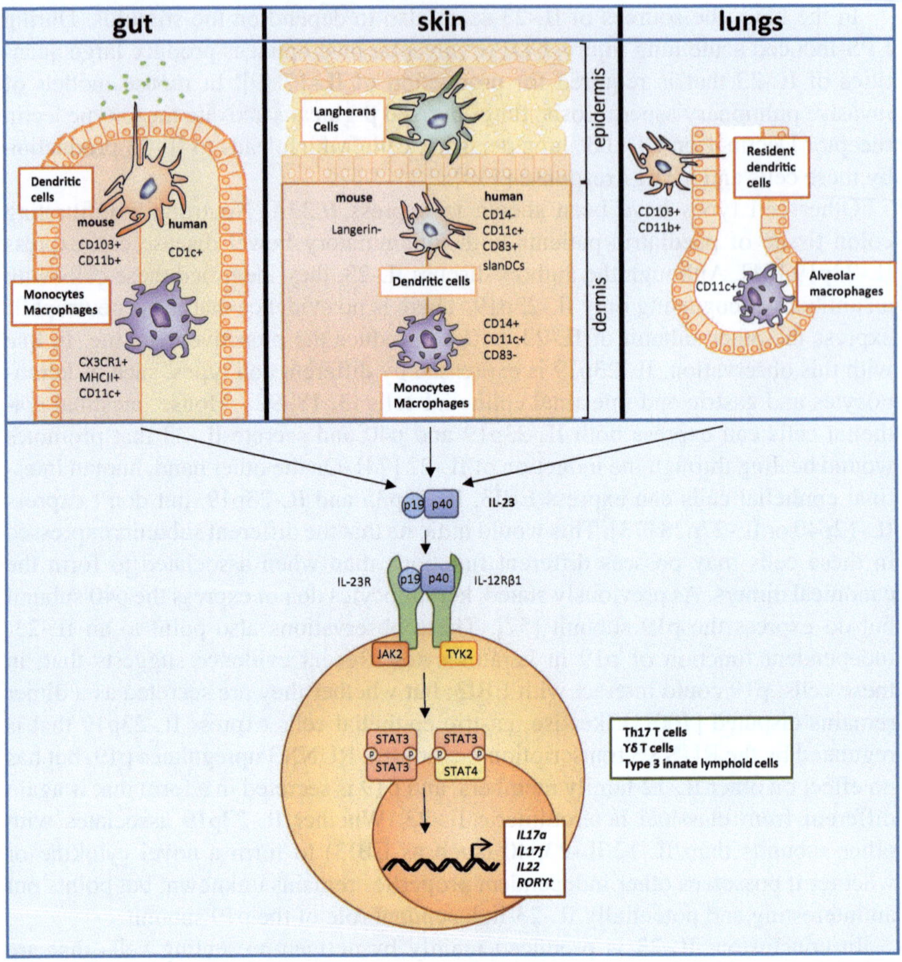

Fig. 8.1 IL-23-secreting cells and their targets. Antigen-presenting cells present in barrier tissues secrete IL-23 in humans and mice. In the gut, mouse CD103$^+$CD11b$^+$ DCs and their equivalent human CD1c$^+$ DCs represent a major source of IL-23. In the skin, mouse Langerhans cells and dermal Langerin$^-$ DCs produce IL-23 in response to TLR ligands that can promote the formation of psoriasis-like skin inflammation. Human mature dermal DCs, inflammatory 6-sulfoLacNAc$^+$ DCs and mature M1 macrophages produce IL-23 and are involved in psoriasis. Lung CD103$^+$CD11b$^+$ DCs and CD11c$^+$ alveolar macrophages can respond to inflammatory stimuli to produce IL-23. The p19/p40 dimers produced by these cells signal via a dimeric IL-23R/IL-12Rβ1 receptor that activates JAK/STAT signalling cascade that *in fine* activates the transcription of "Th17" genes such as *Il17a, Il17f, IL22 and Rorc* in target cells. These target cells include Th17/Tc17 T cells, γδ T cells and type 3 innate lymphoid cells

these loci for future transcription in this particular cell lineage [82, 90]. Sequence-specific transcription factors that are induced by the local microenvironment further shape the enhancer landscapes to provide tissue-specific identity and plasticity [36, 63]. Activation (e.g. through TLR stimulation) allows the subsequent recruitment of transcription factors to these active enhancer/promoter regions, leading to the transcription of specific inflammatory genes in these cells. TLR-induced genes were assigned to multiple classes on the basis of their requirements for new protein synthesis (i.e. primary vs. secondary response genes), kinetics of activation, requirement for the SWI/SNF (SWItch/sucrose non-fermentable) nucleosome remodelling complex for transcriptional activation and other chromatin properties [98]. In macrophages, *Il23a* is a primary response gene (PRG), characterised by rapid inducible expression. However, unlike other PRGs such as *Tnf* or *Cxcl1*, the GC content, basal H3K4me3 marks and RNA polymerase II occupancy at the promoter region are low. Under steady state, NF-κB p50/p50-dependent recruitment of corepressor complexes maintains low transcriptional activity. In response to LPS, several inducible histone modifications occur at the *Il23a* promoter region, including H4K5/8/12 acetylation (possibly mediated by PCAF or GCN5, two histone acetyltransferases) that allows Brd4 (bromodomain-containing protein 4) and P-TEFb (positive transcription elongation factor) recruitment and efficient transcriptional elongation. This makes *Il23a* partially susceptible to synthetic histone mimics that interfere with the recognition of acetylated histones by the bromodomain and extra terminal domain (BET) family of proteins [85]. *Il12b* is a prototypical secondary response gene that requires the activity of the SWI/SNF remodelling complexes [99]. Upon stimulation, a single nucleosome located in the promoter region is selectively displaced [129]. Furthermore, changes in chromatin accessibility are also observed in an enhancer element located 10 KB upstream of the transcription start site. Analysis of chromatin landscape at the genome-wide level indicates that other enhancer elements are likely to participate to *Il12b* expression in macrophages and DC subsets. Hence, dynamic changes in *Il12b* transcription involve both local nucleosomal remodelling and long distance chromatin looping between enhancer-promoter regions. Taken together, these data indicate that *IL23a* and *IL12b* expression follows distinct molecular rules that account for the fact that they can be dissociated.

8.4 Microbial Signals Leading to IL-23 Synthesis

As previously discussed, IL-23 is expressed mainly by dendritic cells and macrophages at barrier tissues, where these cells act as sentinels for the immune system. They express a range of receptors, both intracellular and extracellular, that allow them to sense microbial-derived products. These receptors include Toll-like receptors (TLRs), NOD-like receptors (NLRs), RIG-I-like receptors (RLRs), C-type lectin-like receptors (CLRs) as well as receptors that can sense extracellular metabolites such as ATP. The signalling pathways triggered by these receptors activate downstream transcription factors that are critical for the regulation of both IL-12/IL-23p40 and IL-23p19 subunits. These pathways are summarised in Figs. 8.2 and 8.3.

TLR stimulation is a major inducer of cytokines of the IL-12 family. However, the balance between IL-12 and IL-23 depends on the specific TLR that is engaged

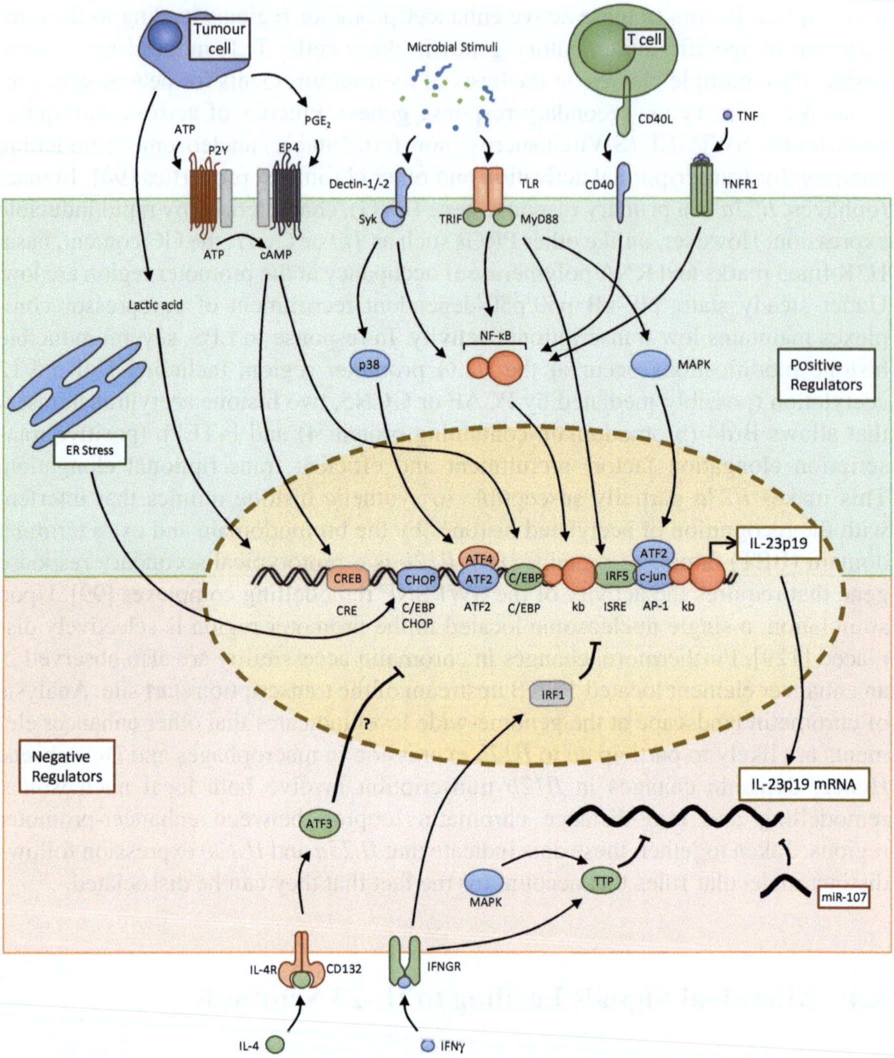

Fig. 8.2 Regulation of IL-23p19 expression. IL-23p19 is both positively regulated and negatively regulated. Positive regulators include microbial stimuli that signal via TLRs or CLRs such as Dectin-1 and Dectin-2. TLR signalling via the adaptors MyD88 or TRIF activates transcription factors of the NF-κB family that bind to at least two different sites in the promoter region. TLRs also activate classical MAP kinase pathways that promote the binding of an ATF2/c-Jun dimer to the AP-1 site. TLR-induced IRF5 activation is also critical for promoting IL-23p19 transcription. Dectin-1 or Dectin-2 primarily signal via the Syk-kinase to activate NF-κB members, but can also promote the binding of an ATF2/ATF4 dimer to the ATF2 site. ATP and tumour-derived PGE2 signal through transmembrane receptors coupled with adenylate cyclases that catalyse the conversion of ATP to cyclic AMP. cAMP signalling regulates the binding of the CREB and c/EBP transcription factors to respectively the CRE and C/EBP sites in the IL-23p19 promoter. Proteins of the TNF receptor superfamily, such as CD40 and TNFR1, activate NF-κB transcription factors to promote IL-23p19 transcription. Tumour-derived lactic acid triggers IL-23p19 expression via the binding of an undetermined transcription factor (but not NF-κB). Unfolded protein responses

and on the subsequent signalling events. IL-12/IL-23p40 is potently induced by different TLR ligands and is usually produced in excess of the IL-12p35 and IL-23p19 subunits that are therefore considered limiting. Activation of human DCs by the TLR4 ligand LPS induces p19, p35 and p40 expression, while a TLR2 agonist, peptidoglycan only increases p19 and p40 mRNA [102]. TLR3 and TLR7 can induce IL-23p19 expression following viral challenge in mouse cells, but while TLR3 knockdown upregulates p40 expression, TLR7 knockdown decreases p40 mRNA in the same model [2]. Different combinations of TLR ligands can also promote IL-23 expression. In human monocyte-derived DCs, a combination of LPS and R848, a TLR7/TLR8 agonist, induces strong expression of p19 and p40, coinciding with secretion of IL-23 [67]. IRF5 is involved in the synergies between TLRs for the induction of p19 and p40. Stimulation of mouse peritoneal macrophages with CpG DNA and poly(I:C), ligands of TLR9 and TLR3, respectively, leads to the synergistic induction of p40 and p19 expression in an IRF5 manner [88]. Additionally, IRF5 directly binds to the promoters of the *IL12B* and *IL23A* genes in human pro-inflammatory M1 macrophages [60]. In line with these observations, IRF5 polymorphisms are associated with an increased risk for Crohn's disease [31].

Transcription factors of the NF-κB family are also involved in TLR-mediated induction of cytokines of the IL-12 family. c-Rel, a member of this family, binds to the promoters of p19, p35 and p40 and regulates their expression [12, 41, 77]. In addition, IL-23p19 mRNA expression is significantly reduced in c-Rel-deficient cells [12]. In agreement with these observations, c-Rel-deficient mice are protected against several autoimmune diseases involving IL-23, such as colitis or experimental autoimmune encephalitis (EAE) [14, 127]. Furthermore, c-Rel alterations are strongly associated with human autoimmune diseases such as RA, coeliac disease or psoriasis (reviewed in [34]).

TLR-mediated signalling events can activate another transcription factor of the NF-κB family, RelB, in DCs. RelB cooperates with c-Rel after stimulation with TLR2 or TLR9 agonists to induce DC maturation and the expression of pro-inflammatory genes such as *Tnf* and *Il23a* [115]. RelB was shown to directly bind the promoter of the *Il23a* gene upon TLR stimulation [115]. Nuclear localisation of RelB and the DNA-binding capacity of the RelB-p50 dimer following TLR4 stimulation are dependent on the kinase RIPK3 (receptor-interacting serine-threonine kinase 3) and RIPK3-induced reactive oxygen species (ROS) production [80].

Transcription factors belonging to the AP-1 family have also been shown to induce p19 expression following TLR stimulation. Ligation of TLR4 by LPS leads

Fig. 8.2 (continued) (also known as ER stress responses) induce the recruitment of CHOP, a transcription factor of the C/EBP family. IL-23p19 can also be negatively regulated, both transcriptionally and post-transcriptionally. The pro-Th1 cytokine IFNγ initiates STAT1 signalling that activates IRF1. This factor interferes with IRF5 binding to the *p19* promoter and inhibits transcription. The pro-Th2 cytokine, IL-4, triggers ATF3 that inhibits IL-23p19 transcription through an unknown mechanism. Multiple pathways, including IFNγ and MAPK signalling, regulate the expression and activity of tristetraprolin (TTP) that binds AU-rich elements in the 3′UTR of IL-23p19 mRNA, thereby promoting its degradation. Finally, the microRNA miR-107 directly targets IL-23p19 mRNA

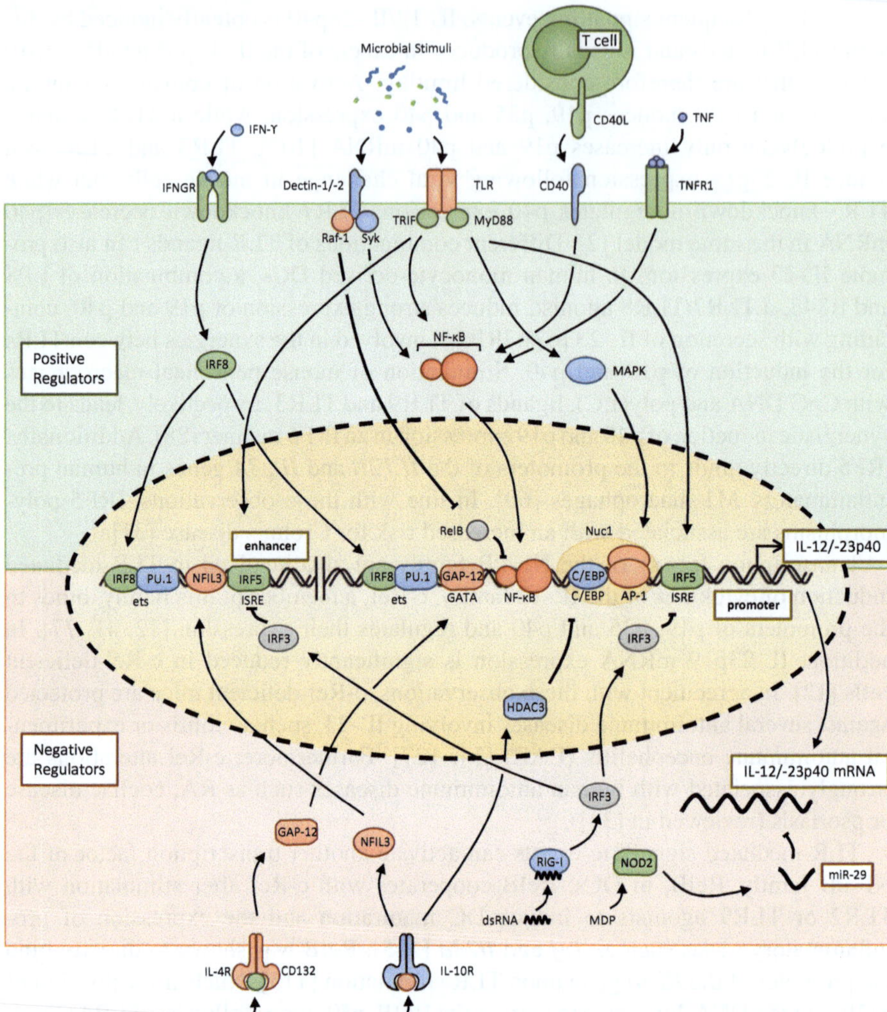

Fig. 8.3 Regulation of IL-12/IL-23p40 expression. *IL12B* is positively and negatively regulated, at the proximal promoter site but also at enhancer sites. Microbial stimuli through TLRs, as well as endogenous inflammatory stimuli such as CD40 ligation or TNF, induce the activation and nuclear translocation of NF-κB dimers that bind the proximal promoter. TLR ligation also promotes MAP kinase activation and ligation of an AP-1 dimer to the proximal promoter. Additionally, TLR signalling triggers the binding of IRF5 to ISREs in both the promoter and enhancer regions of the *Il12b* gene. Contrary to IL-23p19, IFNγ promotes IL-12/IL-23p40 transcription via elements in both the promoter and the enhancer regions. Transcription is associated with local nucleosomal remodelling processes. IL-4 exerts negative regulation through the binding of GAP-12, a putative transcriptional repressor, to a GATA box in the promoter. IL-10 represses IL-12/IL-23p40 expression through multiple mechanisms: NFIL3 binds to the enhancer region and represses transcription, while HDAC3, a histone deacetylase, modifies chromatin accessibility at the promoter. RIG-I, an antiviral double-stranded RNA sensor, induces the nuclear translocation of IRF3 that competes with IRF5 for binding to the ISRE sites in the enhancer and promoter. NOD2 recognises a component of bacteria, muramyl dipeptide (MDP), and induces the expression of a microRNA, miR-29. miR-29 specifically downregulates IL-12/IL-23p40 mRNA expression

to the activation of the ERK, JNK and p38 MAP-kinases which in turn stimulate the binding of c-Jun and ATF2 to an AP-1 site on the *Il23a* promoter [66].

Intracellular pathogen recognition receptors can alter TLR-mediated induction of IL-12 and IL-23. NOD2, a cytoplasmic receptor that recognises muramyl dipeptide (MDP) produced by both Gram-negative and Gram-positive bacteria, can have either a positive or negative effect on IL-12 and IL-23 production when combined with TLR ligation, depending on the context. Prestimulation of human and mouse DCs with MDP activates NOD2 and leads to a decrease in TLR-induced cytokine production, including p40 [128]. This reduction is dependent on the upregulation IRF4, although the exact mechanism still needs to be elucidated. However, when cells are stimulated with MDP and TLR ligands simultaneously, MDP does not inhibit TLR signalling. On the contrary, TLR7/TLR8 ligands combined with MDP induce a strong increase in IL-12 production, and TLR2 ligands + MDP lead to increased production of IL-23 [32, 126]. Thus, NOD2 ligands on their own have a protective effect in intestinal inflammation, but if a pathogen expresses both NOD2 and TLR ligands, production of pro-inflammatory cytokines occurs to preferentially activate the IL-12/Th1 or IL-23/Th17 axis.

Another cytoplasmic pathogen recognition receptor, RIG-I, an RNA helicase that recognises viral double-stranded RNA, can inhibit p40 production induced by TLRs. As previously stated, TLR ligation leads to the activation of IRF5-dependent transcription of p40 expression. Activation of RIG-I leads to the recruitment of another member of the IRF family, IRF3, to the promoter of the *Il12b* gene [83]. IRF3 interferes with IRF5 binding to the promoter and limits the assembly of an activating transcription factor complex, blocking p40 expression and inhibiting IL-12 and IL-23 production. Viral infection can therefore interfere with antibacterial and antifungal immune responses, by inhibiting TH1 or TH17 responses triggered by IL-12 or IL-23 through this mechanism.

The role of IL-17 in response to fungal pathogens is controversial. Whereas most studies have shown that Th17 responses are protective against yeast infections [50, 51], detrimental effects of Th17 responses and IL-23 have been published [68, 140]. Nevertheless, the signalling pathways induced by yeasts leading to IL-23 production are well understood. Among TLRs, TLR2 and TLR4 are the main contributors to yeast recognition. TLR2 recognises β-glucan, the major component of fungal cell walls, and TLR4 recognises *C. albicans*-derived mannan [84]. However, C-type lectin receptors (CLRs), such as Dectin-1 that also recognises β-glucan and Dectin-2 that recognises mannose-based structures, seem to be the major fungal recognition receptors in mice [120] and humans [27]. The balance between IL-12 and IL-23 can be regulated by co-ligation of these different receptors. Initial experiments showed that two forms of yeast can trigger the release of different cytokines. Yeast forms of *C. albicans* can induce both IL-12 and IL-23 production, while tissue-invasive hyphal forms only trigger IL-23 production [1]. This observation may be explained by the recognition of different signals by Dectin-1 and Dectin-2. Dectin-1 can only recognise the yeast form of *Candida*, because β-glucans become masked by mannans in the hyphal form [30]. On the other hand, Dectin-2 recognises mannose structures that are present after hyphal transformation [107]. While Dectin-1 activates several NF-κB subunits in human DCs leading to the production of IL-12 and IL-23, Dectin-2 only activates c-Rel that, as previously stated, is critical for IL-23p19 induction [39]. Dectin-1 and Dectin-2 both signal via Syk (Spleen tyrosine kinase), which drives the formation of a protein complex containing CARD9, BCL-10 and MALT1. This complex then activates c-Rel-dependent transcription of *IL23A*. Dectin-1 however, contrary to Dectin-2, also triggers a Syk-independent, Raf-1-dependent pathway [38]. Raf-1 signalling decreases binding of RelB to the *IL12B* promoter, allowing binding of the activating transcription factor RelA/p65 [38]. Indeed, binding of the non-canonical member of the NF-κB family, RelB, to the *IL12B* promoter has been shown to decrease p40 transcription [105].

Crosstalk between TLRs and CLRs can fine-tune responses to fungal pathogens, modify the IL-12/IL-23 balance and steer the immune system towards Th1 or Th17 responses. Co-stimulation of TLRs and Dectin-1 increases the production of IL-23 and decreases production of IL-12, by modifying the balance between the p19 and p35 transcripts in both murine and human dendritic cells [23, 32]. Raf-1 could be involved in the crosstalk between these receptors via both the regulation of p65 acetylation and the repression of Syk-induced RelB [38]. Simultaneous activation of TLRs and Dectin-1 can activate both NF-κB and ATF-2, which could also explain the specific induction of IL-23 by these two stimuli [104]. Indeed, stimulation of DCs with zymosan triggers Dectin-1 activation of the PLC-ϒ/PKC route. This leads to ATF2 recruitment to the *IL23A* promoter, which induces p19 but not p40 mRNA expression [104]. These observations have possible clinical implications for the treatment of fungal infections. Chromoblastomycosis is a difficult to treat subcutaneous infection caused by fungi such as *Fonsecaea pedrosoi* and has low cure rates [95]. Mouse studies have shown that inflammatory responses to *F. pedrosoi* can be restored by administration of TLR ligands [21]. In line with this observation,topical application of imiquimod, a TLR7 agonist, shows great potential in the treatment of this human disease [118].

Adenosine triphosphate (ATP) is the main intracellular energy carrier for many biological processes. It is present in both prokaryotic and eukaryotic cells but can be released to the extracellular milieu where it acts like a signalling molecule [10]. Extracellular ATP inhibits IL-12 and increases IL-23 production in dendritic cells via the purinergic P2 receptors [106, 110]. Consequently, ATP derived from commensal bacteria can lead to the production of IL-23p19 by intestinal lamina propria cells and drive Th17 responses in mice [5]. ATP and TLR2 signalling can also have a synergistic effect on human monocytes through a P2Y-cAMP-dependent pathway, promoting IL-23 and IL-1β production leading to enhanced Th17 responses [91].

8.5 Control of IL-23 Synthesis by Endogenous Ligands

The previous sections focused mainly on pathogen-derived signals that can stimulate p40 or p19 synthesis, although ATP can be either released from pathogens or from endogenous cells. When released from these cells, it serves as a danger signal and can also activate purinergic receptors on DCs [29]. Cytokines and cell-to-cell interactions also modulate IL-12 or IL-23 production. Original data showed that TNF was critical for the induction of p40 expression in mouse macrophages, as cells from *Tnfr1*$^{-/-}$ mice could not produce this subunit [28]. More recent data show that there is significant crosstalk between TNF and IL-23. Treatment of patients suffering from psoriatic lesions with etanercept, a soluble TNF receptor that blocks TNF signalling, decreases DC activation and IL-23p19 production in these lesions [37]. Etanercept also decreases serum levels of IL-23 in patients suffering from rheumatoid arthritis [52]. Blocking of TNF in mouse models of IL-23-induced epidermal hyperplasia strongly decreases acanthosis [13]. Finally, the absence of TNF signalling decreases IL-23p19 expression in mouse skin lesions [78], indicating that TNF and IL-23 could be part of the same positive feedback loop. However, the situation is not as clear-cut: other data show that TNF signalling via TNFR1 can exert a negative effect on p40 production in human and mouse cells, therefore inhibiting both IL-12 and IL-23 [69, 138]. TNF and p40 can be produced following activation of APCs through similar signals, and IL-23 can induce TNF expression in murine macrophages [20]. TNF could therefore be a key negative feedback regulator for the expression of IL-12 or IL-23, limiting the intensity or duration of Th1 and Th17 responses. Blockade of TNF signalling, either via anti-TNF antibodies or in *Tnfr1*$^{-/-}$ mice, leads to the expansion of Th1 and Th17 cells in mouse collagen-induced arthritis models [86]. Anti-TNF agents are used successfully for the treatment of many inflammatory disorders, but increasing evidence suggests that these therapies can also lead to the development of autoimmune diseases [101]. The absence of TNF-mediated IL-12 and IL-23 inhibition could explain these observations. Depending on the disease model or on the kinetics of expression, TNF can either have a positive or negative regulatory role on IL-12 and IL-23 expression.

Another major regulator of IL-23 production is IL-10. This anti-inflammatory cytokine inhibits both IL-12 and IL-23 production by targeting IL-12/IL-23p40 transcriptional activity. The mechanisms involved are not fully elucidated. Several early events required for the activation of *Il12b* gene, such as nucleosomal remodelling, C/EBP or NF-κB recruitment, were not (or marginally) affected by IL-10 despite robust transcriptional inhibition [142]. IL-10 induces nuclear factor, interleukin 3-regulated (NFIL3), a B-ZIP transcription factor in a STAT3-dependent manner. This transcription factor binds to the −10 kb enhancer region of the *Il12b* gene and represses its activity [117]. Furthermore, Kobayashi et al. showed that IL-10 promotes HDAC3-mediated histone deacetylation of the *Il12b* proximal promoter region [56]. Whether *Il23a* expression is also repressed by IL-10 has not been specifically addressed.

IFN-γ production is a hallmark of Th1 responses and is secreted by T cells in response to IL-12, but it also regulates both IL-12 and IL-23 expression in innate cells. IFN-γ induces both p35 and p40 expression forming a positive feedback loop [70]. LPS and IFN-γ stimulation of murine macrophages leads to the recruitment of IRF1 to an ISRE site at the *Il23a* promoter [113]. This inhibits RelA binding to and histone modification at the *Il23a* promoter [114]. Furthermore, IFN-γ has a protective effect in an IL-10-deficiency experimental model of murine colitis [114].

CD40 is a co-stimulatory molecule found on the surface of antigen-presenting cells. Ligation of CD40L (CD154) present on the surface of T helper cells to CD40 activates APCs and leads to the production of many pro-inflammatory cytokines. Activation of DCs by microbial stimuli upregulates IL-12/IL-23p40 and increases CD40 expression. CD40 ligation by CD40L can induce the production of both IL-12p35 and IL-23p19 by dendritic cells [111, 112]. Stimulation of CD40 signalling in T- and B-cell-deficient mice via the injection of an agonistic CD40 antibody mediates both systemic and intestinal inflammatory responses [125]. Systemic inflammation seems to be dependent on TNF and IL-12, while intestinal inflammation is induced by IL-23 in this model. Furthermore, CD40 ligation induces much higher expression of p19 mRNA in colonic $CD11c^{+}$ cells compared to splenic cells. The type of cell targeted by CD40L or the microenvironment could therefore direct the response through production of IL-12 or IL-23. The CD40-dependent production of IL-23 by DCs appears to rely on prostaglandin E2 (PGE2): blocking of PGE2 signalling inhibits IL-23 production by mouse splenic DCs cultured in vitro [134]. PGE2 signals through the prostaglandin EP4 receptor and triggers cAMP-dependent canonical and non-canonical NF-κB activation that amplifies *Il23a* expression [71].

IL-4 is a cytokine that induces Th2 cell differentiation and IgE isotype switching in B lymphocytes. IL-4 can however also lead to elevated IFN-γ expression in mice and protective Th1 responses against parasites [7, 97]. In addition, IL-4 has a protective role in inflammatory autoimmune diseases, and its mode of action has been recently uncovered [42]. Guenova et al. have shown that IL-4 has opposing effects on IL-12 and IL-23 production. Human and mouse DCs treated with IL-4 produce large amounts of IL-12 and low amounts of IL-23 following LPS stimulation. The mRNAs for IL-12 p35 is upregulated, while the mRNAs for IL-12/IL-23p40 and IL-23p19 are downregulated. This leads to suppression of Th17 responses both in vitro and in vivo. ATF3 is, at least partly, responsible for this repression. ATF3 has been shown to directly regulate *Il12b* transcription [33], and ATF3-deficient BMDCs treated with IL-4 no longer downregulate IL-23 secretion. In line with these observations, IL-4 therapy of psoriasis improves the symptoms of the disease and is correlated with a decrease in IL-23 expression in psoriatic plaques [42].

The endoplasmic reticulum (ER) stress pathway, also known as the unfolded protein response (UPR), is activated when an imbalance between protein synthesis and the folding capacity of the ER occurs. It is involved in a number of pathological processes such as neurodegenerative diseases, obesity, diabetes, atherosclerosis, heart disease, cancer and infection [49]. Components of the ER stress pathway are also activated downstream of TLRs, contributing to inflammatory cytokine production [76]. Activation of the ER stress in conjunction with TLR ligation leads to enhanced IL-23p19 expression and IL-23 production [35]. This is mediated by the UPR-inducible transcription factor CHOP (C/EBP homologous protein) that was shown to be directly recruited to the *IL23A* promoter region. There is strong evidence for crosstalks between pathogen and cellular stress sensors [19]. Activation of DCs by zymosan or hypha forming *Candida* yeast drives the disappearance of nuclear CHOP protein, indicating that this pathway is not involved in IL-23 production under these conditions [104].

The HLA-B27 transgenic rat recapitulates many features of human spondyloarthritis. In this model, very high levels of HLA-B27 are expressed, with consequent misfolding of the B27 heavy chain in the ER. At sites of inflammation, UPR is observed along with upregulation of IL-23 production, suggesting that this specific pathway could be involved in the pathological process [22, 124]. Interestingly, LPS-stimulated macrophages derived from monocytes of HLA-B27$^+$ ankylosing spondylitis patients produce large amounts of IL-23. However, this does not correlate with signs of UPR activation in these cells [141]. Along this line, it was suggested that in *lamina propria* mononuclear cells from ankylosing spondylitis patients, an autophagy-related pathway rather than UPR was associated with excessive IL-23 expression [17]. In conclusion, the relevance of an UPR/IL-23 axis in pathological situations is still unclear.

8.6 Regulation of IL-23 in the Tumour Microenvironment

Inflammation is often associated with the development of solid tumours. In experimental models, IL-23 was shown to promote tumour incidence and growth [62]. IL-23 is thought to favour angiogenesis while reducing CD8 T cell and NK-mediated immunosurveillance [62, 122]. IL-23 allows cancer cells to persist in a state of immune-mediated dormancy by opposing the antitumour effects of IL-12 [123]. This is particularly relevant for colorectal cancers. Indeed, there is a strong upregulation of IL-23 expression by tumour-associated myeloid cells both in mouse models and human samples. It was proposed that adenoma-linked barrier defects lead to abnormal MyD88- and TLR-dependent activation of immune cells by microbial products [40]. Tumour-derived factors were also shown to contribute to IL-23 production. Constitutive STAT3 activation is frequently observed in diverse cancers, both in tumour cells and infiltrating immune cells including DCs and macrophages. Kortylewski and colleagues showed that STAT3 was directly recruited to the *Il23a* promoter region in tumour-associated macrophages. Furthermore, STAT3 also repressed c-Rel-dependent *Il12a* gene expression [58]. Consistent with previous data [54], PGE2 secretion by breast tumour cells was shown to enhance *Il23a* gene expression through a cAMP-response element in the promoter region [93]. Another mechanism could explain tumour cell-induced IL-23 expression in macrophages. Tumour cells favour metabolic pathways such as glycolysis that lead to the production of lactic acid that possesses immunosuppressive and angiogenic activities [16]. Lactic acid secreted by tumour cells can also enhance the transcription of IL-23p19 and lead to increased IL-23 secretion by macrophages [116]. Stress response from the endoplasmic reticulum could also contribute to the propensity of tumour-associated macrophages to produce IL-23. ER stress response in tumour cells has been associated with tumour growth and progression [72]. Activation of this pathway could be transmitted to myeloid cells in the tumour microenvironment, thereby leading to heightened IL-23-dependent pro-inflammatory responses [75]. Along this line, tumour-associated macrophages strongly upregulate the expression of a downstream effector molecular chaperone glycoprotein 96 (gp96) [103]. In the context of experimental colon tumorigenesis model, efficient IL-23 expression was found to be gp96-dependent [79]. As this chaperone is a major effector of ER stress responses and catalyses the folding of most TLRs, this could be secondary to impaired responsiveness of tumour-associated macrophages to microbial stimuli. Taken together, microbial and stress-related signalling pathways seem to converge to promote IL-23 production in the tumour microenvironment, thereby favouring immune evasion.

8.7 Post-transcriptional Regulation of IL-23

In addition to transcriptional regulatory networks, dynamic expression of cytokine and chemokine genes is strongly influenced by mRNA decay [46]. Indeed, 3′ untranslated regions (UTRs) of these genes are enriched in adenine and uridine-rich elements (AREs), sequences that control mRNA stability and translation through recruitment of protein complexes [8]. Tristetraprolin (TTP, encoded by *Zfp36*) is one of the best-characterised ARE-binding protein. In macrophages or DCs, TTP acts as a major regulator of RNA degradation [96] and controls the production of many inflammatory cytokines such as TNF, IL-6 or GM-CSF [59]. Several AREs can be found within the 3′UTR regions of human and murine IL-23p19 mRNAs, and in vitro experiments indicate the direct binding of TTP to these elements. As a consequence, LPS-stimulated *Zfp36*$^{-/-}$ BMDCs cells produce high levels of IL-23. Interestingly, Qian and colleagues observed that inhibition of IL-23 production by IFN-γ treatment was mediated by the induction of TTP and subsequent IL-23p19 mRNA destabilisation [94]. TTP-deficient mice spontaneously develop a complex TNF-dependent inflammatory syndrome that combines cachexia, dermatitis, arthritis, conjunctivitis and myeloid hyperplasia [119]. All these clinical parameters were also found to be dependent on dysregulated IL-23 production and subsequent overproduction of IL-17 by various lymphoid subsets, indicating the importance of this regulatory mechanism in maintaining homeostasis [78].

MicroRNAs (miRNAs) are small, non-coding RNA molecules that also play an important role in the regulation/modulation of inflammatory cytokines. miRNA-induced silencing complex assembles with sequences located mostly in the 3′-UTRs of target mRNAs and induces changes in their subcellular localisation, translation efficiency or stability [44]. IL-23p19 mRNA is a direct target of miR-107. The presence of microbiota downregulates miR-107 expression in intestinal DCs and macrophages, thereby favouring IL-23 expression [133]. In human DCs, miR-29 expression is upregulated in response to NOD2 signals. miR-29 targets IL-12p40 mRNA directly and IL-23p19 indirectly, likely via reduction of ATF2 expression. As NOD2 polymorphisms are associated with Crohn's disease, this NOD2/miR-29 axis might contribute to exacerbated IL-23 in these patients [11].

8.8 Concluding Remarks

The recent clinical results clearly demonstrate that targeting IL-23 has important medical benefit for serious immune-mediated diseases such as psoriasis or ankylosing spondylitis. The data outlined herein indicate that its regulation is highly complex as expression of both subunits follows distinct rules and involves different epigenetic, transcriptional and post-transcriptional mechanisms. The use of novel technological tools such as transgenic reporter mice, transcriptional analysis at the single-cell level and genome-wide profiling of the chromatin state will help define the molecular requirements for bioactive IL-23 production by specific subpopulations. A better understanding of how genetic and environmental factors initiate and maintain excessive IL-23 expression in pathological contexts will help to develop new preventive or therapeutic strategies.

References

1. Acosta-Rodriguez EV, et al. Surface phenotype and antigenic specificity of human interleukin 17-producing T helper memory cells. Nat Immunol. 2007;8:639–46.
2. Al-Salleeh F, Petro TM. TLR3 and TLR7 are involved in expression of IL-23 subunits while TLR3 but not TLR7 is involved in expression of IFN-β by Theiler's virus-infected RAW264.7 cells. Microbes Infect. 2007;9(11):1384–92.
3. Al-Sammak F, et al. Gastric epithelial expression of IL-12 cytokine family in helicobacter pylori infection in human: is it head or tail of the coin? PLoS One. 2013;8(9):e75192.
4. Arnold IC et al. CD11c+ monocyte/macrophages promote chronic Helicobacter hepaticus-induced intestinal inflammation through the production of IL-23. Mucosal Immunol. 2016;9(2):352–63.
5. Atarashi K, et al. ATP drives lamina propria TH17 cell differentiation. Nature. 2008;455(7214):808–12.
6. Awasthi A, et al. Cutting edge: IL-23 receptor gfp reporter mice reveal distinct populations of IL-17-producing cells. J Immunol. 2009;182:5904–8.
7. Biedermann T, et al. IL-4 instructs TH1 responses and resistance to Leishmania major in susceptible BALB/c mice. Nat Immunol. 2001;2(11):1054–60.
8. Blackshear PJ. Tristetraprolin and other CCCH tandem zinc-finger proteins in the regulation of mRNA turnover. Biochem Soc Trans. 2002;30(Pt 6):945–52.
9. Bosmann M, et al. CD11c+ alveolar macrophages are a source of IL-23 during lipopolysaccharide-induced acute lung injury. Shock (Augusta, Ga). 2013;39(5):447–52.
10. Bours MJL, et al. Adenosine 5′-triphosphate and adenosine as endogenous signaling molecules in immunity and inflammation. Pharmacol Ther. 2006;112(2):358–404.
11. Brain O, et al. The intracellular sensor NOD2 induces MicroRNA-29 expression in human dendritic cells to limit IL-23 release. Immunity. 2013;39(3):521–36.
12. Carmody RJ, et al. Essential roles of c-Rel in TLR-induced IL-23 p19 gene expression in dendritic cells. J Immunol. 2007;178(1):186–91.
13. Chan JR, et al. IL-23 stimulates epidermal hyperplasia via TNF and IL-20R2-dependent mechanisms with implications for psoriasis pathogenesis. J Exp Med. 2006;203(12):2577–87.
14. Chen G, et al. The NF-kB transcription factor c-Rel is required for Th17 effector cell development in experimental autoimmune encephalomyelitis. J Immunol (Baltimore, Md : 1950). 2011;187(9):4483–91.

15. Chognard G, et al. The dichotomous pattern of IL-12R and IL-23R expression elucidates the role of IL-12 and IL-23 in inflammation C. Bursill, ed. PLoS One. 2014;9(2):e89092.
16. Choi SYC, et al. Cancer-generated lactic acid: a regulatory, immunosuppressive metabolite? J Pathol. 2013;230(4):350–5.
17. Ciccia F, et al. Evidence that autophagy, but not the unfolded protein response, regulates the expression of IL-23 in the gut of patients with ankylosing spondylitis and subclinical gut inflammation. Ann Rheum Dis. 2014;73(8):1566–74.
18. Ciccia F, et al. Overexpression of interleukin-23, but not interleukin-17, as an immunologic signature of subclinical intestinal inflammation in ankylosing spondylitis. Arthritis Rheum. 2009;60(4):955–65.
19. Cláudio N, et al. Mapping the crossroads of immune activation and cellular stress response pathways. EMBO J. 2013;32(9):1214–24.
20. Cua DJ, et al. Interleukin-23 rather than interleukin-12 is the critical cytokine for autoimmune inflammation of the brain. Nature. 2003;421(6924):744–8.
21. Da Glória Sousa M, et al. Restoration of pattern recognition receptor costimulation to treat chromoblastomycosis, a chronic fungal infection of the skin. Cell Host Microbe. 2011;9(5):436–43.
22. DeLay ML, et al. HLA-B27 misfolding and the unfolded protein response augment interleukin-23 production and are associated with Th17 activation in transgenic rats. Arthritis Rheum. 2009;60(9):2633–43.
23. Dennehy KM, et al. Reciprocal regulation of IL-23 and IL-12 following co-activation of dectin-1 and TLR signaling pathways. Eur J Immunol. 2009;39(5):1379–86.
24. Dillon SM, et al. Human intestinal lamina propria CD1c+ dendritic cells display an activated phenotype at steady state and produce IL-23 in response to TLR7/8 stimulation. J Immunol. 2010;184(12):6612–21.
25. Duerr RH, et al. A genome-wide association study identifies IL23R as an inflammatory bowel disease gene. Science. 2006;314(5804):1461–3.
26. Eken A, et al. IL-23R+ innate lymphoid cells induce colitis via interleukin-22-dependent mechanism. Mucosal Immunol. 2014;7(1):143–54.
27. Ferwerda B, et al. Human dectin-1 deficiency and mucocutaneous fungal infections. N Eng J M. 2009;361(18):1760–7.
28. Flesch IE. Early interleukin 12 production by macrophages in response to mycobacterial infection depends on interferon gamma and tumor necrosis factor alpha. J Exp Med. 1995;181(5):1615–21.
29. Gallucci S, Matzinger P. Danger signals: SOS to the immune system. Curr Opin Immunol. 2001;13(1):114–9.
30. Gantner BN, Simmons RM, Underhill DM. Dectin-1 mediates macrophage recognition of Candida albicans yeast but not filaments. EMBO J. 2005;24(6):1277–86.
31. Gathungu G, et al. A two-marker haplotype in the IRF5 gene is associated with inflammatory bowel disease in a North American cohort. Genes Immun. 2012;13(4):351–5.
32. Gerosa F, et al. Differential regulation of interleukin 12 and interleukin 23 production in human dendritic cells. J Exp Med. 2008;205(6):1447–61.
33. Gilchrist M, et al. Systems biology approaches identify ATF3 as a negative regulator of toll-like receptor 4. Nature. 2006;441(7090):173–8.
34. Gilmore TD, Gerondakis S. The c-Rel transcription factor in development and disease. Genes Cancer. 2011;2(7):695–711.
35. Goodall JC, et al. Endoplasmic reticulum stress-induced transcription factor, CHOP, is crucial for dendritic cell IL-23 expression. Proc Natl Acad Sci U S A. 2010;107(41):17698–703.
36. Gosselin D, et al. Environment drives selection and function of enhancers controlling tissue-specific macrophage identities. Cell. 2014;159(6):1327–40.
37. Gottlieb AB, et al. TNF inhibition rapidly down-regulates multiple proinflammatory pathways in psoriasis plaques. J Immunol. 2005;175(4):2721–9.

38. Gringhuis SI, et al. Dectin-1 directs T helper cell differentiation by controlling noncanonical NF-kB activation through Raf-1 and Syk. Nat Immunol. 2009;10(2):203–13.
39. Gringhuis SI, et al. Selective C-Rel activation via Malt1 controls anti-fungal TH-17 immunity by dectin-1 and dectin-2. PLoS Pathog. 2011;7(1):e1001259.
40. Grivennikov SI, et al. Adenoma-linked barrier defects and microbial products drive IL-23/IL-17-mediated tumour growth. Nature. 2012;491(7423):254–8.
41. Grumont R, et al. c-Rel regulates interleukin 12 p70 expression in CD8(+) dendritic cells by specifically inducing p35 gene transcription. J Exp Med. 2001;194(8):1021–32.
42. Guenova E, et al. IL-4 abrogates T(H)17 cell-mediated inflammation by selective silencing of IL-23 in antigen-presenting cells. Proc Natl Acad Sci U S A. 2015;112(7):2163–8.
43. Guilliams M, et al. Dendritic cells, monocytes and macrophages: a unified nomenclature based on ontogeny. Nat Rev Immunol. 2014;14(8):571–8.
44. Ha M, Kim VN. Regulation of microRNA biogenesis. Nat Rev Mol Cell Biol. 2014;15(8):509–24.
45. Hänsel A, et al. Human slan (6-sulfo LacNAc) dendritic cells are inflammatory dermal dendritic cells in psoriasis and drive strong Th17/Th1 T-cell responses. J Allergy Clin Immunol. 2011;127(3):787–94.
46. Hao S, Baltimore D. The stability of mRNA influences the temporal order of the induction of genes encoding inflammatory molecules. Nat Immunol. 2009;10(3):281–8.
47. Hashimoto D, et al. Tissue-resident macrophages self-maintain locally throughout adult life with minimal contribution from circulating monocytes. Immunity. 2013;38(4):792–804.
48. Hor YT, et al. A role for RUNX3 in inflammation-induced expression of IL23A in gastric epithelial cells. Cell Rep. 2014;8(1):50–8.
49. Hotamisligil GS. Endoplasmic reticulum stress and the inflammatory basis of metabolic disease. Cell. 2010;140(6):900–17.
50. Huang W, et al. Requirement of interleukin-17A for systemic anti-Candida albicans host defense in mice. J Infect Dis. 2004;190(3):624–31.
51. Kagami S, et al. IL-23 and IL-17A, but Not IL-12 and IL-22, are required for optimal skin host defense against candida albicans. J Immunol. 2010;185(9):5453–62.
52. Kageyama Y, et al. Etanercept reduces the serum levels of interleukin-23 and macrophage inflammatory protein-3 alpha in patients with rheumatoid arthritis. Rheumatol Int. 2007;28(2):137–43.
53. Kamada N, et al. Unique CD14+ intestinal macrophages contribute to the pathogenesis of Crohn's disease via IL-23/IFN-γ axis. J Clin Investig. 2008;118(6):2269–80.
54. Khayrullina T, et al. In vitro differentiation of dendritic cells in the presence of prostaglandin E2 alters the IL-12/IL-23 balance and promotes differentiation of Th17 cells. J Immunol (Baltimore, Md : 1950). 2008;181(1):721–35.
55. Kinnebrew MA, et al. Interleukin 23 production by intestinal CD103+CD11b+dendritic cells in response to bacterial flagellin enhances mucosal innate immune defense. Immunity. 2012;36(2):276–87.
56. Kobayashi T, et al. IL-10 regulates Il12b expression via histone deacetylation: implications for intestinal macrophage homeostasis. J Immunol. 2012;189:1792–9.
57. Kopp T, et al. IL-23 production by cosecretion of endogenous p19 and transgenic p40 in keratin 14/p40 transgenic mice: evidence for enhanced cutaneous immunity. J Immunol. 2003;170(11):5438–44.
58. Kortylewski M, et al. Regulation of the IL-23 and IL-12 balance by Stat3 signaling in the tumor microenvironment. Cancer Cell. 2009;15(2):114–23.
59. Kratochvill F, et al. Tristetraprolin-driven regulatory circuit controls quality and timing of mRNA decay in inflammation. Mol Syst Biol. 2011;7(1):560.
60. Krausgruber T, et al. IRF5 promotes inflammatory macrophage polarization and TH1-TH17 responses. Nat Immunol. 2011;12(3):231–8.
61. Kvedaraite, E et al. Tissue-infiltrating neutrophils represent the main source of IL-23 in the colon of patients with IBD. Gut. 2015;pp.gutjnl–2014–309014.

62. Langowski JL, et al. IL-23 promotes tumour incidence and growth. Nature. 2006;442(7101):461–5.
63. Lavin Y, et al. Tissue-resident macrophage enhancer landscapes are shaped by the local microenvironment. Cell. 2014;159(6):1312–26.
64. Lee E, et al. Increased expression of interleukin 23 p19 and p40 in lesional skin of patients with psoriasis vulgaris. J Exp Med. 2004;199(1):125–30.
65. Lee Y, et al. Induction and molecular signature of pathogenic TH17 cells. Nat Immunol. 2012;13(10):991–9.
66. Liu W, et al. AP-1 activated by toll-like receptors regulates expression of IL-23 p19. J Biol Chem. 2009;284(36):24006–16.
67. Lombardi V, et al. Human dendritic cells stimulated via TLR7 and/or TLR8 induce the sequential production of Il-10, IFN-gamma, and IL-17A by naive CD4+ T cells. J Immunol. 2009;182(6):3372.
68. De Luca A, et al. IL-22 defines a novel immune pathway of antifungal resistance. Mucosal Immunol. 2010;3(4):361–73.
69. Ma X, et al. Inhibition of IL-12 production in human monocyte-derived macrophages by TNF. J Immunol. 2000;164(4):1722–9.
70. Ma X, et al. The interleukin 12 p40 gene promoter is primed by interferon gamma in monocytic cells. J Exp Med. 1996;183(1):147–57.
71. Ma, X, Aoki, T, Narumiya, S. Prostaglandin E2-EP4 signaling persistently amplifies CD40-mediated induction of IL-23 p19 expression through canonical and non-canonical NF-kB pathways. Cell Mol Immunol. 2016;13(2):240–50.
72. Ma Y, Hendershot LM. The role of the unfolded protein response in tumour development: friend or foe? Nat Rev Cancer. 2004;4(12):966–77.
73. Maaser C, et al. Expression of Epstein-Barr virus-induced gene 3 and other interleukin-12-related molecules by human intestinal epithelium. Immunology. 2004;112(3):437–45.
74. Macho-Fernandez E, et al. Lymphotoxin beta receptor signaling limits mucosal damage through driving IL-23 production by epithelial cells. Mucosal Immunol. 2014;8(2):403–13.
75. Mahadevan NR, et al. Transmission of endoplasmic reticulum stress and pro-inflammation from tumor cells to myeloid cells. Proc Natl Acad Sci. 2011;108(16):6561–6.
76. Martinon F, et al. TLR activation of the transcription factor XBP1 regulates innate immune responses in macrophages. Nat Immunol. 2010;11(5):411–8.
77. Mise-Omata S, et al. A proximal kappaB site in the IL-23 p19 promoter is responsible for RelA- and c-Rel-dependent transcription. J Immunol. 2007;179(10):6596–603.
78. Molle C, et al. Tristetraprolin regulation of interleukin 23 mRNA stability prevents a spontaneous inflammatory disease. J Exp Med. 2013;210(9):1675–84.
79. Morales C, et al. Immune chaperone gp96 drives the contributions of macrophages to inflammatory colon tumorigenesis. Cancer Res. 2014;74(2):446–59.
80. Moriwaki K, et al. The necroptosis adaptor RIPK3 promotes injury-induced cytokine expression and tissue repair. Immunity. 2014;41(4):567–78.
81. Muranski P, Restifo NP. Essentials of Th17 cell commitment and plasticity. Blood. 2013;121(13):2402–14.
82. Natoli G, Ghisletti S, Barozzi I. The genomic landscapes of inflammation. Genes Dev. 2011;25(2):101–6.
83. Negishi H, et al. Cross-interference of RLR and TLR signaling pathways modulates antibacterial T cell responses. Nat Immunol. 2012;13(7):659–66.
84. Netea MG, et al. An integrated model of the recognition of Candida albicans by the innate immune system. Nat Rev Microbiol. 2008;6(1):67–78.
85. Nicodeme E, et al. Suppression of inflammation by a synthetic histone mimic. Nature. 2010;468(7327):1119–23.
86. Notley CA, et al. Blockade of tumor necrosis factor in collagen-induced arthritis reveals a novel immunoregulatory pathway for Th1 and Th17 cells. J Exp Med. 2008;205(11):2491–7.

87. Oppmann B, et al. Novel p19 protein engages IL-12p40 to form a cytokine, IL-23, with biological activities similar as well as distinct from IL-12. Immunity. 2000;13(5):715–25.
88. Ouyang X, et al. Cooperation between MyD88 and TRIF pathways in TLR synergy via IRF5 activation. Biochem Biophys Res Commun. 2007;354(4):1045–51.
89. Pantelyushin S, et al. Rorγt+ innate lymphocytes and γδ T cells initiate psoriasiform plaque formation in mice. J Clin Invest. 2012;122(6):2252–6.
90. Paul F, Amit I. Plasticity in the transcriptional and epigenetic circuits regulating dendritic cell lineage specification and function. Curr Opin Immunol. 2014;30(1):1–8.
91. Paustian C, et al. Extracellular ATP and toll-like receptor 2 agonists trigger in human monocytes an activation program that favors T helper 17. PLoS One. 2013;8(1):e54804.
92. Piskin G, et al. In vitro and in situ expression of IL-23 by keratinocytes in healthy skin and psoriasis lesions: enhanced expression in psoriatic skin. J Immunol. 2006;176(3):1908–15.
93. Qian X, et al. Increased Th17 cells in the tumor microenvironment is mediated by IL-23 via tumor-secreted prostaglandin E2. J Immunol. 2013;190(11):5894–902.
94. Qian X, et al. Posttranscriptional regulation of IL-23 expression by IFN-gamma through tristetraprolin. J Immunol (Baltimore, Md : 1950). 2011;186(11):6454–64.
95. Queiroz-Telles F, et al. Chromoblastomycosis: an overview of clinical manifestations, diagnosis and treatment. Med Mycol. 2009;47(1):3–15.
96. Rabani M, et al. High-resolution sequencing and modeling identifies distinct dynamic RNA regulatory strategies. Cell. 2014;159(7):1698–710.
97. Racke MK, et al. Cytokine-induced immune deviation as a therapy for inflammatory autoimmune disease. J Exp Med. 1994;180(5):1961–6.
98. Ramirez-Carrozzi VR, et al. A unifying model for the selective regulation of inducible transcription by CpG islands and nucleosome remodeling. Cell. 2009;138(1):114–28.
99. Ramirez-Carrozzi VR. Selective and antagonistic functions of SWI/SNF and Mi-2beta nucleosome remodeling complexes during an inflammatory response. Genes Dev. 2006;20(3):282–96.
100. Ramnath D, et al. TLR3 drives IRF6-dependent IL-23p19 expression and p19/EBI3 heterodimer formation in keratinocytes. Immunol Cell Biol. 2015;93(9):1–9.
101. Ramos-Casals M, et al. Autoimmune diseases induced by TNF-targeted therapies. Best Pract Res Clin Rheumatol. 2008;22(5):847–61.
102. Re F, Strominger JL. Toll-like receptor 2 (TLR2) and TLR4 differentially activate human dendritic cells. J Biol Chem. 2001;276(40):37692–9.
103. Reddy RK, et al. Cancer-inducible transgene expression by the Grp94 promoter: spontaneous activation in tumors of various origins and cancer-associated macrophages. Cancer Res. 2002;62(24):7207–12.
104. Rodriguez M, et al. The unfolded protein response and the phosphorylations of activating transcription factor 2 in the trans-activation of il23a promoter produced by β-Glucans. J Biol Chem. 2014;289(33):22942–57.
105. Saccani S, Pantano S, Natoli G. Modulation of NF-kappaB activity by exchange of dimers. Mol Cell. 2003;11(6):1563–74.
106. La Sala A, et al. Extracellular ATP induces a distorted maturation of dendritic cells and inhibits their capacity to initiate Th1 responses. J Immunol. 2001;166(3):1611–7.
107. Sato K, et al. Dectin-2 is a pattern recognition receptor for fungi that couples with the Fc receptor chain to induce innate immune responses. J Biol Chem. 2006;281(50):38854–66.
108. Satpathy AT, et al. Notch2-dependent classical dendritic cells orchestrate intestinal immunity to attaching-and-effacing bacterial pathogens. Nat Immunol. 2013;14(9):937–48.
109. Schlitzer A, et al. IRF4 transcription factor-dependent CD11b+ dendritic cells in human and mouse control mucosal IL-17 cytokine responses. Immunity. 2013;38(5):970–83.
110. Schnurr M, et al. Extracellular nucleotide signalling via P2 receptors inhibits IL-12 and enhances IL-23 expression in human dendritic cells: a novel role for the cAMP pathway. Gene Expr. 2004;105(4):1582–90.

111. Schulz O, et al. CD40 triggering of heterodimeric IL-12 p70 production by dendritic cells in vivo requires a microbial priming signal. Immunity. 2000;13(4):453–62.
112. Sender LY, et al. CD40 ligand-triggered human dendritic cells mount interleukin-23 responses that are further enhanced by danger signals. Mol Immunol. 2010;47(6):1255–61.
113. Sheikh SZ, et al. Characterization of an Interferon-stimulated Response Element (ISRE) in the Il23a promoter. J Biol Chem. 2011;286(2):1174–80.
114. Sheikh SZ, et al. Cutting edge: IFN-gamma is a negative regulator of IL-23 in murine macrophages and experimental colitis. J Immunol (Baltimore, Md : 1950). 2010;184(8):4069.
115. Shih VF-S, et al. Control of RelB during dendritic cell activation integrates canonical and noncanonical NF-kB pathways. Nat Immunol. 2012;13(12):1162–70.
116. Shime H, et al. Tumor-secreted lactic acid promotes IL-23/IL-17 proinflammatory pathway. J Immunol. 2008;180(11):7175–83.
117. Smith AM, et al. A distal enhancer in Il12b is the target of transcriptional repression by the STAT3 pathway and requires the basic leucine zipper (B-ZIP) protein NFIL3. J Biol Chem. 2011;286(26):23582–90.
118. De Sousa M d GT, et al. Topical application of imiquimod as a treatment for chromoblastomycosis. Clin Infect Dis. 2014;58(12):1734–7.
119. Taylor GA, et al. A pathogenetic role for TNF alpha in the syndrome of cachexia, arthritis, and autoimmunity resulting from tristetraprolin (TTP) deficiency. Immunity. 1996;4(5):445–54.
120. Taylor PR, et al. Dectin-1 is required for β-glucan recognition and control of fungal infection. Nat Immunol. 2007;8(1):31–8.
121. Teng MWL, et al. IL-12 and IL-23 cytokines: from discovery to targeted therapies for immune-mediated inflammatory diseases. Nat Med. 2015;21(7):719–29.
122. Teng MWL, et al. IL-23 suppresses innate immune response independently of IL-17A during carcinogenesis and metastasis. Proc Natl Acad Sci. 2010;107(18):8328–33.
123. Teng MWL, et al. Opposing roles for IL-23 and IL-12 in maintaining occult cancer in an equilibrium state. Cancer Res. 2012;72(16):3987–96.
124. Turner MJ, et al. HLA-B27 misfolding in transgenic rats is associated with activation of the unfolded protein response. J Immunol. 2005;175(4):2438–48.
125. Uhlig HH, et al. Differential activity of IL-12 and IL-23 in mucosal and systemic innate immune pathology. Immunity. 2006;25(2):309–18.
126. Van Beelen AJ, et al. Stimulation of the intracellular bacterial sensor NOD2 programs dendritic cells to promote interleukin-17 production in human memory T cells. Immunity. 2007;27(4):660–9.
127. Wang Y, et al. c-Rel is essential for the development of innate and T cell-induced colitis. J Immunol. 2008;180(12):8118–25.
128. Watanabe T, et al. Muramyl dipeptide activation of nucleotide-binding oligomerization domain 2 protects mice from experimental colitis. J Clin Invest. 2008;118(2):545–59.
129. Weinmann AS, Plevy SE, Smale ST. Rapid and selective remodeling of a positioned nucleosome during the induction of IL-12 p40 transcription. Immunity. 1999;11(6):665–75.
130. Welty NE, et al. Intestinal lamina propria dendritic cells maintain T cell homeostasis but do not affect commensalism. J Exp Med. 2013;210(10):2011–24.
131. Wiekowski MT, et al. Ubiquitous transgenic expression of the IL-23 subunit p19 induces multiorgan inflammation, runting, infertility, and premature death. J Immunol. 2001;166(12):7563–70.
132. Wohn C, et al. Langerin(neg) conventional dendritic cells produce IL-23 to drive psoriatic plaque formation in mice. Proc Natl Acad Sci U S A. 2013;110(26):10723–8.
133. Xue X, et al. Downregulation of microRNA-107 in intestinal CD11c+ myeloid cells in response to microbiota and proinflammatory cytokines increases IL-23p19 expression. Eur J Immunol. 2014;44:673–82.
134. Yao C, et al. Prostaglandin E2–EP4 signaling promotes immune inflammation through TH1 cell differentiation and TH17 cell expansion. Nat Med. 2009;15(6):633–40.

135. Yawalkar N, et al. Increased expression of IL-12p70 and IL-23 by multiple dendritic cell and macrophage subsets in plaque psoriasis. J Dermatol Sci. 2009;54(2):99–105.
136. Yen D, et al. IL-23 is essential for T cell-mediated colitis and promotes inflammation via IL-17 and IL-6. J Clin Invest. 2006;116(5):1310–6.
137. Yoshiki R, et al. IL-23 from Langerhans cells is required for the development of imiquimod-induced psoriasis-like dermatitis by induction of IL-17A-producing γδ T cells. J Invest Dermatol. 2014;134(7):1912–21.
138. Zakharova M, Ziegler HK. Paradoxical anti-inflammatory actions of TNF- : inhibition of IL-12 and IL-23 via TNF receptor 1 in macrophages and dendritic cells. J Immunol. 2005;175(8):5024–33.
139. Zelante T, et al. CD103+ dendritic cells control Th17 cell function in the lung. Cell Rep. 2015;12(11):1789–801.
140. Zelante T, et al. IL-23 and the Th17 pathway promote inflammation and impair antifungal immune resistance. Eur J Immunol. 2007;37(10):2695–706.
141. Zeng L, Lindstrom MJ, Smith JA. Ankylosing spondylitis macrophage production of higher levels of interleukin-23 in response to lipopolysaccharide without induction of a significant unfolded protein response. Arthritis Rheum. 2011;63(12):3807–17.
142. Zhou L, Nazarian AA, Smale ST. Interleukin-10 inhibits interleukin-12 p40 gene transcription by targeting a late event in the activation pathway. Mol Cell Biol. 2004;24(6):2385–96.

Chapter 9
Regulation and Immune Function of IL-27

Qinghong Wang and Jianguo Liu

Abstract IL-27 is a pleiotropic cytokine that has diverse immune regulatory activities under both physiological and pathological conditions. IL-27 enhances the functions of Th1 and CD8⁺ T cells, promotes the development of Tfh and Tr1, and suppresses the functions of Th2, Treg, Th9, and Th17 cells. IL-27 is also involved in regulation of immune responses of B cells, NK, DCs, and macrophages. IL-27 production is strictly regulated at both transcriptional and posttranscriptional levels. Given its broad effects on immune regulation, IL-27 has been implicated in the pathogeneses of autoimmune and infectious diseases as well as cancers. In this chapter, the biological characters of IL-27 have been reviewed, including the molecular mechanisms of IL-27 production and the underlying cellular signaling pathway. The regulatory effects of IL-27 on immune responses in autoimmune disease, infectious disease, and cancer are also discussed.

Keywords IL-27 • T cell function • DC and macrophage • Cancer • Autoimmune diseases • Infectious diseases • Signaling pathway

IL-27 is a member of the IL-12 family and composed of Epstein-Barr virus-induced gene3 (EBI3) and p28 subunits, with the former subunit shared with IL-12 p35 and the latter being the unique subunit for IL 27. IL-27 signals cells through its receptor composed of two intracellular chains, with gp130 shared with IL-6 and a specific chain called IL-27Ra (also known as WSX-1 or TCCR). By engaging its receptor, IL-27 activates Janus kinase (JAK)/signal transducer and activator of transcription (STAT) and mitogen-activated protein kinase (MAPK) signaling to exert diverse immune regulatory activities under different physiological and pathological conditions. In T cells, IL-27 enhances the functions of Th1 and CTL cells and suppresses the functions of Th2, Treg, and Th17 cells. IL-27 also promotes the proliferation

---

Q. Wang • J. Liu (✉)
Division of Infectious Diseases, Allergy and Immunology, Department of Internal Medicine, Saint Louis University School of Medicine, Doisy Research Center, Rm. 811 1100 S. Grand Blvd., St. Louis, MO 63104, USA
e-mail: jliu9@slu.edu

X. Ma (ed.), *Regulation of Cytokine Gene Expression in Immunity and Diseases*, Advances in Experimental Medicine and Biology 941,
DOI 10.1007/978-94-024-0921-5_9

and survival of B cells and suppresses the functions of DCs. IL-27 has been implicated to be involved in the pathogeneses of autoimmune and infectious diseases as well as cancers.

9.1 The Structure and Biological Features of IL-27

EBI3 was first reported in 1996 and its structure resembles type I cytokine receptor family, namely, IL-12 p40 [1]. EBI3 has no membrane-anchoring motif and therefore cannot be secreted effectively by itself. Ectopically expressed EBI3 was accumulated in the endoplasmic reticulum of COS7 cells and could not be secreted as monomer or homodimer [2], suggesting that EBI3 needs partners to provide the membrane-anchoring domain for secretion. p28 is a long-chain protein having four-helix bundle domain and binds to type I cytokine receptor. Through computational analysis and co-immunoprecipitation approaches, p28 has been found to form a heterodimer with EBI3 to compose IL-27 [3].

The major producers of IL-27 are DCs, monocytes, and macrophages. Other types of cells, such as plasma cells, endothelial cells, microglia, placental trophoblasts, and NK cells, also can secrete lower amounts of IL-27. IL-27 production is triggered by various TLR agonists, including LPS, poly(I:C), CPG, and Gram-negative and Gram-positive bacteria [4–8]. LPS induces IL-27 expression, especially the p28 subunit, through MyD88/NF-kB pathway. On the other hand, IFN-γ secreted from T cells and NK cells can induce the expression of IFN-γ regulatory factors (IRFs), including IRF1, IRF3, IRF7, and IRF8, that either alone or together with LPS synergistically enhance IL-27 production. IRF8 promotes p28 transcription through binding to p28 promoter located at -57 to -48 sites. IRF1 is a critical transcription factor for IFN-γ-induced p28 and can cooperate with IRF8 to enhance p28 production. AP-1/c-Fos also can directly bind to p28 promoter and induce p28 gene expression [4, 9, 10]. IFN-γ alone can induce IL-27 production through c-Jun N-terminal kinases (c-JNK), MAPK, and phosphoinositide 3-kinase (PI3K) [11]. IFN-α/IFN-β enhances p28 expression by activating IRF1 through STAT1/STAT2/IRF9 complex [5, 12–15]. It is clear that LPS or IFN-γ activates NF-kB and MAPK through adapter MyD88 to induce transcription factors IRF1, IRF8, or AP-1 that bind to p28 promoter to induce p28 gene expression. Toll/IL-1R-related domain containing adaptor-inducing IFN (TRIF) is another adaptor involved in IL-27 production. p28 and EBI3 expression were increased by activating IRF3 and IRF7 through TLR3/TLR4/TRIF pathway. Though IRF3 is important for p28 expression, it is not essential for the expression of EBI3. So far, much has been learned about the mechanisms of IL-27 production. However, the mechanisms by which IL-27 expression is inhibited are relatively less studied. Osteopontin (OPN) has shown to have suppressive effects on IL-27 production both in vivo and in vitro [16]. p38 MAPK is reported to be able to suppress IL-27 expression through MLK3/MKK3/

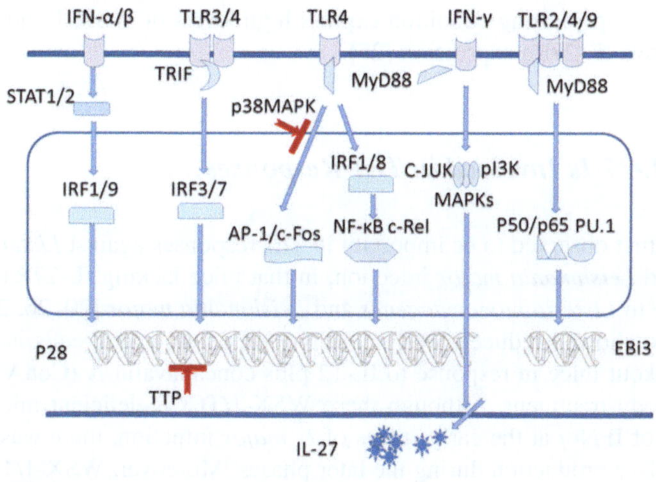

Fig. 9.1 Mechanisms of IL-27 production

MK2 [10]. Tristetraprolin (TTP) is a RNA-binding protein and can directly bind to ARE sites in the 3′UTR of target mRNAs for degradation [17]. We found that TTP inhibits IL-27 production through binding to multiple ARE sites in the 3′UTR of p28 subunit for mRNA degradation. Factors and pathways involved in regulation of IL-27 production are summarized in Fig. 9.1.

IL-27 receptor consists of two chains, WSX-1 and gp130. WSX-1, found in 1998 [18], contains the WSXWS domain. In 2002, WSX-1 was identified as IL-27-specific receptor subunit and later found to partner with gp130 to convey IL-27 signal [3, 19]. WSX-1 is highly expressed in lymphoid tissues, such as spleen, lymph node, and thymus, as well as in several types of tumor cell lines [18, 20–22]. There is a Box1 motif in the cytoplasmic domain of WSX-1, indicating that IL-27/WSX-1 signal can active Janus kinase (JAK) pathway [18]. Studies confirm that IL-27 can activate JAK1, JAK2, and TYK2 [23, 24]. STATs are the downstream of JAKs. IL-27 induces the phosphorylation of STAT1/STAT3/STAT5; activates Blimp-1, T-bet, c-Maf, and Eomes expression; and then exerts its effects on T cells.

9.2 The Effects of IL-27 on T Cell Development and Function

IL-27-specific receptor WSX-1 is preferentially expressed in T cells, particularly in CD4⁺ T cells [19, 20]. Initial studies showed that WSX-1 mRNA levels were highly expressed in naïve CD4⁺ T cells and decreased in activated CD4⁺ T cells upon Th1/Th2 differentiation [20]. Recent studies found that memory and activated CD4⁺ T cells under nonpolarizing condition express high levels of WSX-1 and IL-2 selectively controls IL-27Ra expression [25].

9.2.1 *IL-27 Is Involved in Th1 Responses*

IL-27 was first observed to be important in Th1 responses against *Listeria monocytogenes* and *Leishmania major* infection, in that mice lacking IL-27Ra were more susceptible to *Listeria monocytogenes* and *Leishmania major* [20, 26, 27]. In vitro studies also showed reduced levels of IFN-γ production in T cells isolated from these knockout mice in response to IL-12 plus concanavalin A (ConA) or to anti-CD3 antibody treatment. Although these WSX-1/TCCR-deficient mice produced low levels of IFN-γ at the early stages of *L. major* infection, there was no impairment of IFN-γ production during the later phases. Moreover, WSX-1/TCCR is not required for maintenance of the Th1 response. There studies suggest that there are no intrinsic defects in IFN-γ production in WSX-1 KO mice, rather the ability to prime Th1 response is impaired. IL-27 promotes Th1 differentiation through two different pathways. One is that IL-27 can activate p38 MAPK and upregulate T-bet; the other is to induce the expression of adhesion molecules (ICAM-1/LFA-1) to further activate the ERK1/2. Activation of both pathways leads to upregulation of the IL-12RB which facilitates Th1 differentiation and IFN-γ production during IL-12 stimulation [23, 28].

However, most recent studies found that IL-27 can inhibit Th1 response in vivo. In infection models of *T. gondii*, *T. cruzi*, *P. berghei*, *L. donovani*, and malaria [29–33], IL-27 inhibited IFN-γ production and attenuated disease pathology. Villarino et al. [29] indicated that IL-27 directly suppresses Th1 responses through depleting $CD4^{+}$T cells in the *T. gondii* model. But the underlying mechanisms of suppression remain elusive. IL-27 can indirectly control Th1 responses through suppression of IL-2 via suppressor of cytokine signaling (SOCS)3 [34, 35]. IL-10 is another mediator for IL-27 inhibition of Th1 cells. IL-27 can activate $CD4^{+}$ T cells and $CD8^{+}$ T cells to produce IL-10 that acts through STAT1, STAT3, and Blimp-1 to suppress Th1 responses [36–40].

9.2.2 *IL-27 Inhibits Th2 Responses*

WSX-$1^{-/-}$ mice showed increased resistance to *T. muris* infection. Mesenteric LN cells from WSX-$1^{-/-}$ mice produced significantly more IL-4, IL-5, and IL-13 than WT mice. Administrating anti-IL-4 mAb led to a significant reduction in IL-4 and IL-13 in mesenteric LN. Under Th2 polarization condition, naive WSX-$1^{-/-}$ CD4 T cells secreted higher levels of IL-4, IL-5, and IL-13 [41]. IL-27 inhibits GATA3 expression, while induces T-bet expression in naive $CD4^{+}$ T cells [42, 43]. High levels of Th2 cytokines in the IL-$27R^{-/-}$ or $EBI3^{-/-}$ mice were detected during *T.*

cruzi infection and experimental asthma [30, 43, 44]. Administration of IL-27 attenuated Th2-mediated disease such as *L. major* and *S. venezuelensis* infection and allergic airway inflammation induced by OVA [43]. Th2-related cytokines in pancreatic cancer patients were reduced by IL-27 plus IL-12 treatment [45]. IL-27 also inhibits γδT cells to produce Th2-related cytokines, such as IL-5 and IL-13, through STAT1 and STAT3 [46]. In addition, IL-27 inhibits the development of Th9 cells partially dependent on STAT1 and T-*bet*. Exogenous IL-27 completely abrogated Th9 cells in an EAE model [47].

9.2.3 IL-27 Represses Treg Generation

Treg is a critical regulator for immune responses and homeostasis. It has been reported that 47 % Treg cells express high levels of WSX-1 [25], indicating that IL-27 may regulate the function or phenotype of Treg cells. The effects of IL-27 on Treg were initially shown in vitro experiments [48, 49]. Studies found that IL-27 inhibited the expression of Foxp3, CD25, and CTLA4 through inhibiting STAT3 and therefore suppressed the generation of Treg. In IL-27 transgenic mice, both $CD4^+CD25^+$ T cell and bone marrow transfer models showed a deficiency in Treg differentiation. Since IL-2 levels are decreased in IL-27 Tg mice, it is possible that Treg deficiency may be due to IL-2 reduction. In $CD4^+CD45Rb^{hi}$ transfer colitis model, IL-27Ra$^{-/-}$CD45Rb^{hi} cells contain two to three times more Foxp3$^+$ T cells than IL-27Ra$^{+/+}$CD45Rb^{hi} in the blood, spleen, mesenteric LN, and lamina propria, resulting in attenuation of colitis. In OVA-tolerance model, IL-27 limits Treg conversion from naive $CD4^+$T cells and directly downregulates Foxp3, CD25, and CD69 [50]. All these evidences indicate that IL-27 suppresses the generation and function of Treg. However, the precise underlying mechanisms remain elusive. On the other hand, IL-27 has been reported to promote Treg. Aisling O'Hara Hall et al. found that during *Toxoplasma gondii* infection, IL-27 promoted T-bet and CXCR3 expression in Treg cells through STAT1, STAT3, and STAT5 [51]. IL-27 also activated Treg to express Lag3 and therefore enhanced Treg function. In addition, IL-27 can upregulate IL-10 expression through STAT1, STAT3, ICOS, aryl hydrocarbon receptor (AhR), and c-Maf in Foxp3$^{-/-}$ Treg, resulting in an increase in Tr1 function [37, 52–54].

9.2.4 IL-27 Suppresses Th17 Cell Development and Function

Th17 cells play an important role in the development of inflammatory diseases, such as IBD, EAE, RA, and hepatitis. IL-27 suppresses Th17 cell differentiation in several disease models, such as *Toxoplasma gondii*, *T. Cruzi*, and *L. major* infection as well as multiple sclerosis (MS), pulmonary fibrosis, and IBD [32, 55–63]. We found that IL-27 inhibited severe arthritis mediated by tristetraprolin deficiency via Th17 inhibition (manuscript is under preparation). Both IL-27R$^{-/-}$ and EBI3$^{-/-}$ T cells produce more IL-17 under Th17 polarization condition [32, 64]. IL-27 inhibits naive CD4$^+$ T cells to differentiate into Th17 cells but has little effect on committed Th17 cells [58, 65]. In human non-small cell lung cancer and MS, IL-27 levels in blood negatively correlate with the number of circulating Th17 cells. IL-27 also suppresses the expression of IL-27R and RORrt in naive CD4$^+$T cells [57, 60, 61].

IL-27 binding to IL-27R (WSX-1) phosphorylates JAK1 and then activates STAT1 that plays a critical role in IL-27-mediated inhibition of Th17 cells. STAT1$^{-/-}$ mice produced more IL-17$^+$ cells and were susceptible to EAE development [66, 67]. STAT1$^{-/-}$CD4$^+$ T cells secrete more IL-17 than WT CD4 T cells under Th17 condition in the presence of IL-27 [32, 56, 59], indicating that STAT1 is required for IL-27-mediated Th17 suppression. STAT1 is also involved in IFN-α production by T cells stimulated with IL-27 [23]. How STAT1 contributes to the distinctively different functions of IL-27 remains unknown. The suppression of Th17 cells by IL-27 is independent of T-bet and SOCS3, partially depends on STAT3 [32, 58].

IFN-α inhibits Th17 cell development. IFN-α$^{-/-}$ T cells secrete twice IL-17 than IFN-α$^{+/+}$ T cells. IFN-α inhibits Th17 function, even in STAT1-deficient T cells [56, 59]. More evidence is needed to confirm whether IFN-α directly inhibits Th17 cells independent of STAT1. In addition, IL-27 inhibits Th17 cell differentiation through suppressing IL-6 and IL-23 production in DCs [57]. PD-L1 is another mediator of IL-27-mediated Th17 inhibition. PD-L1 is induced in naïve CD4$^+$ T cells primed by IL-27 through STAT1 but not Foxp3, since PD-L1 expression on naïve CD4$^+$ T cells was completely blocked in STAT1$^{-/-}$ cells. After transferring PD-L1 expressing CD4$^+$ T cells into mice, the Th17 cells were significantly decreased. These mice were resistant to EAE development, further confirming that IL-27 inhibits Th17 cell differentiation through upregulating PD-1 in naive CD4$^+$ T cells [68].

9.2.5 IL-27 Promotes Tfh Cell Development

T follicular helper (Tfh) cells help B cells to differentiate into high-affinity B cells and antibody-producing plasma cells. BCL6 is the master transcription factor for Tfh cell development [69, 70]. Tfh cells express CD44, CXCR5, ICOSL, and PD-1 on the cell surface, and produce IL-4 and IL-21 which then activates Tfh cell differentiation as a feedback loop. IL-27 increases IL-21 production in naïve T cells stimulated by anti-CD3/CD28 antibody, dependent on STAT3 but not STAT1 [52, 60]. IL-27 derived of DC-SIGN-primed DCs induces Tfh cell polarization [71]. CD4$^+$ T cells from IL-27Ra$^{-/-}$ mice immunized by OVA/CFA secreted less IL-21 than T cells from WT mice. IL-27Ra$^{-/-}$ mice showed intrinsic defect in Tfh development, resulting in prevention of the development of murine lupus [60]. However, the molecular mechanisms by which IL-27 inhibits Tfh development and function still need further study.

9.2.6 *IL-27 Enhances CD8 T Cell Development and Function*

Under physiological condition, 22 % of $CD8^+$ T cells express WSX-1. The percentages of WSX-1^+ $CD8^+$ T cells were increased during *T. gondii* infection [25]. IL-27 activates STAT1, STAT2, STAT3, STAT4, and STAT5 in naïve $CD8^+$ T cells, induces T-bet and Eomes expression, and then enhances the levels of granzyme B, perforin, and IL-12Rβ2. Among them, STAT1 plays a critical role in IL-27-mediated T-bet, granzyme B, and IL-12Rβ2 expression, since these molecules were greatly diminished in naïve $CD8^+$ T cells deficient of STAT1 [72]. In addition, IL-27 increases IFN-γ production in $CD8^+$ T cells in the presence of IL-12. IL-27Ra was also detected in human $CD8^+$ T cells. IL-27 rapidly activates STAT1 and STAT3 and then induces T-box transcription, leading to enhanced cell proliferation, IFN-γ, and granzyme B production in human $CD8^+$ T cells. IL-27 also can increase IL-21 and granzyme B levels in human $CD8^+$ T cells [73, 74].

IL-27 is important for CTL generation and function. In HLA-A*0201 transgenic mice, administration with IL-27 and IL-12, in a prime-boost immunization model, increased HCV-specific CTL generation [75]. During subsequent immunization, T cells require IL-27/IL-27R signal to elicit responses [76]. In tumors, IL-27 promotes CTL to suppress tumor growth and metastasis. There are at least four mechanisms involved in the development of IL-27-mediated tumor-specific CTL. Firstly, IL-27 directly induces T-bet and Eomes expression in $CD8^+$ T cells, augments granzyme B and IFN-γ production, and then increases cytotoxic function. Secondly, IL-27 induces IL-21 production in $CD8^+$ T cells. IL-21 plays an important role in antitumor immune responses and in enhancing proliferation, survival, and cytotoxic function of tumor-specific CTL. Thirdly, IL-27 facilitates infiltration of $CD8^+$ T cells into tumors. There are dramatically more $CD8^+$ T cells in TBJ-IL-27 tumors than TBJ-FLAG or TBJ parental tumors. Finally, IL-27 regulates the intake and processing of antigens in DCs, which makes DCs more efficient to activate $CD8^+$ T cells [77–79]. Collectively, IL-27 promotes Th1, Tfh, and $CD8^+$ T cells, and inhibits Th2, Th17, Th9, and Treg T cells (Fig. 9.2).

9.3 The Effects of IL-27 on B Cell Development and Function

The initial study found lower total serum IgG2a in $TCCR^{-/-}$ mice than WT mice, while other immunoglobulin isotypes are compatible. $TCCR^{-/-}$ mice challenged by OVA had much lower OVA-specific IgG2a [20]. Further study discovered that WSX-1 expression in B cells is independent of anti-CD40 stimulation. In in vitro experiments, IL-27 induces IgG2a in primary B cells dependent on STAT1, independent of IFN-γ, and represses IgG1 class switching induced by IL-4. T-bet is required for IL-27-induced IgG2a class switching in response to LPS stimulation but not required to anti-CD40 stimulation [80]. In humans, IL-27R and gp130 are

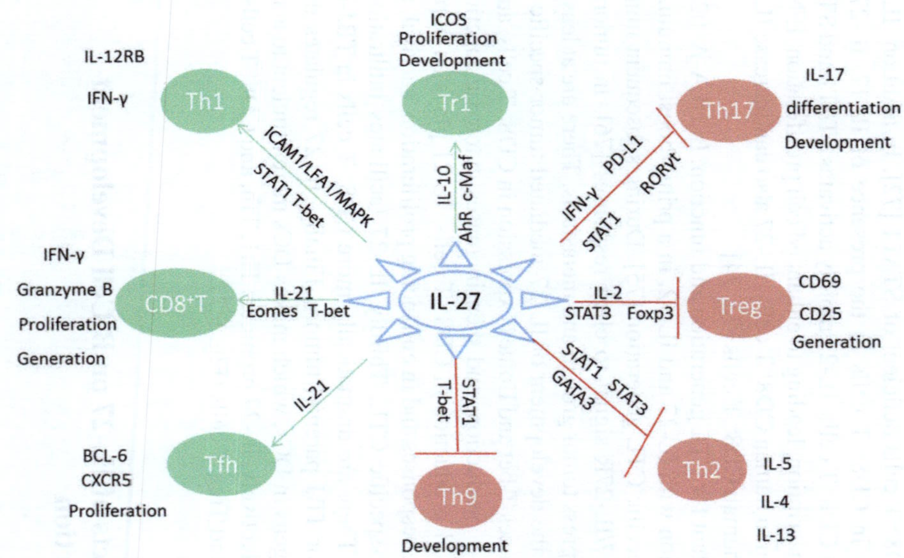

Fig. 9.2 The effects of IL-27 on T cells

detected in B cells, and the expression of IL-27R is variable depending on the differentiation stages of B cells [19, 81, 82]. High levels of IL-27R were detected on both IgD$^+$CD38$^-$ naïve B cells and IgD$^-$CD38$^-$ memory B cells, but very little on IgD$^-$CD38$^+$GC B cells. IL-27 activates STAT1, STAT3, and T-bet and enhances IL-12Rβ2 surface expression in B cells. IL-27 also induces ICAM-1, CD95, CD86, and CXCL9 expression on B cells and promotes B cell proliferation [81, 82]. In addition, IL-27 indirectly regulates B cell functions through IL-21, IL-10, and Tfh [39, 60, 71]. IL-27 induces IL-21 production in CD4$^+$ T cells, which increases the expression of Blim-1 and Bcl-6 in B cells and then promotes plasma cell differentiation and antibody production. Tfh cells are critical for B cell development. IL-27 directly or indirectly induces Tfh cells and then enhances IgG production and survival of B cells. IL-27 may also promote IL-10$^+$ B cell (B10) generation and effect function.

9.4 The Effects of IL-27 on DC and Macrophage Function

IL-27Ra is expressed on activated DCs. Upon LPS stimulation, higher levels of CD80/CD86 and Th1-related molecules were induced in WSX-1$^{-/-}$ DCs than WT DCs. DCs from WSX deficient mice promote T and NK cells to produce higher IFN-γ than WT DCs [83]. Treatment of DCs with IL-27 prior to LPS stimulation reduced the levels of CD40, CD86, and MCH II compared with DCs activated without IL-27, which may be due to increased CD39 expression by IL-27. CD39 limits ATP concentration and downregulates NLRP3 inflammasome activation [84]. DCs with high levels of CD39 suppress pathogenic T cell responses and autoimmune disease development. In humans, both immature and mature DCs express functional IL-27R. IL-27 upregulates B7-H1 in DCs, diminishes APM, decreases HLA restricted antigen presentation, and inhibits proliferation and cytokine production of allogeneic T cells [85, 86]. Taken together, IL-27 inhibits DC functions in both mice and humans. However, in cord blood dendritic cells (CB DCs), IL-27 augments CXCL10, CCR1, IRF8, IL-8, and TNF-α expression, suggesting that IL-27 can also enhance the function of DCs [87]. Whether this is unique to CB DCs needs further study.

IL-27 has minimal effects on murine macrophages. However, in human resting monocytes or macrophages, IL-27 significantly promotes differentiation of monocytes into macrophages and enhances pro-inflammatory cytokine production as well as antiviral activity. IL-27 increases inflammatory cytokine production in monocytes dependent on STAT1, while inhibits IL-10 induced by TLR [88]. Monocytes pretreated by IL-27 increase TLR4 expression through STAT3 and NF-kB and further enhance NF-kB activation induced by LPS, leading to higher levels of IL-6, TNF-α, MIP-1α, and MIP-1β [89]. IL-27 suppresses spectrin β nonerythrocyte 1(SPTBN1) through the TAK-1/MAPK pathway and prevents HIV-1 infection. IL-27 also upregulates HLA-E in human monocytes [90]. The pro-inflammatory effects of IL-27 were rapidly reversed by LPS stimulation in resting macrophages.

IL-27 has been found to inhibit macrophage functions through TNF-α and IL-1β [91], indicating that the effects of IL-27 on macrophages depend on different stimuli.

9.5 The Effects of IL-27 on NK Cell Development and Function

Resting NK cells express high levels of WSX-1. Its expression is reduced once NK cells are activated [25]. IL-27 promotes phosphorylation of STAT1 and STAT3, induces T-bet and granzyme B expression, and increases cytotoxic function of NK cells both in vitro and in vivo [92]. IL-27 also indirectly activates and recruits NK cells through CXCL10 [93]. In human NK cells, IL-27 increases IFN-γ production in the presence of IL-2 and IL-12 [3, 94]. IL-27 not only enhances IFN-γ production but also upregulates CD25, CD69, and T-bet expression in NK cells. IL-27 triggers NK cell cytotoxicity through upregulation of perforin and granule exocytosis via NKp46. On the other hand, IL-27 inhibits the activity of regulatory NK subsets (CD56bright) [95]. Taken together, these studies suggest that IL-27 promotes NK cell cytotoxic activity.

9.6 The Effects of IL-27 on Autoimmune Diseases

IL-27 plays pleiotropic roles in regulation of immune responses in autoimmune diseases. IL-27 exerts pro-inflammatory roles through promoting Th1, CD8, NK, Tfh, and B cell proliferation/function and through inhibiting Treg and Th2 cell generation. On the other hand, IL-27 also plays an anti-inflammatory role through enhancing IL-10 production and inhibiting DC function. High levels of IL-27 are existed in synovial fluid of RA patients. The increased IL-27 is positively correlated with IFN-γ levels, whereas negatively correlated with IL-17 expression. IL-27 is produced by synovial monocytes and macrophages [25, 96, 97]. The first arthritis animal model indicated the involvement of IL-27 is adjuvant-induced arthritis in Lewis rats. In this model, antibody against IL-27 p28 quickly suppressed ongoing disease through regulating the polarization of naïve $CD4^{+}$T cells and through affecting the proliferation and cytokine production from Ag-specific Th1 cells [98]. Later on, it was found that IL-$27^{-/-}$ mice delayed to develop the proteoglycan-induced rheumatoid arthritis compared with IL-$27^{+/+}$ mice, due to reduced IFN-γ production [99]. These studies indicate that IL-27 plays a pro-inflammatory role in arthritis development. However, in collagen-induced arthritis model, if IL-27 was administrated in short-term at the onset of disease, the disease severity was significantly attenuated compared with untreated mice. IL-27 suppresses IL-17 production and increases IL-10 and Foxp3 expression in CD4⁺T cells. So, the effects of IL-27 on arthritis depend on different models and administration time.

The role of IL-27 in colitis is complicated. Low IL-27 p28 gene expression is observed at the early stage in IBD patients. In active CD (but not UC) patients, both IL-27 p28 and EBI3 levels are increased [100, 101]. IL-27Ra and gp130 are expressed in intestinal epithelial cells and upregulated during inflammation and bacterial infection. IL-27 induces ERK/p38 MAPK, STAT1, STAT3, and STAT6 and increases the proliferation and restitution of IEC [102]. In colitis mouse model, IL-27 inhibits inflammatory responses in the colon and prevents colitis development. Both IL-27Ra$^{-/-}$ and IL-27Ra$^{-/-}$/Rag$^{-/-}$ were susceptible to high-dose (5–10 %) and short-time (3 days) DSS-induced colitis through increasing Th17 cells [103]. Another DSS model showed that the severity of colitis in WSX-1$^{-/-}$mice was decreased and the cytokines, such as IFN-γ, IL-6, and TNF-α, were reduced compared with WT mice [104]. Treatment with exogenous IL-27 either at the onset or active stage of colitis in mice fed with 0.25 % DSS or TNBS prevented mice from developing colitis. However, there are reports showing that IL-27 promotes colitis. In the low-dose (0.5 %) and long-time (14 days) DSS and *Helminth* infection model, IL-27Ra$^{-/-}$mice are resistant to develop colitis [104, 105]. In T cell transfer model, IL-27 was found to promote colitis [106–108]. IL-27Ra$^{-/-}$CD4^{+}T cells did not induce exacerbated colitis in SCID mice due to high levels of Treg and low levels of IFN-γ. In Visperas's T cell transfer model, IL-27 was found to promote colitis through indirectly inducing Th17 cells dependent on APCs. It seems that the effects of IL-27 on colitis depend on different mouse models. IL-27 also involves in the pathogeneses of EAE, SLE, psoriasis, type I diabetes mellitus, and uveitis through regulating immune responses. Strategies blocking or promoting IL-27 pathway to prevent or dampen autoimmune diseases need to be carefully considered on different diseases.

9.7 The Effects of IL-27 on Infectious Diseases

The roles of IL-27 in parasitic infections are broadly elaborated. During *L. major*, *L. donovani*, *Plasmodium berghei*, *T. gondii*, *Trichuris muris*, and *Trypanosoma cruzi* infection, IL-27 signal pathways are impaired; while Th1 response is normal, Th2 and Th17 responses are increased [29, 30, 32, 41, 83, 109]. In *L. monocytogenes* infection, initial study showed that there were defective Th1 cells in IL-27Ra$^{-/-}$mice, but later studies showed that Th1 cells were normal, whereas Th17 cells were increased and IL-10 decreased [20, 39, 110]. IL-27 is involved in *Mycobacterium tuberculosis* (Mtb) infection. Both *M. tuberculosis* and BCG can induce IL-27 production in humans. High levels of IL-27 were found in granulomas of TB patients and in active TB patients [111–114]. Mice lacking IL-27Ra showed lower burden of *M. tuberculosis* in the lung and increased inflammatory responses. The possible reason is that IL-27R$^{-/-}$ T cells have reduced T-bet expression, decreased cell death markers and increased IL-2 production, resulting in T cell activation and Mtb repression. Therefore, during *M. tuberculosis* infection, IL-27 plays a detrimental role in protective immune responses. IL-27 also has antiviral effects. Studies showed that IL-27 prevents HIV-1 replication in PBMCs, $CD4^+$ T cells, macrophages, and dendritic cells [64, 115, 116]. IL-27 activates STAT1 and antiviral factor protein kinase R (PKR) phosphorylation and then dampens influenza A virus replication. In hepatitis B patients, there were high levels of IL-27 in plasma. IL-27 has also been shown to be involved in resisting CMV, HSV-2, SIV, and KSHV infection [117–121]. Therefore, IL-27 is a promising antiviral cytokine.

9.8 The Effects of IL-27 on Antitumor Immune Responses

IL-27 has complex effects on antitumor immunity. Firstly, IL-27 acts as an antitumor cytokine through enhancing CTL function. CTL is the most important cells in antitumor immune response [41]. IL-27 promotes the generation, proliferation, and survival of CTL and increases IFN-γ, perforin, and granzyme B production through STAT1/T-bet. Overexpression of IL-27 in Colon26 (CT26) tumors induced CTL development with augmented IFN-γ production [78, 79]. Local delivery of IL-27 to tumors significantly decreased the metastasis and growth of neuroblastoma through inducing tumor-specific CTL [122]. Mice bearing CT26 tumors treated with IL-27 intramuscularly induced lots of tumor-specific IFN-γ^+CD8$^+$ T cells and promoted tumor eradication [123]. CTL functions and recall responses were enhanced by IL-27-mediated IL-10 production [38, 124, 125]. NK cells are one of the key antitumor innate immune cells. IL-27 upregulates IFN-γ production and cytotoxic activity of NK cells [92]. In B16 melanoma, IL-27 derived of DCs promotes the recruitment and activation of NK cells. Reconstitution of IL-27 in local tumor sites restores the number and activity of NK cells and represses tumor growth [93]. In addition, IL-27 also battles with tumors through inhibiting Treg, Th2, and Th17 cells. Treg cells promote tumor growth and metastasis through repressing antitumor immune response and inducing tumor-promoting factors, such as COX-2, PGE2, and VEGF. Th2 cells are well known as an immune suppressor, and Th17 cells promote tumor growth and angiogenesis mediated through IL-17. IL-27 dampens both Th2 and Th17 cells directly or indirectly and therefore exerts its antitumor effects.

Secondly, IL-27R is expressed in several types of tumors originated from epithelial cells, such as colon, breast, and from melanocytes [22, 126, 127]. IL-27 activates STAT1 and STAT3 to inhibit melanoma cell growth. Overexpressing WSX-1 in B16F10 increases MHC I levels and induces IRF1 and IRF8 expression through STAT1 upon IL-27 treatment. Overexpression of IL-27 in B16F10 cells markedly inhibits the neovascularization induced by tumors and suppresses lung metastasis and angiogenesis [127, 128]. In leukemia, B-ALL cells treated by IL-27 decreased miR-155 expression and promoted apoptosis. IL-27 significantly dampens leukemia growth and spreading in vivo [129, 130]. Studies indicated that IL-27 also inhibited the proliferation of tumors and angiogenesis in multiple myeloma and lymphoma [131, 132]. In summary, IL-27 shows a direct antitumor characteristic.

Finally, IL-27 may indirectly promote tumor growth through IL-10 production. IL-27 induces IL-10 production from CD4$^+$T cells in vivo and in vitro. Moreover, IL-27R$^{-/-}$ mice secreted lower IL-10 than WT in toxoplasmic encephalitis or EAE model [36, 37]. IL-27 also stimulates CD8$^+$ T cells and B cells to produce more IL-10 and induces IL-10$^+$ Tr1 cells [38, 124, 133, 134]. IL-21 mediates IL-10 production in response to IL-27. Induction of IL-21 by IL-27 can also induce more IL-10 [52, 135]. IL-10 contributes to the immunosuppressive environment and facilitates tumor escape from immune elimination [136]. In addition, IL-27 also inhibits the functions of DCs by suppressing co-stimulatory molecular expression and cytokine production related to Th1, CD8$^+$ T cells, and NK cell function as discussed previously. Furthermore, IL-27 induces PD-L1 expression on DCs that also mediates tumor immune escape [137]. A few studies also showed that IL-27 inhibits GM-CSF production and Th9 generation [47, 63], with both being reported to be able to suppress tumor progression [138].

9.9 Conclusion

IL-27 is a pleiotropic cytokine having both anti-inflammatory and pro-inflammatory effects. As an immune regulator, IL-27 suppresses the occurrence and development of autoimmune diseases; while as an immune activator, IL-27 prevents infection and cancer development. In general, IL-27 is an attractive candidate for immune therapies in treating autoimmune and infectious diseases and cancers. Prior to clinical application, more work needs to be done, including the following: (1) the mechanisms and functions of endogenous IL-27 in different diseases need to be elaborated; (2) the doses of IL-27 need to be figured out in different diseases; (3) the time starting IL-27 administration needs to be determined; (4) last but not the least, the effective mechanisms of exogenous IL-27 need to be investigated. Overall, IL-27 is a promising cytokine for treating different human diseases.

References

1. Devergne O, Hummel M, Koeppen H, Le Beau MM, Nathanson EC, Kieff E, Birkenbach M. A novel interleukin-12 p40-related protein induced by latent Epstein-Barr virus infection in B lymphocytes. J Virol. 1996;70(2):1143–53.
2. Devergne O, Birkenbach M, Kieff E. Epstein-Barr virus-induced gene 3 and the p35 subunit of interleukin 12 form a novel heterodimeric hematopoietin. Proc Natl Acad Sci U S A. 1997;94(22):12041–6.
3. Pflanz S, Timans JC, Cheung J, Rosales R, Kanzler H, Gilbert J, Hibbert L, Churakova T, Travis M, Vaisberg E, et al. IL-27, a heterodimeric cytokine composed of EBI3 and p28 protein, induces proliferation of naive CD4+ T cells. Immunity. 2002;16(6):779–90.

4. Liu J, Guan X, Ma X. Regulation of IL-27 p28 gene expression in macrophages through MyD88- and interferon-gamma-mediated pathways. J Exp Med. 2007;204(1):141–52.
5. Wirtz S, Becker C, Fantini MC, Nieuwenhuis EE, Tubbe I, Galle PR, Schild HJ, Birkenbach M, Blumberg RS, Neurath MF. EBV-induced gene 3 transcription is induced by TLR signaling in primary dendritic cells via NF-kappa B activation. J Immunol. 2005;174(5):2814–24.
6. Smits HH, van Beelen AJ, Hessle C, Westland R, de Jong E, Soeteman E, Wold A, Wierenga EA, Kapsenberg ML. Commensal Gram-negative bacteria prime human dendritic cells for enhanced IL-23 and IL-27 expression and enhanced Th1 development. Eur J Immunol. 2004;34(5):1371–80.
7. Veckman V, Miettinen M, Pirhonen J, Siren J, Matikainen S, Julkunen I. Streptococcus pyogenes and Lactobacillus rhamnosus differentially induce maturation and production of Th1-type cytokines and chemokines in human monocyte-derived dendritic cells. J Leukoc Biol. 2004;75(5):764–71.
8. Hashimoto SI, Suzuki T, Nagai S, Yamashita T, Toyoda N, Matsushima K. Identification of genes specifically expressed in human activated and mature dendritic cells through serial analysis of gene expression. Blood. 2000;96(6):2206–14.
9. Zhang J, Qian X, Ning H, Yang J, Xiong H, Liu J. Activation of IL-27 p28 gene transcription by interferon regulatory factor 8 in cooperation with interferon regulatory factor 1. J Biol Chem. 2010;285(28):21269–81.
10. Zhang J, Qian X, Ning H, Eickhoff CS, Hoft DF, Liu J. Transcriptional suppression of IL-27 production by Mycobacterium tuberculosis-activated p38 MAPK via inhibition of AP-1 binding. J Immunol. 2011;186(10):5885–95.
11. Blahoianu MA, Rahimi AA, Kozlowski M, Angel JB, Kumar A. IFN-gamma-induced IL-27 and IL-27p28 expression are differentially regulated through JNK MAPK and PI3K pathways independent of Jak/STAT in human monocytic cells. Immunobiology. 2014;219(1):1–8.
12. Molle C, Nguyen M, Flamand V, Renneson J, Trottein F, De Wit D, Willems F, Goldman M, Goriely S. IL-27 synthesis induced by TLR ligation critically depends on IFN regulatory factor 3. J Immunol. 2007;178(12):7607–15.
13. Molle C, Goldman M, Goriely S. Critical role of the IFN-stimulated gene factor 3 complex in TLR-mediated IL-27p28 gene expression revealing a two-step activation process. J Immunol. 2010;184(4):1784–92.
14. Remoli ME, Gafa V, Giacomini E, Severa M, Lande R, Coccia EM. IFN-beta modulates the response to TLR stimulation in human DC: involvement of IFN regulatory factor-1 (IRF-1) in IL-27 gene expression. Eur J Immunol. 2007;37(12):3499–508.
15. van Seventer JM, Nagai T, van Seventer GA. Interferon-beta differentially regulates expression of the IL-12 family members p35, p40, p19 and EBI3 in activated human dendritic cells. J Neuroimmunol. 2002;133(1–2):60–71.
16. Shinohara ML, Kim JH, Garcia VA, Cantor H. Engagement of the type I interferon receptor on dendritic cells inhibits T helper 17 cell development: role of intracellular osteopontin. Immunity. 2008;29(1):68–78.
17. Brooks SA, Blackshear PJ. Tristetraprolin (TTP): interactions with mRNA and proteins, and current thoughts on mechanisms of action. Biochim Biophys Acta. 2013;1829(6–7):666–79.
18. Sprecher CA, Grant FJ, Baumgartner JW, Presnell SR, Schrader SK, Yamagiwa T, Whitmore TE, O'Hara PJ, Foster DF. Cloning and characterization of a novel class I cytokine receptor. Biochem Biophys Res Commun. 1998;246(1):82–90.
19. Pflanz S, Hibbert L, Mattson J, Rosales R, Vaisberg E, Bazan JF, Phillips JH, McClanahan TK, de Waal MR, Kastelein RA. WSX-1 and glycoprotein 130 constitute a signal-transducing receptor for IL-27. J Immunol. 2004;172(4):2225–31.
20. Chen Q, Ghilardi N, Wang H, Baker T, Xie MH, Gurney A, Grewal IS, de Sauvage FJ. Development of Th1-type immune responses requires the type I cytokine receptor TCCR. Nature. 2000;407(6806):916–20.

21. Pradhan A, Lambert QT, Reuther GW. Transformation of hematopoietic cells and activation of JAK2-V617F by IL-27R, a component of a heterodimeric type I cytokine receptor. Proc Natl Acad Sci U S A. 2007;104(47):18502–7.
22. Dibra D, Cutrera JJ, Xia X, Birkenbach MP, Li S. Expression of WSX1 in tumors sensitizes IL-27 signaling-independent natural killer cell surveillance. Cancer Res. 2009;69(13):5505–13.
23. Takeda A, Hamano S, Yamanaka A, Hanada T, Ishibashi T, Mak TW, Yoshimura A, Yoshida H. Cutting edge: role of IL-27/WSX-1 signaling for induction of T-bet through activation of STAT1 during initial Th1 commitment. J Immunol. 2003;170(10):4886–90.
24. Kamiya S, Owaki T, Morishima N, Fukai F, Mizuguchi J, Yoshimoto T. An indispensable role for STAT1 in IL-27-induced T-bet expression but not proliferation of naive CD4+ T cells. J Immunol. 2004;173(6):3871–7.
25. Villarino AV, Larkin 3rd J, Saris CJ, Caton AJ, Lucas S, Wong T, de Sauvage FJ, Hunter CA. Positive and negative regulation of the IL-27 receptor during lymphoid cell activation. J Immunol. 2005;174(12):7684–91.
26. Yoshida H, Hamano S, Senaldi G, Covey T, Faggioni R, Mu S, Xia M, Wakeham AC, Nishina H, Potter J, et al. WSX-1 is required for the initiation of Th1 responses and resistance to L. major infection. Immunity. 2001;15(4):569–78.
27. Artis D, Johnson LM, Joyce K, Saris C, Villarino A, Hunter CA, Scott P. Cutting edge: early IL-4 production governs the requirement for IL-27-WSX-1 signaling in the development of protective Th1 cytokine responses following Leishmania major infection. J Immunol. 2004;172(8):4672–5.
28. Owaki T, Asakawa M, Fukai F, Mizuguchi J, Yoshimoto T. IL-27 induces Th1 differentiation via p38 MAPK/T-bet- and intercellular adhesion molecule-1/LFA-1/ERK1/2-dependent pathways. J Immunol. 2006;177(11):7579–87.
29. Villarino A, Hibbert L, Lieberman L, Wilson E, Mak T, Yoshida H, Kastelein RA, Saris C, Hunter CA. The IL-27R (WSX-1) is required to suppress T cell hyperactivity during infection. Immunity. 2003;19(5):645–55.
30. Hamano S, Himeno K, Miyazaki Y, Ishii K, Yamanaka A, Takeda A, Zhang M, Hisaeda H, Mak TW, Yoshimura A, et al. WSX-1 is required for resistance to Trypanosoma cruzi infection by regulation of proinflammatory cytokine production. Immunity. 2003;19(5):657–67.
31. Rosas LE, Satoskar AA, Roth KM, Keiser TL, Barbi J, Hunter C, de Sauvage FJ, Satoskar AR. Interleukin-27R (WSX-1/T-cell cytokine receptor) gene-deficient mice display enhanced resistance to leishmania donovani infection but develop severe liver immunopathology. Am J Pathol. 2006;168(1):158–69.
32. Findlay EG, Greig R, Stumhofer JS, Hafalla JC, de Souza JB, Saris CJ, Hunter CA, Riley EM, Couper KN. Essential role for IL-27 receptor signaling in prevention of Th1-mediated immunopathology during malaria infection. J Immunol. 2010;185(4):2482–92.
33. Villegas-Mendez A, de Souza JB, Lavelle SW, Gwyer Findlay E, Shaw TN, van Rooijen N, Saris CJ, Hunter CA, Riley EM, Couper KN. IL-27 receptor signalling restricts the formation of pathogenic, terminally differentiated Th1 cells during malaria infection by repressing IL-12 dependent signals. PLoS Pathog. 2013;9(4):e1003293.
34. Villarino AV, Stumhofer JS, Saris CJ, Kastelein RA, de Sauvage FJ, Hunter CA. IL-27 limits IL-2 production during Th1 differentiation. J Immunol. 2006;176(1):237–47.
35. Owaki T, Asakawa M, Kamiya S, Takeda K, Fukai F, Mizuguchi J, Yoshimoto T. IL-27 suppresses CD28-mediated [correction of medicated] IL-2 production through suppressor of cytokine signaling 3. J Immunol. 2006;176(5):2773–80.
36. Fitzgerald DC, Zhang GX, El-Behi M, Fonseca-Kelly Z, Li H, Yu S, Saris CJ, Gran B, Ciric B, Rostami A. Suppression of autoimmune inflammation of the central nervous system by interleukin 10 secreted by interleukin 27-stimulated T cells. Nat Immunol. 2007;8(12):1372–9.

37. Stumhofer JS, Silver JS, Laurence A, Porrett PM, Harris TH, Turka LA, Ernst M, Saris CJ, O'Shea JJ, Hunter CA. Interleukins 27 and 6 induce STAT3-mediated T cell production of interleukin 10. Nat Immunol. 2007;8(12):1363–71.
38. Sun J, Dodd H, Moser EK, Sharma R, Braciale TJ. CD4+ T cell help and innate-derived IL-27 induce Blimp-1-dependent IL-10 production by antiviral CTLs. Nat Immunol. 2011;12(4):327–34.
39. Batten M, Kljavin NM, Li J, Walter MJ, de Sauvage FJ, Ghilardi N. Cutting edge: IL-27 is a potent inducer of IL-10 but not FoxP3 in murine T cells. J Immunol. 2008;180(5):2752–6.
40. Neumann C, Heinrich F, Neumann K, Junghans V, Mashreghi MF, Ahlers J, Janke M, Rudolph C, Mockel-Tenbrinck N, Kuhl AA, et al. Role of Blimp-1 in programming Th effector cells into IL-10 producers. J Exp Med. 2014;211(9):1807–19.
41. Artis D, Villarino A, Silverman M, He W, Thornton EM, Mu S, Summer S, Covey TM, Huang E, Yoshida H, et al. The IL-27 receptor (WSX-1) is an inhibitor of innate and adaptive elements of type 2 immunity. J Immunol. 2004;173(9):5626–34.
42. Lucas S, Ghilardi N, Li J, de Sauvage FJ. IL-27 regulates IL-12 responsiveness of naive CD4+ T cells through Stat1-dependent and -independent mechanisms. Proc Natl Acad Sci U S A. 2003;100(25):15047–52.
43. Yoshimoto T, Yoshimoto T, Yasuda K, Mizuguchi J, Nakanishi K. IL-27 suppresses Th2 cell development and Th2 cytokines production from polarized Th2 cells: a novel therapeutic way for Th2-mediated allergic inflammation. J Immunol. 2007;179(7):4415–23.
44. Dokmeci E, Xu L, Robinson E, Golubets K, Bottomly K, Herrick CA. EBI3 deficiency leads to diminished T helper type 1 and increased T helper type 2 mediated airway inflammation. Immunology. 2011;132(4):559–66.
45. Tassi E, Braga M, Longhi R, Gavazzi F, Parmiani G, Di Carlo V, Protti MP. Non-redundant role for IL-12 and IL-27 in modulating Th2 polarization of carcinoembryonic antigen specific CD4 T cells from pancreatic cancer patients. PLoS One. 2009;4(10):e7234.
46. Morandi F, Prigione I, Airoldi I. Human TCRgammadelta+ T cells represent a novel target for IL-27 activity. Eur J Immunol. 2012;42(6):1547–52.
47. Murugaiyan G, Beynon V, Pires Da Cunha A, Joller N, Weiner HL. IFN-gamma limits Th9-mediated autoimmune inflammation through dendritic cell modulation of IL-27. J Immunol. 2012;189(11):5277–83.
48. Huber M, Steinwald V, Guralnik A, Brustle A, Kleemann P, Rosenplanter C, Decker T, Lohoff M. IL-27 inhibits the development of regulatory T cells via STAT3. Int Immunol. 2008;20(2):223–34.
49. Wojno ED, Hosken N, Stumhofer JS, O'Hara AC, Mauldin E, Fang Q, Turka LA, Levin SD, Hunter CA. A role for IL-27 in limiting T regulatory cell populations. J Immunol. 2011;187(1):266–73.
50. Cox JH, Kljavin NM, Ramamoorthi N, Diehl L, Batten M, Ghilardi N. IL-27 promotes T cell-dependent colitis through multiple mechanisms. J Exp Med. 2011;208(1):115–23.
51. Hall AO, Beiting DP, Tato C, John B, Oldenhove G, Lombana CG, Pritchard GH, Silver JS, Bouladoux N, Stumhofer JS, et al. The cytokines interleukin 27 and interferon-gamma promote distinct Treg cell populations required to limit infection-induced pathology. Immunity. 2012;37(3):511–23.
52. Pot C, Jin H, Awasthi A, Liu SM, Lai CY, Madan R, Sharpe AH, Karp CL, Miaw SC, Ho IC, et al. Cutting edge: IL-27 induces the transcription factor c-Maf, cytokine IL-21, and the costimulatory receptor ICOS that coordinately act together to promote differentiation of IL-10-producing Tr1 cells. J Immunol. 2009;183(2):797–801.
53. Xu J, Yang Y, Qiu G, Lal G, Wu Z, Levy DE, Ochando JC, Bromberg JS, Ding Y. c-Maf regulates IL-10 expression during Th17 polarization. J Immunol. 2009;182(10):6226–36.
54. Pot C, Apetoh L, Awasthi A, Kuchroo VK. Induction of regulatory Tr1 cells and inhibition of T(H)17 cells by IL-27. Semin Immunol. 2011;23(6):438–45.
55. Sasaoka T, Ito M, Yamashita J, Nakajima K, Tanaka I, Narita M, Hara Y, Hada K, Takahashi M, Ohno Y, et al. Treatment with IL-27 attenuates experimental colitis through the suppression

of the development of IL-17-producing T helper cells. Am J Physiol Gastrointest Liver Physiol. 2011;300(4):G568–76.
56. Villarino AV, Gallo E, Abbas AK. STAT1-activating cytokines limit Th17 responses through both T-bet-dependent and -independent mechanisms. J Immunol. 2010;185(11):6461–71.
57. Murugaiyan G, Mittal A, Lopez-Diego R, Maier LM, Anderson DE, Weiner HL. IL-27 is a key regulator of IL-10 and IL-17 production by human CD4+ T cells. J Immunol. 2009;183(4):2435–43.
58. Yoshimura T, Takeda A, Hamano S, Miyazaki Y, Kinjyo I, Ishibashi T, Yoshimura A, Yoshida H. Two-sided roles of IL-27: induction of Th1 differentiation on naive CD4+ T cells versus suppression of proinflammatory cytokine production including IL-23-induced IL-17 on activated CD4+ T cells partially through STAT3-dependent mechanism. J Immunol. 2006;177(8):5377–85.
59. Batten M, Li J, Yi S, Kljavin NM, Danilenko DM, Lucas S, Lee J, de Sauvage FJ, Ghilardi N. Interleukin 27 limits autoimmune encephalomyelitis by suppressing the development of interleukin 17-producing T cells. Nat Immunol. 2006;7(9):929–36.
60. Batten M, Ramamoorthi N, Kljavin NM, Ma CS, Cox JH, Dengler HS, Danilenko DM, Caplazi P, Wong M, Fulcher DA, et al. IL-27 supports germinal center function by enhancing IL-21 production and the function of T follicular helper cells. J Exp Med. 2010;207(13):2895–906.
61. Liu L, Meng J, Zhang C, Duan Y, Zhao L, Wang S, Shan B. Effects on apoptosis and cell cycle arrest contribute to the antitumor responses of interleukin-27 mediated by retrovirus in human pancreatic carcinoma cells. Oncol Rep. 2012;27(5):1497–503.
62. Dong Z, Lu X, Yang Y, Zhang T, Li Y, Chai Y, Lei W, Li C, Ai L, Tai W. IL-27 alleviates the bleomycin-induced pulmonary fibrosis by regulating the Th17 cell differentiation. BMC Pulm Med. 2015;15:13.
63. Young A, Linehan E, Hams E, O'Hara Hall AC, McClurg A, Johnston JA, Hunter CA, Fallon PG, Fitzgerald DC. Cutting edge: suppression of GM-CSF expression in murine and human T cells by IL-27. J Immunol. 2012;189(5):2079–83.
64. Dai L, Lidie KB, Chen Q, Adelsberger JW, Zheng X, Huang D, Yang J, Lempicki RA, Rehman T, Dewar RL, et al. IL-27 inhibits HIV-1 infection in human macrophages by down-regulating host factor SPTBN1 during monocyte to macrophage differentiation. J Exp Med. 2013;210(3):517–34.
65. El-behi M, Ciric B, Yu S, Zhang GX, Fitzgerald DC, Rostami A. Differential effect of IL-27 on developing versus committed Th17 cells. J Immunol. 2009;183(8):4957–67.
66. Bettelli E, Sullivan B, Szabo SJ, Sobel RA, Glimcher LH, Kuchroo VK. Loss of T-bet, but not STAT1, prevents the development of experimental autoimmune encephalomyelitis. J Exp Med. 2004;200(1):79–87.
67. Harrington LE, Hatton RD, Mangan PR, Turner H, Murphy TL, Murphy KM, Weaver CT. Interleukin 17-producing CD4+ effector T cells develop via a lineage distinct from the T helper type 1 and 2 lineages. Nat Immunol. 2005;6(11):1123–32.
68. Hirahara K, Ghoreschi K, Yang XP, Takahashi H, Laurence A, Vahedi G, Sciume G, Hall AO, Dupont CD, Francisco LM, et al. Interleukin-27 priming of T cells controls IL-17 production in trans via induction of the ligand PD-L1. Immunity. 2012;36(6):1017–30.
69. Nurieva RI, Chung Y, Hwang D, Yang XO, Kang HS, Ma L, Wang YH, Watowich SS, Jetten AM, Tian Q, et al. Generation of T follicular helper cells is mediated by interleukin-21 but independent of T helper 1, 2, or 17 cell lineages. Immunity. 2008;29(1):138–49.
70. Nurieva RI, Chung Y, Martinez GJ, Yang XO, Tanaka S, Matskevitch TD, Wang YH, Dong C. Bcl6 mediates the development of T follicular helper cells. Science. 2009;325(5943):1001–5.
71. Gringhuis SI, Kaptein TM, Wevers BA, van der Vlist M, Klaver EJ, van Die I, Vriend LE, de Jong MA, Geijtenbeek TB. Fucose-based PAMPs prime dendritic cells for follicular T helper cell polarization via DC-SIGN-dependent IL-27 production. Nat Commun. 2014;5:5074.

72. Morishima N, Owaki T, Asakawa M, Kamiya S, Mizuguchi J, Yoshimoto T. Augmentation of effector CD8+ T cell generation with enhanced granzyme B expression by IL-27. J Immunol. 2005;175(3):1686–93.
73. Mittal A, Murugaiyan G, Beynon V, Hu D, Weiner HL. IL-27 induction of IL-21 from human CD8+ T cells induces granzyme B in an autocrine manner. Immunol Cell Biol. 2012;90(8):831–5.
74. Schneider R, Yaneva T, Beauseigle D, El-Khoury L, Arbour N. IL-27 increases the proliferation and effector functions of human naive CD8+ T lymphocytes and promotes their development into Tc1 cells. Eur J Immunol. 2011;41(1):47–59.
75. Matsui M, Moriya O, Belladonna ML, Kamiya S, Lemonnier FA, Yoshimoto T, Akatsuka T. Adjuvant activities of novel cytokines, interleukin-23 (IL-23) and IL-27, for induction of hepatitis C virus-specific cytotoxic T lymphocytes in HLA-A*0201 transgenic mice. J Virol. 2004;78(17):9093–104.
76. Pennock ND, Gapin L, Kedl RM. IL-27 is required for shaping the magnitude, affinity distribution, and memory of T cells responding to subunit immunization. Proc Natl Acad Sci U S A. 2014;111(46):16472–7.
77. Hisada M, Kamiya S, Fujita K, Belladonna ML, Aoki T, Koyanagi Y, Mizuguchi J, Yoshimoto T. Potent antitumor activity of interleukin-27. Cancer Res. 2004;64(3):1152–6.
78. Chiyo M, Shimozato O, Yu L, Kawamura K, Iizasa T, Fujisawa T, Tagawa M. Expression of IL-27 in murine carcinoma cells produces antitumor effects and induces protective immunity in inoculated host animals. Int J Cancer. 2005;115(3):437–42.
79. Salcedo R, Stauffer JK, Lincoln E, Back TC, Hixon JA, Hahn C, Shafer-Weaver K, Malyguine A, Kastelein R, Wigginton JM. IL-27 mediates complete regression of orthotopic primary and metastatic murine neuroblastoma tumors: role for CD8+ T cells. J Immunol. 2004;173(12):7170–82.
80. Yoshimoto T, Okada K, Morishima N, Kamiya S, Owaki T, Asakawa M, Iwakura Y, Fukai F, Mizuguchi J. Induction of IgG2a class switching in B cells by IL-27. J Immunol. 2004;173(4):2479–85.
81. Larousserie F, Charlot P, Bardel E, Froger J, Kastelein RA, Devergne O. Differential effects of IL-27 on human B cell subsets. J Immunol. 2006;176(10):5890–7.
82. Cocco C, Di Carlo E, Zupo S, Canale S, Zorzoli A, Ribatti D, Morandi F, Ognio E, Airoldi I. Complementary IL-23 and IL-27 anti-tumor activities cause strong inhibition of human follicular and diffuse large B-cell lymphoma growth in vivo. Leukemia. 2012;26(6):1365–74.
83. Wang S, Miyazaki Y, Shinozaki Y, Yoshida H. Augmentation of antigen-presenting and Th1-promoting functions of dendritic cells by WSX-1(IL-27R) deficiency. J Immunol. 2007;179(10):6421–8.
84. Mascanfroni ID, Yeste A, Vieira SM, Burns EJ, Patel B, Sloma I, Wu Y, Mayo L, Ben-Hamo R, Efroni S, et al. IL-27 acts on DCs to suppress the T cell response and autoimmunity by inducing expression of the immunoregulatory molecule CD39. Nat Immunol. 2013;14(10):1054–63.
85. Karakhanova S, Bedke T, Enk AH, Mahnke K. IL-27 renders DC immunosuppressive by induction of B7-H1. J Leukoc Biol. 2011;89(6):837–45.
86. Morandi F, Di Carlo E, Ferrone S, Petretto A, Pistoia V, Airoldi I. IL-27 in human secondary lymphoid organs attracts myeloid dendritic cells and impairs HLA class I-restricted antigen presentation. J Immunol. 2014;192(6):2634–42.
87. Birkholz J, Doganci A, Darstein C, Gehring S, Zepp F, Meyer CU. IL-27 improves migrational and antiviral potential of CB dendritic cells. Hum Immunol. 2014;75(6):584–91.
88. Kalliolias GD, Ivashkiv LB. IL-27 activates human monocytes via STAT1 and suppresses IL-10 production but the inflammatory functions of IL-27 are abrogated by TLRs and p38. J Immunol. 2008;180(9):6325–33.

89. Guzzo C, Ayer A, Basta S, Banfield BW, Gee K. IL-27 enhances LPS-induced proinflammatory cytokine production via upregulation of TLR4 expression and signaling in human monocytes. J Immunol. 2012;188(2):864–73.
90. Morandi F, Airoldi I, Pistoia V. IL-27 driven upregulation of surface HLA-E expression on monocytes inhibits IFN-gamma release by autologous NK cells. J Immunol Res. 2014;2014:938561.
91. Kalliolias GD, Gordon RA, Ivashkiv LB. Suppression of TNF-alpha and IL-1 signaling identifies a mechanism of homeostatic regulation of macrophages by IL-27. J Immunol. 2010;185(11):7047–56.
92. Matsui M, Kishida T, Nakano H, Yoshimoto K, Shin-Ya M, Shimada T, Nakai S, Imanishi J, Yoshimoto T, Hisa Y, et al. Interleukin-27 activates natural killer cells and suppresses NK-resistant head and neck squamous cell carcinoma through inducing antibody-dependent cellular cytotoxicity. Cancer Res. 2009;69(6):2523–30.
93. Wei J, Xia S, Sun H, Zhang S, Wang J, Zhao H, Wu X, Chen X, Hao J, Zhou X, et al. Critical role of dendritic cell-derived IL-27 in antitumor immunity through regulating the recruitment and activation of NK and NKT cells. J Immunol. 2013;191(1):500–8.
94. Ziblat A, Domaica CI, Spallanzani RG, Iraolagoitia XL, Rossi LE, Avila DE, Torres NI, Fuertes MB, Zwirner NW. IL-27 stimulates human NK-cell effector functions and primes NK cells for IL-18 responsiveness. Eur J Immunol. 2015;45(1):192–202.
95. Laroni A, Gandhi R, Beynon V, Weiner HL. IL-27 imparts immunoregulatory function to human NK cell subsets. PLoS One. 2011;6(10):e26173.
96. Baek SH, Lee SG, Park YE, Kim GT, Kim CD, Park SY. Increased synovial expression of IL-27 by IL-17 in rheumatoid arthritis. Inflamm Res. 2012;61(12):1339–45.
97. Tanida S, Yoshitomi H, Ishikawa M, Kasahara T, Murata K, Shibuya H, Ito H, Nakamura T. IL-27-producing CD14(+) cells infiltrate inflamed joints of rheumatoid arthritis and regulate inflammation and chemotactic migration. Cytokine. 2011;55(2):237–44.
98. Goldberg R, Wildbaum G, Zohar Y, Maor G, Karin N. Suppression of ongoing adjuvant-induced arthritis by neutralizing the function of the p28 subunit of IL-27. J Immunol. 2004;173(2):1171–8.
99. Cao Y, Doodes PD, Glant TT, Finnegan A. IL-27 induces a Th1 immune response and susceptibility to experimental arthritis. J Immunol. 2008;180(2):922–30.
100. Imielinski M, Baldassano RN, Griffiths A, Russell RK, Annese V, Dubinsky M, Kugathasan S, Bradfield JP, Walters TD, Sleiman P, et al. Common variants at five new loci associated with early-onset inflammatory bowel disease. Nat Genet. 2009;41(12):1335–40.
101. Schmidt C, Giese T, Ludwig B, Mueller-Molaian I, Marth T, Zeuzem S, Meuer SC, Stallmach A. Expression of interleukin 12-related cytokine transcripts in inflammatory bowel disease: elevated interleukin-23p19 and interleukin-27p28 in Crohn's disease but not in ulcerative colitis. Inflamm Bowel Dis. 2005;11(1):16–23.
102. Diegelmann J, Olszak T, Goke B, Blumberg RS, Brand S. A novel role for interleukin-27 (IL-27) as mediator of intestinal epithelial barrier protection mediated via differential signal transducer and activator of transcription (STAT) protein signaling and induction of antibacterial and anti-inflammatory proteins. J Biol Chem. 2011;287(1):286–98.
103. Troy AE, Zaph C, Du Y, Taylor BC, Guild KJ, Hunter CA, Saris CJ, Artis D. IL-27 regulates homeostasis of the intestinal CD4+ effector T cell pool and limits intestinal inflammation in a murine model of colitis. J Immunol. 2009;183(3):2037–44.
104. Honda K, Nakamura K, Matsui N, Takahashi M, Kitamura Y, Mizutani T, Harada N, Nawata H, Hamano S, Yoshida H. T helper 1-inducing property of IL-27/WSX-1 signaling is required for the induction of experimental colitis. Inflamm Bowel Dis. 2005;11(12):1044–52.
105. Villarino AV, Artis D, Bezbradica JS, Miller O, Saris CJ, Joyce S, Hunter CA. IL-27R deficiency delays the onset of colitis and protects from helminth-induced pathology in a model of chronic IBD. Int Immunol. 2008;20(6):739–52.
106. Visperas A, Do JS, Bulek K, Li X, Min B. IL-27, targeting antigen-presenting cells, promotes Th17 differentiation and colitis in mice. Mucosal Immunol. 2013;7(3):625–33.

107. Kim G, Shinnakasu R, Saris CJ, Cheroutre H, Kronenberg M. A novel role for IL-27 in mediating the survival of activated mouse CD4 T lymphocytes. J Immunol. 2013;190(4):1510–8.
108. Cox JH, Kljavin NM, Ramamoorthi N, Diehl L, Batten M, Ghilardi N. IL-27 promotes T cell-dependent colitis through multiple mechanisms. J Exp Med. 2010;208(1):115–23.
109. Stumhofer JS, Laurence A, Wilson EH, Huang E, Tato CM, Johnson LM, Villarino AV, Huang Q, Yoshimura A, Sehy D, et al. Interleukin 27 negatively regulates the development of interleukin 17-producing T helper cells during chronic inflammation of the central nervous system. Nat Immunol. 2006;7(9):937–45.
110. Yang J, Yang M, Htut TM, Ouyang X, Hanidu A, Li X, Sellati R, Jiang H, Zhang S, Li H, et al. Epstein-Barr virus-induced gene 3 negatively regulates IL-17, IL-22 and RORgamma t. Eur J Immunol. 2008;38(5):1204–14.
111. Lim J, Derrick SC, Kolibab K, Yang AL, Porcelli S, Jacobs WR, Morris SL. Early pulmonary cytokine and chemokine responses in mice immunized with three different vaccines against Mycobacterium tuberculosis determined by PCR array. Clin Vaccine Immunol: CVI. 2009;16(1):122–6.
112. Wu B, Huang C, Garcia L, Ponce De Leon A, Osornio JS, Bobadilla-del-Valle M, Ferreira L, Canizales S, Small P, Kato-Maeda M, et al. Unique gene expression profiles in infants vaccinated with different strains of Mycobacterium bovis bacille Calmette-Guerin. Infect Immun. 2007;75(7):3658–64.
113. Larousserie F, Pflanz S, Coulomb-L'Hermine A, Brousse N, Kastelein R, Devergne O. Expression of IL-27 in human Th1-associated granulomatous diseases. J Pathol. 2004;202(2):164–71.
114. Torrado E, Fountain JJ, Liao M, Tighe M, Reiley WW, Lai RP, Meintjes G, Pearl JE, Chen X, Zak DE, et al. Interleukin 27R regulates CD4+ T cell phenotype and impacts protective immunity during Mycobacterium tuberculosis infection. J Exp Med. 2015;212(9):1449–63.
115. Fakruddin JM, Lempicki RA, Gorelick RJ, Yang J, Adelsberger JW, Garcia-Pineres AJ, Pinto LA, Lane HC, Imamichi T. Noninfectious papilloma virus-like particles inhibit HIV-1 replication: implications for immune control of HIV-1 infection by IL-27. Blood. 2007;109(5):1841–9.
116. Swaminathan S, Dai L, Lane HC, Imamichi T. Evaluating the potential of IL-27 as a novel therapeutic agent in HIV-1 infection. Cytokine Growth Factor Rev. 2013;24(6):571–7.
117. Liu L, Cao Z, Chen J, Li R, Cao Y, Zhu C, Wu K, Wu J, Liu F, Zhu Y. Influenza A virus induces interleukin-27 through cyclooxygenase-2 and protein kinase A signaling. J Biol Chem. 2012;287(15):11899–910.
118. Shin MS, Lee JS, Lee N, Lee WW, Kim SH, Kang I. Maintenance of CMV-specific CD8+ T cell responses and the relationship of IL-27 to IFN-gamma levels with aging. Cytokine. 2013;61(2):485–90.
119. Zhu C, Zhang R, Liu L, Rasool ST, Mu Y, Sun W, Hao Q, Liu F, Zhu Y, Wu J. Hepatitis B virus enhances interleukin-27 expression both in vivo and in vitro. Clin Immunol. 2009;131(1):92–7.
120. Wang HL, Zhang HY, Zhai ZL, Zhou X. The correlation between hepatitis B virus infection and IL-27. Biomed Mater Eng. 2012;22(1–3):187–93.
121. Swaminathan S, Hu X, Zheng X, Kriga Y, Shetty J, Zhao Y, Stephens R, Tran B, Baseler MW, Yang J, et al. Interleukin-27 treated human macrophages induce the expression of novel microRNAs which may mediate anti-viral properties. Biochem Biophys Res Commun. 2013;434(2):228–34.
122. Salcedo R, Hixon JA, Stauffer JK, Jalah R, Brooks AD, Khan T, Dai RM, Scheetz L, Lincoln E, Back TC, et al. Immunologic and therapeutic synergy of IL-27 and IL-2: enhancement of T cell sensitization, tumor-specific CTL reactivity and complete regression of disseminated neuroblastoma metastases in the liver and bone marrow. J Immunol. 2009;182(7):4328–38.
123. Zhu S, Lee DA, Li S. IL-12 and IL-27 sequential gene therapy via intramuscular electroporation delivery for eliminating distal aggressive tumors. J Immunol. 2010;184(5):2348–54.

124. Perona-Wright G, Kohlmeier JE, Bassity E, Freitas TC, Mohrs K, Cookenham T, Situ H, Pearce EJ, Woodland DL, Mohrs M. Persistent loss of IL-27 responsiveness in CD8+ memory T cells abrogates IL-10 expression in a recall response. Proc Natl Acad Sci U S A. 2012;109(45):18535–40.
125. Liu Z, Liu JQ, Talebian F, Wu LC, Li S, Bai XF. IL-27 enhances the survival of tumor antigen-specific CD8+ T cells and programs them into IL-10-producing, memory precursor-like effector cells. Eur J Immunol. 2012;43(2):468–79.
126. Dibra D, Cutrera J, Xia X, Li S. WSX1 expression in tumors induces immune tolerance via suppression of effector immune cells. PLoS One. 2011;6(4):e19072.
127. Yoshimoto T, Morishima N, Mizoguchi I, Shimizu M, Nagai H, Oniki S, Oka M, Nishigori C, Mizuguchi J. Antiproliferative activity of IL-27 on melanoma. J Immunol. 2008;180(10):6527–35.
128. Shimizu M, Shimamura M, Owaki T, Asakawa M, Fujita K, Kudo M, Iwakura Y, Takeda Y, Luster AD, Mizuguchi J, et al. Antiangiogenic and antitumor activities of IL-27. J Immunol. 2006;176(12):7317–24.
129. Canale S, Cocco C, Frasson C, Seganfreddo E, Di Carlo E, Ognio E, Sorrentino C, Ribatti D, Zorzoli A, Basso G, et al. Interleukin-27 inhibits pediatric B-acute lymphoblastic leukemia cell spreading in a preclinical model. Leukemia. 2011;25(12):1815–24.
130. Zorzoli A, Di Carlo E, Cocco C, Ognio E, Ribatti D, Ferretti E, Dufour C, Locatelli F, Montagna D, Airoldi I. Interleukin-27 inhibits the growth of pediatric acute myeloid leukemia in NOD/SCID/Il2rg-/- mice. Clin Cancer Res. 2012;18(6):1630–40.
131. Cocco C, Giuliani N, Di Carlo E, Ognio E, Storti P, Abeltino M, Sorrentino C, Ponzoni M, Ribatti D, Airoldi I. Interleukin-27 acts as multifunctional antitumor agent in multiple myeloma. Clin Cancer Res. 2010;16(16):4188–97.
132. Cocco C, Pistoia V, Airoldi I. Anti-leukemic properties of IL-12, IL-23 and IL-27: differences and similarities in the control of pediatric B acute lymphoblastic leukemia. Crit Rev Oncol Hematol. 2012;83(3):310–8.
133. Zhu C, Sakuishi K, Xiao S, Sun Z, Zaghouani S, Gu G, Wang C, Tan DJ, Wu C, Rangachari M, et al. An IL-27/NFIL3 signalling axis drives Tim-3 and IL-10 expression and T-cell dysfunction. Nat Commun. 2015;6:6072.
134. Mishima Y, Liu B, Hansen JJ, Sartor RB. Resident bacteria-stimulated IL-10-secreting B cells ameliorate T cell-mediated colitis by inducing Tr-1 cells that require IL-27-signaling. Cell Mol Gastroenterol Hepatol. 2015;1(3):295–310.
135. Ansari NA, Kumar R, Gautam S, Nylen S, Singh OP, Sundar S, Sacks D. IL-27 and IL-21 are associated with T cell IL-10 responses in human visceral leishmaniasis. J Immunol. 2011;186(7):3977–85.
136. Mannino MH, Zhu Z, Xiao H, Bai Q, Wakefield MR, Fang Y. The paradoxical role of IL-10 in immunity and cancer. Cancer Lett. 2015;367(2):103–7.
137. Lipson EJ, Forde PM, Hammers HJ, Emens LA, Taube JM, Topalian SL. Antagonists of PD-1 and PD-L1 in cancer treatment. Semin Oncol. 2015;42(4):587–600.
138. Eager R, Nemunaitis J. GM-CSF gene-transduced tumor vaccines. Mol Ther. 2005;12(1):18–27.

Chapter 10
The Immunobiology of Interleukin-35 and Its Regulation and Gene Expression

Mei Song and Xiaojing Ma

Abstract Interleukin-35 (IL-35) is the latest addition to the IL-12 family of heterodimeric cytokines, consisting of IL-12 p35 subunit and IL-27β subunit Epstein–Barr virus induced 3 (EBI3). Since its discovery, IL-35 has been shown to exhibit immunosuppressive activities which are distinct from other members of IL-12 family. IL-35 is also unique in that it is expressed primarily by regulatory T-cells (T_{regs}) rather than by antigen-presenting cells (APCs). IL-35 can directly suppress effector T-cell proliferation and function and inhibit the differentiation of Th17 cells. It is also able to expand regulatory responses to promote tolerance to infections by generating a potent population of IL-35-producing inducible Tregs (iTr35). As the new cellular sources of IL-35 such as $CD8^{+}T_{regs}$, B_{regs}, and tolerogenic dendritic cells DCs (tolDCs) are identified, more immunoregulatory functions of this cytokine are explored. IL-35 has been shown to be associated with a range of autoimmune diseases and cancer models. It appears to be a promising diagnostic biomarker. A greater understanding of the expression and regulatory mechanisms of IL-35 will be beneficial to the development of novel immune therapies.

Keywords Interleukin-35 • EBI3 • Induced regulatory T-cells • IL-35 signaling • Breg

---

The authors declare no conflict of interest.

M. Song
Department of Microbiology and Immunology, Weill Cornell Medical College,
1300 York Avenue, New York, NY 10065, USA

X. Ma (✉)
State Key Laboratory of Microbial Metabolism, Sheng Yushou Center of Cell Biology and Immunology, School of Life Science and Biotechnology, Shanghai Jiaotong University, 800 Dongchuan Road, Shanghai 200240, China

Department of Microbiology and Immunology, Weill Cornell Medical College,
1300 York Avenue, New York, NY 10065, USA
e-mail: xim2002@sjtu.edu.cn; xim2002@med.cornell.edu

X. Ma (ed.), *Regulation of Cytokine Gene Expression in Immunity and Diseases*,
Advances in Experimental Medicine and Biology 941,
DOI 10.1007/978-94-024-0921-5_10

10.1 IL-35 Composition, Expression, Receptors, and Target Cells

IL-35 belongs to the IL-12 family of cytokines, which is the only heterodimeric cytokine family consisting of an α chain (p19, p28, or p35) and a β chain (p40 or Ebi3) [6]. Chain pairing in various combinations to form IL-12 [18], IL-23 [31], IL-27 [32], and IL-35 [7], as illustrated in Fig. 10.1, IL-35, the newest member of this family, was identified as an inhibitory cytokine produced by natural, thymus-derived regulatory T-cell (nT_{reg}) populations [7]. In contrast to all other known IL-12 family members, which are primarily produced by activated APCs [17, 23, 24], IL-35 appears to be primarily produced by non-stimulated mouse T_{reg} cells but not detected in non-stimulated human T_{regs} [1]. Recently, IL-35 was shown to be expressed in regulatory B-cells [36, 42] and DCs [12], to a lesser extent, in endothelial cells, smooth muscle cells, and monocytes [3]. Mice in which only B-cells did not express IL-35 lose their ability to recover from the T-cell-mediated demyelinating autoimmune disease experimental autoimmune encephalomyelitis (EAE) and are more susceptible to Gram-negative bacterium *Salmonella enterica* serovar Typhimurium. In addition, some tumor cells seem to produce IL35 [22, 43].

IL-35 is the newest member in IL-12 family. It is composed of the IL-12α chain p35 and the IL-27β chain Epstein–Barr virus induced 3 (EBI3) and utilizes IL-12Rβ2 and gp130 heterodimers/homodimers to transduce signals.

IL-35 is composed of the IL-12α chain p35 and the IL-27β chain Epstein–Barr virus induced 3 (EBI3). Despite their shared components, they possess contrasting biological functions: IL-35 is so far believed to be strictly immunosuppressive, while IL-12, IL-23, and IL-27 all have immunostimulatory or proinflammatory functions. IL-35 mediates regulatory B- [36] and T-cell function [7, 26] and converts naïve T-cells into IL-35-producing induced regulatory T-cells (iTr35) [4], as well as inhibits effector T-cell proliferation and Th17 differentiation in vitro [26].

Using cells deficient in the IL-12 family receptor chains, it was demonstrated that IL-35 signals through IL12Rβ2 and gp130 in T-cells [5], which are also

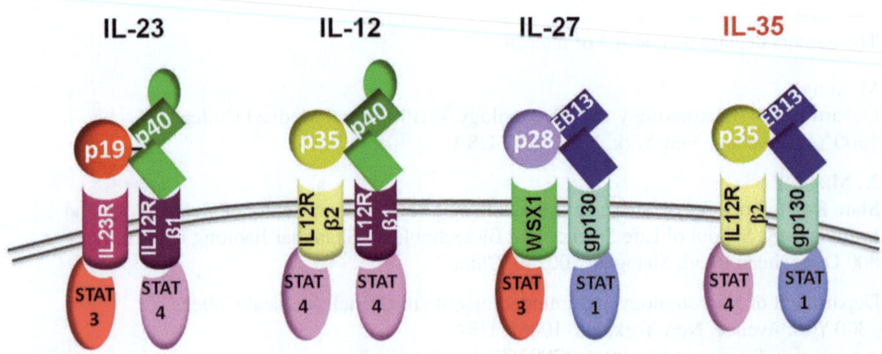

Fig. 10.1 Four known members of the IL-12 cytokine family and their downstream signaling pathways

associated with the IL-12 and IL-27 receptors, respectively. IL-35 was also demonstrated to signal through IL12Rβ2 homodimers and gp130 homodimers in T-cells [5]. These homodimeric receptors can only activate one branch of its signaling pathway (either STAT1 or STAT4 for gp130–gp130 or IL12Rβ2–IL12Rβ2 homodimers, respectively), which are able to suppress T-cell proliferation, but cannot induce iTr35 conversion (Fig. 10.1). In contrast, T-cell suppression and iTr35 induction can be mediated by IL12Rβ2–gp130 heterodimeric receptor. Signaling through this fully functional receptor is transduced through STAT1 and STAT4, which then form a unique heterodimer and promote the expression of target genes *Il12α* and *Ebi3*, thus resulting in a feedback loop promoting IL-35 expression. Recent studies have also suggested that IL-35 signaling in B-cells is mediated through an IL-12Rβ2–WSX-1 heterodimer and induces STAT1 and STAT3 [36]. Gp130 is expressed in most cell types [34], whereas IL-12Rβ2 is expressed predominantly on activated T-cells, NK cells, and to a lesser extent DCs and B-cells [39]. In addition, IL-12Rβ2 is undetectable on most resting T-cells, but rapidly induced in the presence of cytokines (such as IL-2, IFN-γ, IL-12, IL-27) [5]. WSX1 (IL-27Rα) is expressed in many immune cell types (such as monocytes, DCs, T- and B-cells, NK cells, and master cells) [2]. The use of these receptors may shed some light onto the targets of IL-35 signaling.

10.2 Immunoregulatory Activities of IL-35

The predominant mechanism of suppression associated with the activity of IL-35 is its ability to suppress T-cell proliferation and effector functions. IL-35 is required for maximal suppressive activity of T_{regs}, as genetic deletion (*Ebi3*$^{-/-}$ and *Il12α*$^{-/-}$) significantly reduces Treg's proliferation and regulatory activities in vitro and fails to cure inflammatory bowel disease (IBD) in vivo [7]. Ectopically expressed EbI3–p35–Fc fusion protein suppresses $CD4^+CD25^-$ effector T-cell proliferation and IFN-γ production [26]. The suppressive activity of IL-35 is not restricted to $CD4^+$ T_{regs}, as a novel population of $CD8^+$ $CTLA4^+$ IL-35-secreting tumor Ag-specific T_{regs} in prostate cancer patients was found to suppress the proliferation of autologous T-cells by an IL-35-dependent mechanism [29]. In addition, IL-35 inhibits the differentiation of Th17 cells in vitro and effectively attenuated established collagen-induced arthritis in vivo [26]. EBI3-deficient mice have a significant increase in the production of IL-17 [20, 38]. Intratracheal delivery of single-chain recombinant IL-35 significantly attenuated allergic airway inflammation by reducing the Th2 effector and humoral response [16]. Recombinant IL-35 treatment induced IL-10-secreting B_{regs}, which suppressed proliferation of CD19 B-cells. IL-35 also converted B-cells into a B_{reg}-cell population that secreted IL-35(IL-35B_{regs}). Treatment of mice with IL-35 or IL-35B_{regs} conferred protection from experimental autoimmune uveitis (EAU). This IL-35 $^+$ B_{reg}-mediated protection was dependent on increased induction and expansion of endogenous B_{regs} and $Foxp3^+$ T_{regs} and inhibition of pathogenic Th1 and Th17 effector cells [42].

Furthermore, IL-35 seems to mediate infectious tolerance, similar to IL-10 and TGFβ [13]. Naïve human or mouse T-cells treated with IL-35 are not only suppressed but also converted into a IL-35-producing (but not IL-10 or TGFβ) regulatory population, called iTr35 cells [4]. Mouse iTreg35 cells suppress T-cell proliferation in a contact-independent fashion through the activity of IL-35. The stability and efficiency of iTr35 are relative to TGFβ-iTr and IL-10-iTr cells. iTr35 cells were almost as efficient as nT_{regs} in restoring homeostasis and preventing autoimmunity in $Foxp3^{-/-}$ mice, limiting T-cell proliferation in lymphopenic settings, curtailing inflammation in experimental autoimmune encephalitis (EAE) and colitis models, and promoting tumor growth in the B16 melanoma model. iTr35 cells function independently of Foxp3, IL-10, and TGFβ and exhibit tolerance-promoting abilities in several tumor mouse models. R-DC (DC activated by Human rhinoviruses)-induced T_{reg} produced and released IL-35, which is responsible for the inhibitory effect of these iTr35 [35].

Inhibition of effector T-cell proliferation, modulating T-cell differentiation, and induction of iTr35 seem to be the major mechanisms for IL-35-mediated regulatory functions.

Several diseases have been shown to be associated with IL-35, including multiple inflammatory diseases (IBD, EAE, EAU) [7, 36, 42], coronary artery disease [26], and cancer. In acute myeloid leukemia [47] patients, the development of disease has been shown to be associated with elevated plasma levels of IL-35. Evaluated Ebi3 levels were found in lung cancer patients with malignancy [28]. In murine models of melanoma and colorectal carcinoma, the establishment of tumors leads to increased IL-35 expression in $CD4^+$ tumor-infiltrating lymphocytes [4]. In the tumor microenvironment, $Foxp3^+$ T_{regs} are considered to be the primary IL-35 producer [46]. T_{reg}-derived IL-35 has been found to potently inhibit antitumor T-cell responses [4]. In addition, iTr35 was shown to contribute to the regulatory milieu in the tumor microenvironment, thus confirming a critical role for regulatory triad of IL-35 and iTr35 cells in tumor progression [4]. IL-35 can be induced in human cancer cell lines following TNF-α and IFN-γ stimulation [22]. Tumor-derived IL-35 promoted tumor growth through the accumulation of suppressive myeloid cell populations (MDSCs) and increasing angiogenesis [43]. IL-35 over expression in tumor cells can also induce cell cycle arrest at the G1 phase and sensitize tumor cells to serum starvation-induced apoptosis [22].

10.3 IL-35 Coding Genes

The two IL-35-encoding genes *IL12A* (for p35) and *EBI3* (for IL-27 β) are located on separate chromosomes (3p12-q13.2 and 19p13.3, respectively, in humans; chromosomes 6 and 17qD, respectively, in the mouse) [11]. Sequence comparisons showed that p35 is related in sequence to ligands of the cytokine receptor family, most closely to IL-6 and GCSF [14].

EBI3 was first identified from a subtractive hybridization screen of genes in B lymphocytes in response Epstein–Barr virus (EBV) infection. It encodes a 34-kDa (229 amino acids) glycoprotein which is 30 % identical to the ciliary neurotrophic factor receptor (CNTFR) and 27 % identical to the p40 unit of IL-12 [30].

Coimmunoprecipitation analysis showed that EBI3 can specifically be associated with the p35 in lysates and culture media of transfected EBV-negative Burkitt lymphoma cells and COS-7 cells [8]. Co-expression of EBI3 and p35 mutually facilitates their secretion. EBI/p35 heterodimer was also coimmunoprecipitated from trophoblasts of a human full-term normal placenta. Covalently linked EBI3 with p35–Fc protein was found in human and mouse to form the heterodimeric protein IL-35 [26]. The mRNA of *IL12A* is ubiquitously and constitutively expressed in many cell types at low levels [44], while *EBI3* is expressed in EBV-transformed B lymphocytes, spleen, tonsil, placental trophoblasts, and activated APCs in vivo [10, 11, 32].

10.4 Transcriptional Regulation of *EBI3*

EBI3 was initially discovered as a transcriptionally activated gene in Epstein–Barr virus-infected human B lymphocytes. While typically low or absent in resting cells, EBI3 expression is constitutive in several human tumor cells, including Hodgkin lymphoma cells [27], and acute myeloid leukemia cells [33]. Immunohistochemical analysis revealed that EBI3 is highly expressed in tumor tissues originated from lung cancer, colon cancer, esophageal carcinoma, hepatocellular carcinoma, and cervical carcinoma [22]. It can be induced by pathogen- and host-derived inflammatory stimuli: In B lymphocytes, EBI3 mRNA expression was directly induced by EBV latent membrane protein 1(LMP1) [9]; in CD40L/IFN-γ-stimulated immature human DCs, IFN-β enhanced EBI3 mRNA levels [41]. It is also highly expressed in LPS-stimulated mature and activated DCs (MADCs) [15]. Influenza A virus (IAV) infection enhanced EBI3 promoter activity through a PGE_2-PKA-CREB signaling pathway [21].

Consistent with the finding that the TLR2, TLR4, and TLR9 agonists induced EBI mRNA production in murine macrophages and B-cell as well as mouse BMDC, the adapter protein MyD88 was found to be required for these events [45]. EBI3 gene expression was reduced or abrogated in TLR2-/TLR4, TLR9, and MyD88 knockout mice. In addition, the NF-κB family members p50 and p65 directly bind to the EBI3 promoter in bone marrow-derived dendritic cells (BMDCs) (Fig. 10.2). The absence of p50 resulted in decreased EBI3 gene expression in response to LPS; p50/p65 was able to synergize with the Ets transcription factor PU.1 to increase EBI3 transcription [45]. In contrast, EBI3 levels were virtually unaffected in B-cells from NF-κB p50 knockout mice, suggesting the existence of p50-independent pathways for the regulation of EBI3 expression in B lymphocytes. Besides MyD88, another adaptor protein linked to TLR signaling – TRIF – was shown to be required for EBI3 expression [25]. TRIF is associated with TLR3 and TLR4 and serves to

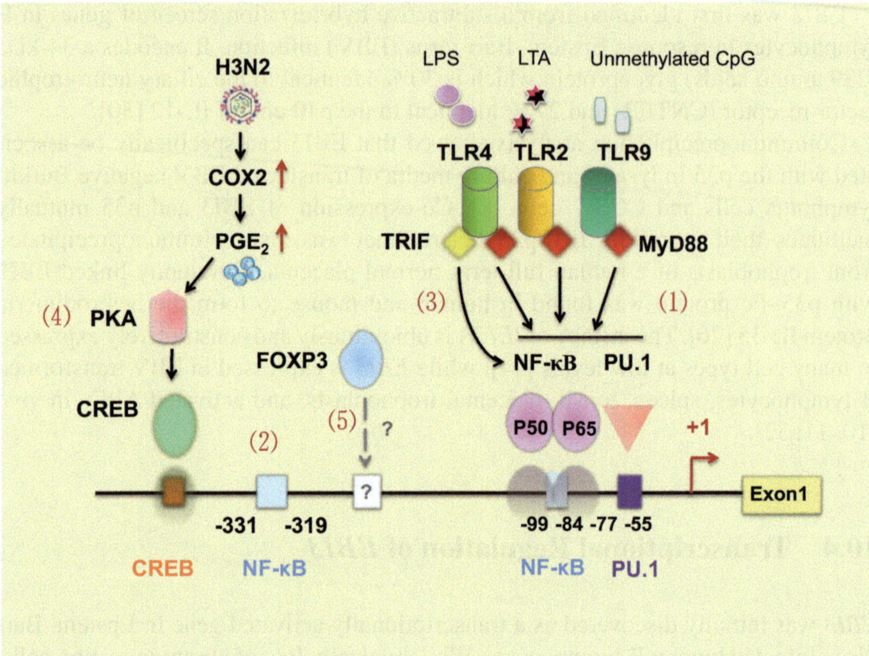

Fig. 10.2 Transcriptional regulation of *EBI3*

activate IFN regulatory factor 3 (IRF3). BMDCs from TRIF-deficient mice stimulated with LPS cannot upregulate EBI3 gene expression [48]. As illustrated in Fig. 10.2, bioinformatics analysis determined that there are several consensus *cis*-elements within EBI3 promoter, including bind sites for NF-κB, CREB, and AP1 [21, 33]. IAV (H3N2) infection significantly activated EBI3 promoter activity by enhancing CREB binding to the specific enhancer element site within the EBI3 promoter. This may be achieved by activated PKA-CREB signaling, which is the consequence of triggered COX-2 expression and PGE2 accumulation after IAV infection. Thus EBI is significantly upregulated in A549 cells and peripheral blood mononuclear cells (PBMCs) [21]. In mice, while p35 mRNA was comparable in $Foxp3^-CD4^+$ and $Foxp3^+CD4^+$ thymocytes, EBI3 mRNA was present in $Foxp3^+$ $CD4^+$ thymocytes, but was essentially absent in $CD4^+CD8^+$ and $CD4^+Foxp3^-$ thymocytes [7]. Intracellular Ebi3 expression was detected in resting wild-type T_{reg} cells, but not wild-type T_{eff} or *Ebi3*$^{-/-}$ T_{reg} cells. *Foxp3*-transduced T_{eff} cells exhibited increased levels of *EBI3* transcripts (Foxp3-GFP encoding gene on retroviral vector) compared with the GFP-alone controls, suggesting that *EBI3* is a downstream target of Foxp3. Foxp3 is a transcription factor which is required for T_{reg}-cell development, homeostasis, and function [37]. This may partially explain the preferential expression of IL-35 by T_{reg} cells.

(1) EBI3 expression in DCs is transcriptionally regulated by TLR signaling via MyD88, NF-κB, and PU.1. NF-κB and Ets binding sites were found at −99 and −77

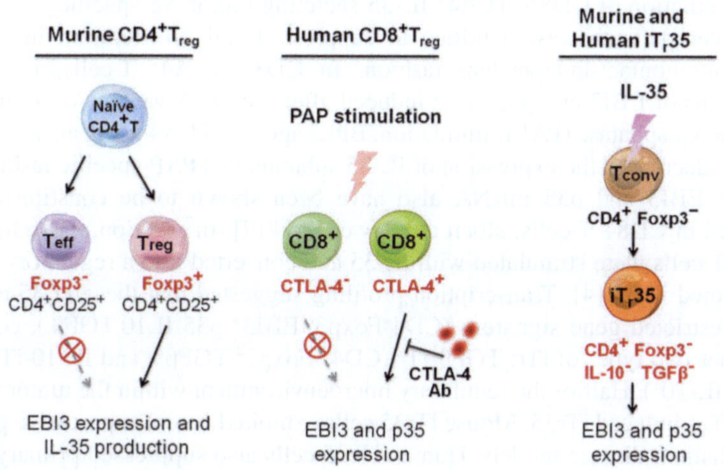

Fig. 10.3 IL-35 expression in T-cells

bp upstream of the transcriptional state site, respectively. This −99 to −84 NF-κB binding site is not present in the murine EBI3 promoter. (2) Another conserved NF-κB binding site was found in the region between −331 and −319, a region of high homology between the murine and the human EBI3 promoter [33]. (3). Besides MyD88, TRIF was found to be another TLR4 downstream mediator to regulate EBI3 mRNA expression. As EBI mRNA levels upon LPS stimulation were decreased in BMDCs from both MyD88- and TRIF-deficient mice. (4) EBI3 expression was upregulated in A549 human lung epithelial cells and human peripheral blood mononuclear cells infected with IAV (H3N2). IAV activated EBI3 promoter activity by enhancing CREB binding to the specific enhancer elements site within the EBI3 promoter through PKA-CREB signaling, which was mediated by COX2 derived PGE_2. (5) EBI3 was identified as a downstream target of FOXP3 in murine T-cells, whereas the regulatory mechanism and binding site for FOXP3 are not clear.

The transcriptome of EBI3 and p35 in different cell types determine the production and function of IL-35. It was also reported that p35 needs to be part of a heterodimer in order to be secreted [40]. Indeed, IL-35 is only upregulated when the expressions of both subunits are induced [19].

EBI3 and p35 are constitutively secreted by T_{reg} ($CD4^+CD25^+$) but not T_{eff} cells ($CD4^+CD25^-$). T_{reg}-cell restriction of IL-35 occurs because EBI3 is a downstream target of Foxp3 [7]. Foxp3 is an exclusive marker of murine T_{reg} cells because of its strictly restricted expression in T_{reg} cells [50]. $EBI3^{-/-}$ and $p35^{-/-}$ T_{reg} cells had significantly reduced regulatory activity in vitro and failed to control homeostatic proliferation and to cure inflammatory bowel disease in vivo. In contrast to murine T_{reg}, neither $CD4^+CD25^+$ $Foxp3^+$ nor $CD8^+CD25^+$ $Fox3^+$ regulatory thymocytes express EBI3, suggesting that in humans, Foxp3 expression is not sufficient to induce significant EBI3 expression [1]. As shown in Fig. 10.3, besides $CD4^+$ T_{reg}, a novel population of $CD8^+CTLA4^+$ IL-35-secreting tumor Ag-specific T_{regs} in prostate cancer patients was identified to suppress T-cell proliferation in a IL-35-dependent, contact-independent fashion. In $CD8^+CTLA4^+$ T-cells, the mRNA expressions of EBI3 and p35 were induced after the DNA vaccine encoding prostatic acid phosphatase (PAP) stimulation. Blockage of CTLA4 using an anti-CTLA4 antibody decreased the expression of IL-35 subunits in a PAP-specific fashion [29]. In mice, EBI3 and p35 mRNA also have been shown to be constitutively co-expressed in CD8+ T-cells, albeit at a lower level [7]. In addition, naïve human or murine T cells were stimulated with IL-35 and converted into a regulatory population, termed iT_r35 [4]. Transcription profiling suggested that these iT_r35 exhibit a highly restricted gene signature ($CD4^+Foxp3^-$ $EBI3^+$ $p35^+IL10^-TGF\beta^-$), compared with other two types of iTr: TGFβ-iTr ($CD4^+$ $Foxp3^+$ $TGF\beta^+$) and IL-10-iTr ($CD4^+$ $Foxp3^-$ IL-10^+). Half of the regulatory microenvironment within the tumor is mediated by T_{reg}-induced iTr35. Mouse iTr35 cells exhibited potent suppressive potential in five distinct disease models. Human iTr35 cells also suppressed primary human T-cell proliferation through IL-35 [35].

(1) IL-35 is constitutively secreted by mouse T_{reg} but not T_{eff} cells. T_{reg}-cell restriction of this cytokine occurs because *Ebi3* is a downstream target of Foxp3. In contrast, EBI3 (as well as IL-35) is not expressed in human T_{reg} cells, constitutively or upon stimulation. (2) Antigen-specific $CD8^+CTLA4^+$ T_{regs} in human prostate cancer produce IL-35. This induction depends on CTLA4, as blockage of CTLA4 using specific antibody decreased the expression of IL-35 subunits. In addition, antigen-specific induction of IL-35 was not observed in $CD8^+CTLA4^-$ T-cells. Naïve human or murine T cells were stimulated with IL-35 and converted into a regulatory population – iTr35 cells – which exhibit a highly restricted gene signature ($CD4^+Foxp3^-$ $EBI3^+$ $p35^+IL10^-TGF\beta^-$).

Transcriptome analysis in B-cells showed that EBI3 was one of the differentially regulated genes in B-cell activated by TLR4 plus CD40 relative to TLR4 priming alone [36]. B-cells constitutively express the IL-35 alpha chain, p35, but do not express p40. Thus, B-cell receptor for antigen (BCR) engagement along with TLR4 and CD40 increased EBI3 and p35 transcription and IL-35 secretion, confirming B-cells as a novel source of IL-35. The regulatory potential of B-cell-derived IL-35 was further confirmed and complemented by another study: Recombinant IL-35 (rIL-35) treatment promoted the generation and conversion of IL-10 secreting B_{reg}. It also mediated conversion of B-cells into a B_{reg} subset that produces IL-35 (IL-35^+B_{reg}) [42]. rIL-35 induced the binding of STAT1 to p35 and EBI3 proximal promoter in B-cells and upregulated p35 and EBI mRNA expression and secretion of IL-35 by B_{reg} cells. In B-cells, IL-35 signals through IL-12Rβ/IL-27Rα (WSX1) and mediates its biological effects by activating STAT1 and to a lesser extent, STAT3, as IL-12Rβ2- or IL-27Rα-deficient B-cells could not inhibit proliferation of B-cells or induce IL-35^+ B_{regs} in the presence of rIL-35. IL-12Rβ/IL-27Rα-STAT1/3 pathway is different from the IL-35 signal in T-cells (Fig. 10.4).

IL-35 mediates conversion of B-cells into a B_{reg} subset that produces IL-35 (IL-35^+B_{reg}). B-cells respond to IL-35 through IL-12Rβ2/IL-27Rα (WSX1) heterodimers and STAT1/STAT3 pathways, which are required for the generation of

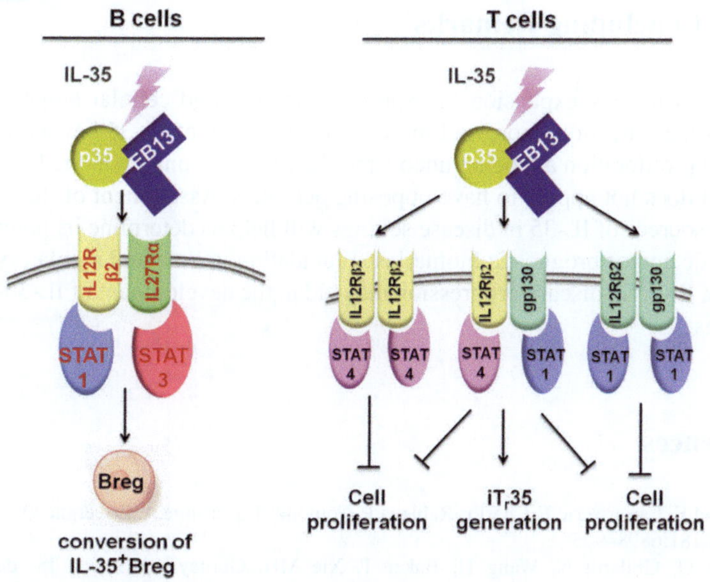

Fig. 10.4 The expression and signaling pathway of IL-35 in B-cells

IL-35-producing B_{regs}. In T-cells, IL-35 signals through gp130 and IL-12Rβ2 homodimers or gp130/IL-12Rβ2 heterodimer.

Distinct from other members of the IL-12 family, the expression of IL-35 has not been described in antigen-presenting cells (APCs). Recently, IL-35 was shown to be produced in human tolerogenic dendritic cells (tolDCs) [12]. TolDCs don't express IL-12p40 transcript, but maintain mRNA and protein expression of p35 and EBI3. EBI3 was strongly upregulated by IFN-γ alone or plus LPS, while IL-12p40 was not expressed using any stimuli in tolDCs, suggesting that IL-35 producing in tolDCs may due to the lack of competition between EBI3 and IL-12p40. In addition, p35 produced in tolDCs is required for their full tolerogenic potential, thereby implying a role of IL-35 in tolDCs. In addition, heat-killed *C. albicans* induced EBI3 expression at a high level in type2 anti-inflammatory macrophages (M2 macrophages) [49]. EBI3 served as a suppressor for LPS-induced M1 to M2 transition and IL-12p70 production. However, IL-35 production was not detected in M2 macrophages.

Gene expression analysis has revealed that IL-35 may have much broader tissue distribution [19]. Based on experimental evidence, the authors demonstrated that IL-35 is a responsive anti-inflammatory cytokine that could be induced by proinflammatory cytokines in non-T-cells (intestinal epithelial cells, primary aortic smooth muscle cells, and intestinal microvascular endothelial cells).

10.5 Concluding Remarks

With an increasing expansion of expression sources and cellular targets of IL-35 since its identification, more and more novel functions of IL-35 beyond inhibition of T-cell proliferation are being uncovered. IL-35 seems unique in the IL-12 family in that it does not appear to have opposing activities. Assessment of the tissue and cellular sources of IL-35 in disease settings will help us determine its potential as a diagnostic and/or prognostic biomarker. Elucidating its role and regulatory mechanisms of IL-35 in disease progression may aid in the development of IL-35-directed therapies.

References

1. Bardel E, Larousserie F, Charlot-Rabiega P, Coulomb-L'Hermine A, Devergne O. J Immunol. 2008;181:6898–905.
2. Chen Q, Ghilardi N, Wang H, Baker T, Xie MH, Gurney A, Grewal IS, de Sauvage FJ. Development of Th1-type immune responses requires the type I cytokine receptor TCCR. Nature. 2000;407:916–20.
3. Choi J, Leung PS, Bowlus C, Gershwin ME. Macrophage transformation. Clin Rev Allergy Immunol 2015;49:327–332.
4. Collison LW, Chaturvedi V, Henderson AL, Giacomin PR, Guy C, Bankoti J, Finkelstein D, Forbes K, Workman CJ, Brown SA, et al. IL-35-mediated induction of a potent regulatory T cell population. Nat Immunol. 2010;11:1093–101.
5. Collison LW, Delgoffe GM, Guy CS, Vignali KM, Chaturvedi V, Fairweather D, Satoskar AR, Garcia KC, Hunter CA, Drake CG, et al. The composition and signaling of the IL-35 receptor are unconventional. Nat Immunol. 2012;13:290–9.
6. Collison LW, Vignali DA. Interleukin-35: odd one out or part of the family? Immunol Rev. 2008;226:248–62.
7. Collison LW, Workman CJ, Kuo TT, Boyd K, Wang Y, Vignali KM, Cross R, Sehy D, Blumberg RS, Vignali DA. The inhibitory cytokine IL-35 contributes to regulatory T-cell function. Nature. 2007;450:566–9.
8. Devergne O, Birkenbach M, Kieff E. Epstein-Barr virus-induced gene 3 and the p35 subunit of interleukin 12 form a novel heterodimeric hematopoietin. Proc Natl Acad Sci U S A. 1997;94:12041–6.
9. Devergne O, Cahir McFarland ED, Mosialos G, Izumi KM, Ware CF, Kieff E. Role of the TRAF binding site and NF-kappaB activation in Epstein-Barr virus latent membrane protein 1-induced cell gene expression. J Virol. 1998;72:7900–8.
10. Devergne O, Coulomb-L'Hermine A, Capel F, Moussa M, Capron F. Expression of Epstein-Barr virus-induced gene 3, an interleukin-12 p40-related molecule, throughout human pregnancy: involvement of syncytiotrophoblasts and extravillous trophoblasts. Am J Pathol. 2001;159:1763–76.
11. Devergne O, Hummel M, Koeppen H, Le Beau MM, Nathanson EC, Kieff E, Birkenbach M. A novel interleukin-12 p40-related protein induced by latent Epstein-Barr virus infection in B lymphocytes. J Virol. 1996;70:1143–53.
12. Dixon KO, van der Kooij SW, Vignali DA, van Kooten C. Human tolerogenic dendritic cells produce IL-35 in the absence of other IL-12 family members. Eur J Immunol. 2015;45:1736–47.

13. Gravano DM, Vignali DA. The battle against immunopathology: infectious tolerance mediated by regulatory T cells. Cell Mol Life Sci. 2012;69:1997–2008.
14. Gubler U, Chua AO, Schoenhaut DS, Dwyer CM, McComas W, Motyka R, Nabavi N, Wolitzky AG, Quinn PM, Familletti PC, et al. Coexpression of two distinct genes is required to generate secreted bioactive cytotoxic lymphocyte maturation factor. Proc Natl Acad Sci U S A. 1991;88:4143–7.
15. Hashimoto SI, Suzuki T, Nagai S, Yamashita T, Toyoda N, Matsushima K. Identification of genes specifically expressed in human activated and mature dendritic cells through serial analysis of gene expression. Blood. 2000;96:2206–14.
16. Huang CH, Loo EX, Kuo IC, Soh GH, Goh DL, Lee BW, Chua KY. Airway inflammation and IgE production induced by dust mite allergen-specific memory/effector Th2 cell line can be effectively attenuated by IL-35. J Immunol. 2011;187:462–71.
17. Kastelein RA, Hunter CA, Cua DJ. Discovery and biology of IL-23 and IL-27: related but functionally distinct regulators of inflammation. Annu Rev Immunol. 2007;25:221–42.
18. Kobayashi M, Fitz L, Ryan M, Hewick RM, Clark SC, Chan S, Loudon R, Sherman F, Perussia B, Trinchieri G. Identification and purification of natural killer cell stimulatory factor (NKSF), a cytokine with multiple biologic effects on human lymphocytes. J Exp Med. 1989;170:827–45.
19. Li X, Mai J, Virtue A, Yin Y, Gong R, Sha X, Gutchigian S, Frisch A, Hodge I, Jiang X, et al. IL-35 is a novel responsive anti-inflammatory cytokine – a new system of categorizing anti-inflammatory cytokines. PLoS One. 2012;7:e33628.
20. Liu JQ, Liu Z, Zhang X, Shi Y, Talebian F, Carl Jr JW, Yu C, Shi FD, Whitacre CC, Trgovcich J, Bai XF. Increased Th17 and regulatory T cell responses in EBV-induced gene 3-deficient mice lead to marginally enhanced development of autoimmune encephalomyelitis. J Immunol. 2012;188:3099–106.
21. Liu L, Cao Z, Chen J, Li R, Cao Y, Zhu C, Wu K, Wu J, Liu F, Zhu Y. Influenza A virus induces interleukin-27 through cyclooxygenase-2 and protein kinase A signaling. J Biol Chem. 2012;287:11899–910.
22. Long J, Zhang X, Wen M, Kong Q, Lv Z, An Y, Wei XQ. IL-35 over-expression increases apoptosis sensitivity and suppresses cell growth in human cancer cells. Biochem Biophys Res Commun. 2013;430:364–9.
23. Ma X, Trinchieri G. Regulation of interleukin-12 production in antigen-presenting cells. Adv Immunol. 2001;79:55–92.
24. Molle C, Goldman M, Goriely S. Critical role of the IFN-stimulated gene factor 3 complex in TLR-mediated IL-27p28 gene expression revealing a two-step activation process. J Immunol. 2010;184:1784–92.
25. Molle C, Nguyen M, Flamand V, Renneson J, Trottein F, De Wit D, Willems F, Goldman M, Goriely S. IL-27 synthesis induced by TLR ligation critically depends on IFN regulatory factor 3. J Immunol. 2007;178:7607–15.
26. Niedbala W, Wei XQ, Cai B, Hueber AJ, Leung BP, McInnes IB, Liew FY. IL-35 is a novel cytokine with therapeutic effects against collagen-induced arthritis through the expansion of regulatory T cells and suppression of Th17 cells. Eur J Immunol. 2007;37:3021–9.
27. Niedobitek G, Pazolt D, Teichmann M, Devergne O. Frequent expression of the Epstein-Barr virus (EBV)-induced gene, EBI3, an IL-12 p40-related cytokine, in Hodgkin and Reed-Sternberg cells. J Pathol. 2002;198:310–6.
28. Nishino R, Takano A, Oshita H, Ishikawa N, Akiyama H, Ito H, Nakayama H, Miyagi Y, Tsuchiya E, Kohno N, et al. Identification of Epstein-Barr virus-induced gene 3 as a novel serum and tissue biomarker and a therapeutic target for lung cancer. Clin Cancer Res. 2011;17:6272–86.
29. Olson BM, Jankowska-Gan E, Becker JT, Vignali DA, Burlingham WJ, McNeel DG. Human prostate tumor antigen-specific CD8+ regulatory T cells are inhibited by CTLA-4 or IL-35 blockade. J Immunol. 2012;189:5590–601.

30. Omata F, Birkenbach M, Matsuzaki S, Christ AD, Blumberg RS. The expression of IL-12 p40 and its homologue, Epstein-Barr virus-induced gene 3, in inflammatory bowel disease. Inflamm Bowel Dis. 2001;7:215–20.
31. Oppmann B, Lesley R, Blom B, Timans JC, Xu Y, Hunte B, Vega F, Yu N, Wang J, Singh K, et al. Novel p19 protein engages IL-12p40 to form a cytokine, IL-23, with biological activities similar as well as distinct from IL-12. Immunity. 2000;13:715–25.
32. Pflanz S, Timans JC, Cheung J, Rosales R, Kanzler H, Gilbert J, Hibbert L, Churakova T, Travis M, Vaisberg E, et al. IL-27, a heterodimeric cytokine composed of EBI3 and p28 protein, induces proliferation of naive CD4+ T cells. Immunity. 2002;16:779–90.
33. Poleganov MA, Bachmann M, Pfeilschifter J, Muhl H. Genome-wide analysis displays marked induction of EBI3/IL-27B in IL-18-activated AML-derived KG1 cells: critical role of two kappaB binding sites in the human EBI3 promotor. Mol Immunol. 2008;45:2869–80.
34. Saito M, Yoshida K, Hibi M, Taga T, Kishimoto T. Molecular cloning of a murine IL-6 receptor-associated signal transducer, gp130, and its regulated expression in vivo. J Immunol. 1992;148:4066–71.
35. Seyerl M, Kirchberger S, Majdic O, Seipelt J, Jindra C, Schrauf C, Stockl J. Human rhinoviruses induce IL-35-producing Treg via induction of B7-H1 (CD274) and sialoadhesin (CD169) on DC. Eur J Immunol. 2010;40:321–9.
36. Shen P, Roch T, Lampropoulou V, O'Connor RA, Stervbo U, Hilgenberg E, Ries S, Dang VD, Jaimes Y, Daridon C, et al. IL-35-producing B cells are critical regulators of immunity during autoimmune and infectious diseases. Nature. 2014;507:366–70.
37. Shevach EM, DiPaolo RA, Andersson J, Zhao DM, Stephens GL, Thornton AM. The lifestyle of naturally occurring CD4+ CD25+ Foxp3+ regulatory T cells. Immunol Rev. 2006;212:60–73.
38. Tong H, Miyazaki Y, Yamazaki M, Hara H, Waldmann H, Hori S, Yoshida H. Exacerbation of delayed-type hypersensitivity responses in EBV-induced gene-3 (EBI-3)-deficient mice. Immunol Lett. 2010;128:108–15.
39. Trinchieri G. Interleukin-12 and the regulation of innate resistance and adaptive immunity. Nat Rev Immunol. 2003;3:133–46.
40. Trinchieri G, Pflanz S, Kastelein RA. The IL-12 family of heterodimeric cytokines: new players in the regulation of T cell responses. Immunity. 2003;19:641–4.
41. van Seventer JM, Nagai T, van Seventer GA. Interferon-beta differentially regulates expression of the IL-12 family members p35, p40, p19 and EBI3 in activated human dendritic cells. J Neuroimmunol. 2002;133:60–71.
42. Wang RX, Yu CR, Dambuza IM, Mahdi RM, Dolinska MB, Sergeev YV, Wingfield PT, Kim SH, Egwuagu CE. Interleukin-35 induces regulatory B cells that suppress autoimmune disease. Nat Med. 2014;20:633–41.
43. Wang Z, Liu JQ, Liu Z, Shen R, Zhang G, Xu J, Basu S, Feng Y, Bai XF. Tumor-derived IL-35 promotes tumor growth by enhancing myeloid cell accumulation and angiogenesis. J Immunol. 2013;190:2415–23.
44. Watford WT, Hissong BD, Bream JH, Kanno Y, Muul L, O'Shea JJ. Signaling by IL-12 and IL-23 and the immunoregulatory roles of STAT4. Immunol Rev. 2004;202:139–56.
45. Wirtz S, Becker C, Fantini MC, Nieuwenhuis EE, Tubbe I, Galle PR, Schild HJ, Birkenbach M, Blumberg RS, Neurath MF. EBV-induced gene 3 transcription is induced by TLR signaling in primary dendritic cells via NF-kappa B activation. J Immunol. 2005;174:2814–24.
46. Wolf D, Wolf AM, Rumpold H, Fiegl H, Zeimet AG, Muller-Holzner E, Deibl M, Gastl G, Gunsilius E, Marth C. The expression of the regulatory T cell-specific forkhead box transcription factor FoxP3 is associated with poor prognosis in ovarian cancer. Clin Cancer Res. 2005;11:8326–31.
47. Wu H, Li P, Shao N, Ma J, Ji M, Sun X, Ma D, Ji C. Aberrant expression of Treg-associated cytokine IL-35 along with IL-10 and TGF-beta in acute myeloid leukemia. Oncol Lett. 2012;3:1119–23.

48. Yamamoto M, Sato S, Hemmi H, Hoshino K, Kaisho T, Sanjo H, Takeuchi O, Sugiyama M, Okabe M, Takeda K, Akira S. Role of adaptor TRIF in the MyD88-independent toll-like receptor signaling pathway. Science. 2003;301:640–3.
49. Zheng XF, Hong YX, Feng GJ, Zhang GF, Rogers H, Lewis MA, Williams DW, Xia ZF, Song B, Wei XQ. Lipopolysaccharide-induced M2 to M1 macrophage transformation for IL-12p70 production is blocked by Candida albicans mediated up-regulation of EBI3 expression. PLoS One. 2013;8:e63967.
50. Ziegler SF. FOXP3: of mice and men. Annu Rev Immunol. 2006;24:209–26.

MIX
Papier aus verantwortungsvollen Quellen
Paper from responsible sources
FSC® C105338

If you have any concerns about our products,
you can contact us on
ProductSafety@springernature.com

In case Publisher is established outside the EU,
the EU authorized representative is:
Springer Nature Customer Service Center GmbH
Europaplatz 3, 69115 Heidelberg, Germany

Printed by Libri Plureos GmbH
in Hamburg, Germany